GLOBAL PERSPECTIVES ON SUSTAINABLE FOREST MANAGEMENT

Edited by **Clement Akais Okia**

Global Perspectives on Sustainable Forest Management
Edited by Clement Akais Okia

CBS, Edition **2016**

Published by InTech

Janeza Trdine 9, 51000 Rijeka, Croatia

Notice

Publishing Process Manager Ana Skalamera

Technical Editor Teodora Smiljanic

Cover Designer InTech Design Team

Additional hard copies can be obtained from orders@intechopen.com

Global Perspectives on Sustainable Forest Management, Edited by Clement Akais Okia
p. cm.
ISBN 978-953-51-0569-5

Contents

Preface   IX

Section 1   Forest Management   1

Chapter 1   **Deforestation: Causes, Effects and Control Strategies   3**
Sumit Chakravarty, S. K. Ghosh, C. P. Suresh,
A. N. Dey and Gopal Shukla

Chapter 2   **Depletion of Oak (*Quercus* spp.) Forests in the Western
Himalaya: Grazing, Fuelwood and Fodder Collection   29**
Gajendra Singh and G. S. Rawat

Chapter 3   **Nepal's Community Forestry:
Need of Better Governance   43**
Ridish K. Pokharel, Santosh Rayamajhi
and Krishna R. Tiwari

Chapter 4   **The Influence of Decisional Autonomy on
Performance and Strategic Choices – The Case
of Subcontracting SMEs in Logging Operations   59**
Etienne St-Jean and Luc LeBel

Chapter 5   **Member, Owner, Customer, Supplier? –
The Question of Perspective on Membership and
Ownership in a Private Forest Owner Cooperative   75**
Gun Lidestav and Ann-Mari Arvidsson

Chapter 6   **Economic Feasibility of an
Eucalyptus Agroforestry System in Brazil   95**
Álvaro Nogueira de Souza, Humberto Ângelo,
Maísa Santos Joaquim, Sandro Nogueira de Souza
and John Edward Belknap

Chapter 7   **Effects of a Unilateral Tariff Liberalisation
on Forestry Products and Trade in Australia:
An Economic Analysis Using the GTAP Model   107**
Luz Centeno Stenberg and Mahinda Siriwardana

Chapter 8   **Strategic Cost Management Practices Adopted by Segments of Brazilian Agribusiness 121**
Marcos Antonio Souza and Kátia Arpino Rasia

Chapter 9   **Forestry and Life Cycle Assessment 139**
Andreja Bosner, Tomislav Poršinsky and Igor Stankić

Chapter 10   **Wildlife Forestry 161**
Daniel J. Twedt

Chapter 11   **Fire Ignition Trends in Durango, México 191**
Marín Pompa-García and Eusebio Montiel-Antuna

Chapter 12   **Remote Sensing of Forestry Studies 205**
Kai Wang, Wei-Ning Xiang, Xulin Guo and Jianjun Liu

Chapter 13   **GIS for Operative Support 217**
Gunnar Bygdén

Section 2   **Forest Utilization 223**

Chapter 14   **Adding Value Prior to Pulping: Bioproducts from Hemicellulose 225**
Lew Christopher

Chapter 15   **Incorporating Carbon Credits into Breeding Objectives for Plantation Species in Australia 247**
Miloš Ivković, Keryn Paul and Tony McRae

Chapter 16   **Irreversibility and Uncertainty in Multifunctional Forest Management Allocation 263**
Jens Abildtrup, Jacques Laye, Maximilien Laye and Anne Stenger

Chapter 17   **Ecoefficient Timber Forwarding on Lowland Soft Soils 275**
Tomislav Poršinsky, Tibor Pentek, Andreja Bosner and Igor Stankić

Chapter 18   **Comparison of Timber Consumption in U.S. and China 289**
Lanhui Wang and Huiwen Wang

Preface

Forests are an integral part of global sustainable development. The World Bank estimates more than 1.6 billion people to be dependent on forests for their livelihoods with some 300 million living in them. The forest product industry is a source of economic growth and employment, with global forest products traded internationally is estimated at $327 billion. The Food and Agriculture Organization (FAO) estimates that every year 130,000 km² of the world's forests are lost due to deforestation. Sustainable forest management aims to ensure that the goods and services derived from the forests meet present-day needs while at the same time securing their continued availability and contribution to long-term development. There is worldwide evidence that forest resources can provide long-term national economic benefits. The year 2011 was therefore designated as 'The International Year of Forests' mainly to draw greater attention to the forests worldwide.

This book is therefore dedicated to global perspectives on sustainable forest management. It focuses on a need to move away from purely protective management of forests to innovative approaches for multiple use and management of forest resources. The book is divided into two sections; the first section, with thirteen chapters deals with the forest management aspects while the second section, with five chapters is dedicated to forest utilization. The first two chapters set the scene with chapter one highlighting the causes, effects and control strategies for deforestation and chapter two presenting depletion of Oak forests in the western Himalaya. It is observed that by destroying the forests, all potential future revenues and future employment that could be derived from their sustainable management for timber and non-timber products will disappear.

Chapter three presents Nepal's experiences as one of the leading countries in community-based forest management. Unlike in other countries where forest authorities have not yet trusted local communities in forest management, Nepal has used this approach in rejuvenating forests in denuded hills and naked areas. The success revolves around empowerment of Community Forest User Group (CFUG) that are legal entities with autonomy in decision making such as access rules, forest products prices, mechanism for allocation of forest products and user fees. The influence of decisional autonomy on performance and strategic choices for sub-contracting SMEs in logging operations is discussed in chapter four. The authors note that in a situation of commercial dependency, some contractors use their power over the SME to interfere in the management of their business, thereby curtailing their decisional autonomy. It is concluded that, businesses which find themselves in a situation where they have to obey the orders of a contractor may have limited resources to innovate or develop new and promising niche markets.

Chapter five highlights the perspective on membership and ownership in a private forest owner cooperative in a Swedish private forest owner cooperative. It looks at the mismatch between goal of the cooperative and that of the members, to illustrate how the members, the inspectors and the managers look upon the private forest owner in terms of identity, benefits, and agreement. It is suggested that improvements can be achieved by applying a new institutional way of organization, which will make it easier to create a meaningful communication. Chapter six evaluates the economic feasibility of an eucalyptus agroforestry system in Brazil. It is noted that in decision making between projects which have different time horizons, the Equivalent Periodic Benefits (BPE) method has priority over the Net Present Value (VPL). Since each different age of eucalyptus represents an alternative project with different planning horizons, the alternative with a greater BPE must be chosen.

Chapter seven analyses the effects of a unilateral tariff liberalisation on forestry products in Australia, amongst the leading exporters and importers of forest products, in particular, as well as global merchandise, in general. The Global Trade Analysis Project (GTAP) model and its database, version 7 with 2004 data are utilized. Given that forest products only comprise a small proportion of world merchandise trade, it is expected that trade liberalisation would cause small changes in terms of trade, real GDP, production, consumption and prices of forest products in most countries. Chapter eight is dedicated to strategic cost management practices adopted by segments of Brazilian agribusiness. It focus on investigating the use, perceived benefits and difficulties in implementation by agribusiness companies of strategic cost management practices, treated in the literature as the most appropriate ones to help managers in the management process of organizations. The findings of this exploratory study provide the foundation for conducting further research in the form of multiple case studies and cross-case analyses.

Chapter nine is about forestry and life cycle assessment (LCA). LCA helps in measuring environmental aspects and potential impacts of a product through its entire life-cycle from raw material acquisition (beginning of the life-cycle) to manufacturing, use, recycling, re-use and final disposal (end of the product's life-cycle). Though LCA is particularly difficult in forestry due to the long rotation period, it remains a good tool especially today when there is increased awareness of environmentally friendly technologies, renewable energy and eco-efficiency. The greatest use of LCA tool in forestry, from the timber harvesting aspect, is noted to be in comparing different harvesting systems for selecting environmentally friendly versions.

Chapter ten presents an exciting concept - wildlife forestry. It goes beyond the traditional production forestry which is limited to a few favoured tree species. The concept of desired landscape conditions with regard to wildlife forestry is well-explained using the Lower Mississippi Valley Joint Venture Forest Resource Conservation Working Group model. Wildlife forestry silviculture exploits the versatility of legacy retention by combining specific retention or removal of canopy to achieved habitat conditions that are deemed suitable for priority wildlife species. Fire ignition trends in Durango, Mexico in covered in Chapter eleven. The study assessed wildfire ignition history in an ecologically heterogeneous landscape over a decade in order to develop clustering maps that would be easy for managers to understand in setting up a strategy for fighting forest fires. While the fires exhibited a clustered spatial distribution, there is need for further studies to thoroughly analyse the underlying mechanisms responsible for the spatially clustered fires. Nevertheless, the assessment of fire clustering can provide useful new information to researchers in predicting potential fire dangers and behaviours across the landscape.

Chapter twelve is an introductory review on the use of remote sensing in forestry studies. It introduces the basics of remote sensing, summarizes the recent developments, and elucidates several typical applications of remote sensing in forestry studies. A number of case studies are used to illustrate the application of application of remote sensing in forestry including; species composition, forest ecophysiology, forest ecosystem ecology and measuring and monitoring forest resources among others. The use of GIS for operative support in forestry is highlighted in Chapter thirteen. The authors contend that the use of GIS in forestry planning should be extended beyond giving directives for the machine operators where to cut the trees and marking areas with high environmental values. It should as well incorporate soil information maps as an overlay to give information on where less good trafficability is to be expected, especially after a heavy rain. Such information can improve forest operations and save both time and money.

In chapter fourteen, adding value prior to pulping focusing on bio-products from hemicelluloses in the USA is presented. Hemicellulose (the second most abundant polysaccharide after cellulose) is available in large quantities in USA, unfortunately, it's not efficiently utilized and often discarded as industrial waste with limited usage. A number of bio-products such as bio-chemicals and bio-materials can be produced from biomass hemicelluloses. Such value addition will enhance the value extracted from wood fibre and improve the process economics in a forest-based biorefinery, thus ensuring that forest-based industries to remain competitive. Chapter fifteen examines the incorporation of carbon credits into breeding objectives for plantation species in Australia. It is concluded that the revenue from sales of carbon credits significantly increases the profitability of plantation species on a per hectare basis. However, the economic weights for breeding objective traits should be calculated as an average of those for the production systems that are receiving and those that are not receiving carbon credits, weighted by the respective projected plantation areas.

Irreversibility and uncertainty in multifunctional forest management allocation is considered in chapter sixteen. A simple model is formulated to derive general decision rules which can be applied in a given situation defined by the production function and uncertainty about future changes in policy. The authors observe that when a manager of two forest areas targets a multifunctional use of the forest in the context of irreversibility among regimes and uncertainty about the forthcoming information on environmental policy affecting the forest areas, two alternatives must be compared in terms of specialization of the areas and implementation of mixed regime in both forest areas. Chapter seventeen presents results of a study on eco-efficient timber forwarding on lowland soft soils in Croatia. The authors assessed the environmental soundness of forest vehicles, whose contact with the soil is likely to cause soil damage basing on trafficking and soil compaction. The findings indicate that using wide tires on the front wheels of the vehicle is advantageous as it leads to a decrease in the exceeded nominal pressure with respect to the allowed value. From the environmental soundness aspect, the solution of the overloaded front axle of the six-wheel forwarder is an eight-wheel forwarder.

Finally, chapter eighteen presents a comparison of timber consumption in U.S. and China. It is observed that in view of the security of timber resources and promotion of environmental consciousness, many countries have tightened or are trying to tighten the timber trade policy. It is hard for China to source the timber overseas, thus the fast growing plantations are a feasible choice. Compared with U.S, China's timber industry needs to enlarge the forest investment to guarantee sustainable timber supply.

This book will fill the existing gaps in the knowledge about emerging perspectives on sustainable forest management. It will be an interesting and helpful resource to managers, specialists and students in the field of forestry and natural resources management.

Dr. Clement Akais Okia
Makerere University, Kampala
Uganda

Section 1

Forest Management

Deforestation: Causes, Effects and Control Strategies

Sumit Chakravarty[1], S. K. Ghosh[2], C. P. Suresh[2],
A. N. Dey[1] and Gopal Shukla[3]
[1]Department of Forestry
[2]Pomology & Post Harvest Technology, Faculty of Horticulture
Uttar Banga Krishi Viswavidyalaya, Pundibari
[3]ICAR Research Complex for Eastern Region, Research Center, Plandu Ranchi
India

1. Introduction

The year 2011 is 'The International Year of Forests'. This designation has generated momentum bringing greater attention to the forests worldwide. Forests cover almost a third of the earth's land surface providing many environmental benefits including a major role in the hydrologic cycle, soil conservation, prevention of climate change and preservation of biodiversity (Sheram, 1993). Forest resources can provide long-term national economic benefits. For example, at least 145 countries of the world are currently involved in wood production (Anon., 1994a). Sufficient evidence is available that the whole world is facing an environmental crisis on account of heavy deforestation. For years remorseless destruction of forests has been going on and we have not been able to comprehend the dimension until recently. Nobody knows exactly how much of the world's rainforests have already been destroyed and continue to be razed each year. Data is often imprecise and subject to differing interpretations. However, it is obvious that the area of tropical rainforest is diminishing and the rate of tropical rain forest destruction is escalating worldwide, despite increased environmental activism and awareness.

Deforestation is the conversion of forest to an alternative permanent non-forested land use such as agriculture, grazing or urban development (van Kooten and Bulte, 2000). Deforestation is primarily a concern for the developing countries of the tropics (Myers, 1994) as it is shrinking areas of the tropical forests (Barraclough and Ghimire, 2000) causing loss of biodiversity and enhancing the greenhouse effect (Angelsen *et al.*, 1999). FAO considers a plantation of trees established primarily for timber production to be forest and therefore does not classify natural forest conversion to plantation as deforestation (but still records it as a loss of natural forests). However, FAO does not consider tree plantations that provide non-timber products to be forest although they do classify rubber plantations as forest. Forest degradation occurs when the ecosystem functions of the forest are degraded but where the area remains forested rather cleared (Anon., 2010).

Thirty per cent of the earth's land area or about 3.9 billion hectares is covered by forests. It was estimated that the original forest cover was approximately six billion hectares (Bryant *et al.*, 1997). The Russian Federation, Brazil, Canada, the United States of America and China were the most forest rich countries accounting to 53 per cent of the total forest area of the globe. Another 64 countries having a combined population of two billions was reported to have forest on less than ten per cent of their total land area and unfortunately ten of these countries have no forest at all. Among these countries 16 are such which had relatively substantial forest areas of more 1than one million hectares each and three of these countries namely Chad, the Islamic Republic of Iran and Mongolia each had more than ten million hectares of forest. The forest area remained fairly stable in North and Central America while it expanded in Europe during the past decade. Asian continent especially in India and China due to their large scale afforestation programme in the last decade registered a net gain in forest area. Conversely the South America, Africa and Oceania had registered the net annual loss of forest area (Anon., 2010; 2011$_a$).

2. World deforestation

According to Professor Norman Myers, one of the foremost authorities on rates of deforestation in tropical forests, "the annual destruction rates seems set to accelerate further and could well double in another decade" (Myers, 1992). Mostly deforestation has occurred in the temperate and sub-tropical areas. Deforestation is no longer significant in the developed temperate countries now and in fact many temperate countries now are recording increases in forest area (Anon., 1990$_a$; 2010). In most instances developed nations are located in temperate domains and developing nations in tropical domains. However deforestation was significantly less in tropical moist deciduous forest in 1990-2000 than 1980-1990 but using satellite imagery it was found that FAO overestimated deforestation of tropical rainforests by 23 per cent (Anon., 2001$_{a;\ b}$). However the definition of what is and what is not forest remains controversial. The tropical rainforests capture most attention but 60 per cent of the deforestation that occurred in tropical forests during 1990-2010 was in moist deciduous and dry forests.

However extensive tropical deforestation is a relatively modern event that gained momentum in the 20th century and particularly in the last half of the 20th century. The FAO FRA 2001 and 2010 reports indicate considerable deforestation in the world during 1990-2010 but this was almost entirely confined to tropical regions (Anon., 2001$_a$; 2010). A summary of deforestation during the decades 1990-2010 is given in tables 1 and 2. These tables show there was considerable deforestation in the world during 1990-2010 but this was almost entirely confined to tropical regions. Rowe *et al.* (1992) estimated that 15 per cent of the world's forest was converted to other land uses between 1850 and 1980. Deforestation occurred at the rate of 9.2 million hectares per annum from 1980-1990, 16 million hectares per annum from 1990-2000 and decreased to 13 million hectares per annum from 2000-2010. The net change in forest area during the last decade was estimated at -5.2 million hectares per year, the loss area equivalent to the size of Costa Rica or 140 km^2 of forest per day, was however lesser than that reported during 1990-2000 which was 8.3 million hectares per year equivalent to a loss of 0.20 per cent of the remaining forest area each year. The current annual net loss is 37 per cent lower than that in the 1990s and equals a loss of 0.13 per cent of the remaining forest area each year during this period. By contrast some smaller countries have very high losses per year and they are in risk of virtually losing all their forests within the next decade if current rates of deforestation are maintained. Indeed some 31 countries do not even make the list because they have already removed most of their forests and even if that remain are seriously fragmented and degraded. The changes in area of forest by region and subregion are shown in table 1.

Region/subregion	1990-2000		2000-2010	
	1 000 ha/year	%	1 000 ha/year	%
Eastern and Southern Africa	-1841	-0.62	-1839	-0.66
Northern Africa	-590	-0.72	-41	-0.05
Western and Central Africa	-1637	-0.46	-1535	-0.46
Total Africa	**-4067**	**-0.56**	**-3414**	**-0.49**
East Asia	1762	0.81	2781	1.16
South and Southeast Asia	-2428	-0.77	-677	-0.23
Western and Central Asia	72	0.17	131	0.31
Total Asia	**-595**	**-0.10**	**2235**	**0.39**
Russian Federation (RF)	32	n.s.	-18	n.s.
Europe excluding RF	845	0.46	694	0.36
Total Europe	**877**	**0.09**	**676**	**0.07**
Caribbean	53	0.87	50	0.75
Central America	-374	-1.56	-248	-1.19
North America	32	n.s.	188	0.03
Total North and Central America	**-289**	**-0.04**	**-10**	**0.00**
Total Oceania	**-41**	**-0.02**	**-700**	**-0.36**
Total South America	**-4213**	**-0.45**	**-3997**	**-0.45**
World	**-8327**	**-0.20**	**-5211**	**-0.13**

Table 1. Annual change in forest area by region and subregion, 1990-2010
(Source: Anon., 2010)

South America with about four million hectares per year suffered the largest net loss of forests during the last decade followed by Africa with 3.4 million hectares annually and the least Oceania with seven lakh hectares annually. Oceania suffered mainly due to Australia where severe drought and forest fires from 2000 AD had exacerbated their loss. Both Brazil and Indonesia had the highest net loss of forest during the decade of 1990 but has significantly reduced their rate of loss after this decade. Brazil and Indonesia dominate accounting for almost 40 per cent of net forest loss over the decade of 1990s. Even though Brazil was the top deforesting country by area, the forests in Brazil are so extensive that this represents a loss of 0.4 per cent per year. The forest area in North and Central America remained stable during the past decade. The forest area in Europe continued to expand although at a slower rate of seven lakh hectare per year during the last decade than in the 1990s with nine lakh hectares per year. Asia lossed some six lakh hectares annually during 1990s but gained more than 2.2 million hectares per year during the last decade. The ten countries with the largest net loss per year in the period 1990-2000 AD had a combined net loss of forest area of 7.9 million hectares per year. In the period 2000-2010 AD this was reduced to six million hectares per year as a result of reductions in Indonesia, Sudan, Brazil and Australia (table 1). There were 28 countries and areas which have an estimated net loss of one per cent or more of their forest area per year. The five countries with the largest annual net loss for 2000-2010 AD were Comoros (-9.3 per cent), Togo (-5.1 per cent), Nigeria (-3.7 per cent), Mauritania (-2.7 per cent) and Uganda (-2.6 per cent). The area of other wooded land globally decreased by about 3.1 million hectares per year during 1990-2000 AD and by about 1.9 million hectares per year during the last decade. The area of other wooded land also decreased during the past two decades in Africa, Asia and South America.

Country	Annual change 1990-2000		Country	Annual change 1990-2000	
	1 000 ha/year	%		1 000 ha/year	%
Brazil	-2890	-0.51	Brazil	-2642	-0.49
Indonesia	-1914	-1.75	Australia	-562	-0.37
Sudan	-589	-0.80	Indonesia	-498	-0.51
Myanmar	-435	-1.17	Nigeria	-410	-3.67
Nigeria	-410	-2.68	Tanzania	-403	-1.13
Tanzania	-403	-1.02	Zimbabwe	-327	-1.88
Mexico	-354	-0.52	the Congo	-311	-0.20
Zimbabwe	-327	-1.58	Myanmar	-310	-0.93
Congo	-311	-0.20	Bolivia	-290	-0.49
Argentina	-293	-0.88	Venezuela	-288	-0.60
Total	**-7926**	**-0.71**	**Total**	**-6040**	**-0.53**

Table 2. Countries with largest annual net loss of forest area, 1990-2010 (Source: Anon., 2010)

3. The causes of deforestation

As Myers pointed out, "we still have half of all tropical forests that ever existed" (Myers, 1992). The struggle to save the world's rainforests and other forests continues and there is a growing worldwide concern about the issue. In order to save forests, we need to know why they are being destroyed. Distinguishing between the agents of deforestation and its causes is very important in order to understand the major determinants of deforestation. The agents of deforestation are those slash and burn farmers, commercial farmers, ranchers, loggers, firewood collectors, infra-structure developers and others who are cutting down the forests. Causes of deforestation are the forces that motivate the agents to clear the forests. However, most of the existing literature typically distinguishes between two levels of specific factors: direct and indirect causes of deforestation. Direct agents and causes of deforestation, also typically referred to as sources of deforestation, first level or proximate causes (Panayotou, 1990; Barbier *et al.*, 1994; Caviglia, 1999) are relatively easy to identify but the indirect causes which are usually the main divers of deforestation are the ones that cause most disagreement and the ones that are hardest to quantify (Bhatnagar, 1991; Mather, 1991; Humphreys, 2006; Sands, R. 2005).

Similarly, Pearce and Brown (1994) identified two main forces affecting deforestation. They are:

- Competition between humans and other species for the remaining ecological niches on land and in coastal regions. This factor is substantially demonstrated by the conversion of forest land to other uses such as agriculture, infrastructure, urban development, industry and others.

\- Failure in the working of the economic systems to reflect the true value of the environment. Basically, many of the functions of tropical forests are not marketed and as such are ignored in decision making. Additionally, decisions to convert tropical forests are themselves encouraged by fiscal and other incentives.

The former can be regarded as the direct and latter as indirect cause of deforestation.

3.1 Direct causes

3.1.1 Expansion of farming land

About 60 per cent of the clearing of tropical moist forests is for agricultural settlement (Myers, 1994; Anon., 1991) with logging and other reasons like roads, urbanization and Fuelwood accounting for the rest (Anon., 1994b). Tropical forests are one of the last frontiers in the search for subsistence land for the most vulnerable people worldwide (Myers, 1992). Millions of people live on the tropical forest with less than a dollar a day where a third of a billion are estimated to be foreign settlers. However, as the land degrades people are forced to migrate, exploring new forest frontiers increasing deforestation (Wilkie *et al.*, 2000; Amor, 2008; Amor and Pfaff, 2008). Deforestation is proxied by the expansion of agricultural land. This is because agricultural land expansion is generally viewed as the main source of deforestation contributing around 60 per cent of total tropical deforestation.

Shifting agriculture also called slash and burn agriculture is the clearing of forested land for raising or growing the crops until the soil is exhausted of nutrients and/or the site is overtaken by weeds and then moving on to clear more forest. It is been often reported as the main agent of deforestation. Smallholder production in deforestation and the growing number of such producers notably shifting cultivators were the main cause of deforestation (Anon., 1990b; c; Dick, 1991; Anon., 1992a; b; Barbier *et al.*, 1993; Ascher, 1993; Dove, 1993; 1996; Dauvergne, 1994; Porter, 1994; Thiele, 1994; Anon., 1994c; Angelsen, 1995; Ross, 1996). Mostly all reports indicate shifting agriculture as responsible for about one half of tropical deforestation and some put it up to two-thirds. Shifting agriculture was greatest in Asia (about 30 per cent) but only about 15 per cent over the whole tropical world. It appears that the proportion of direct conversion of forest to agriculture is increasing and the proportion of shifting agriculture is decreasing with time.

3.1.2 Forest and other plantations

Plantations are a positive benefit and should assist in reducing the rate of deforestation. The fact that plantations remove the timber pressure on natural forests does not translate eventually into less, but rather into more deforestation. Indeed, it is feared that agricultural expansion which is the main cause of deforestation in the tropics might replace forestry in the remaining natural forests (Anon., 2002; Cossalter and Pye-Smith, 2003; Anon., 2005). The impact of timber plantations could thus turn out to be quite detrimental to tropical forest ecosystems (Kartodihardjo and Supriono, 2000). Tree crops and rubber in particular plays a more important role in deforestation in Indonesia than subsistence-oriented shifting cultivation (Chomitz and Griffiths, 1996). Unfortunately about one-half of the plantations in the tropics are established on native forest cleared for the purpose. Moreover plantations can promote deforestation by constructing roads that improve access of the shifting cultivators and others to the forest frontier.

3.1.3 Logging and fuel wood

Logging does not necessarily cause deforestation. However, logging can seriously degrade forests (Putz *et al.*, 2001). Logging in Southeast Asia is more intensive and can be quite destructive. However, logging provides access roads to follow-on settlers and log scales can help finance the cost of clearing remaining trees and preparing land for planting of crops or pasture. Logging thus catalyzes deforestation (Chomitz *et al.*, 2007).

Fuelwood gathering is often concentrated in tropical dry forests and degraded forest areas (Repetto, 1988; 1990; Rowe *et al.*, 1992; Anon., 1994$_a$). Fuelwood is not usually the major cause of deforestation in the humid tropics although it can be in some populated regions with reduced forest area such as in the Philippines, Thailand and parts of Central America. Fuelwood gathering was considered to be the main cause of deforestation and forest degradation in El Salvador (Repetto, 1990). In the drier areas of tropics, Fuelwood gathering can be a major cause of deforestation and degradation.

3.1.4 Overgrazing

Overgrazing is more common in drier areas of the tropics where pastures degraded by overgrazing are subject to soil erosion. Stripping trees to provide fodder for grazing animals can also be a problem in some dry areas of the tropics but is probably not a major cause of deforestation. Clear cutting and overgrazing have turned large areas of Qinghai province in China into a desert. Overgrazing are causing large areas of grasslands north of Beijing and in Inner Mongolia and Qinghai province to turn into a desert. One man who lived in a village on the eastern edge of the Qinghai-Tibet plateau that was being swallowed up by sand told the New York Times, "The pasture here used to be so green and rich. But now the grass is disappearing and the sand is coming." Huge flocks of sheep and goats strip the land of vegetation. In Xillinggol Prefecture in Inner Mongolia, for example, the livestock population increased from 2 million in 1977 to 18 million in 2000, turning one third of the grassland area to desert. Unless something is done the entire prefecture could be uninhabitable by 2020. Overgrazing is exacerbated by sociological phenomena called "the tragedy of the common." People share land but raises animals for themselves and try to enrich them by rising as many as they can. This leads to more animals than the land can support. Grassland in Qinghai that can support 3.7 million sheep had 5.5 million sheep in 1997. Animals remove the vegetation and winds finished the job by blowing away the top soil, transforming grasslands into desert. When a herder was asked why he was grazing goats next to a sign that said "Protect vegetation, no grazing," he said, "The lands are too infertile to grow crops—herding is the only way for us to survive." (Hays, 2008 web page).

3.1.5 Fires

Fires are a major tool used in clearing the forest for shifting and permanent agriculture and for developing pastures. Fire is a good servant but has a poor master. Fire used responsibly can be a valuable tool in agricultural and forest management but if abused it can be a significant cause of deforestation (Repetto, 1988; Rowe *et al.*, 1992). Based on the data available from 118 countries representing 65 per cent of the global forest area, an average of 19.8 million hectares or one per cent of all forests were reported to be significantly affected each year by forest fires (Anon., 2010). Deforestation due to road pavements in Brazil had also lead to higher incidences of forest fires (Carvalho *et al.*, 2001; Nepstad *et al.*, 2001).

3.1.6 Mining

Mining is very intensive and very destructive (Mather, 1991; Sands, 2005). The area of land involved is quite small and it is not seen as a major cause of primary deforestation. Mining is a lucrative activity promoting development booms which may attract population growth with consequent deforestation. The deforestation rate due to mining activities in Guyana from 2000 to 2008 increased 2.77 times according to an assessment by the World Wildlife Fund-Guianas (Staff, 2010). Similarly, in the Philippines, mining, along with logging, has been among the forces behind the country's loss of forest cover: from 17 million hectares in 1934 to just three million in 2003 or an 82 per cent decline (Docena, 2010). Nearly 2,000 hectares of tropical forest in the Municipality of Coahuayana in the State of Michoacán (south-western Mexico) will completely be destroyed by mining iron minerals planned by the Italo-Argentine mining company TERNIUM (Anonymous, 2008). Similarly, Nyamagari hills in Orissa India currently threatened by Vedanta Aluminum Corporation's plan to start bauxite mining will destroy 750 hectares of reserved forest (Griffiths and Hirvelä, 2008). Massive and unchecked mining of coal, iron ore and bauxite in Jharkhand, India has caused large scale deforestation and created a huge water scarcity (Anon., 2011b). In return for US$3.8 billion of investment, the agreements between the State government of Jharkhand, India and mining companies, there will be a massive land acquisition which will deforest no less than 57,000 hectares of forest and displace 9,615 families, many of them located in legally protected Scheduled Areas set aside for indigenous peoples in the State (Mullick and Griffiths. 2007). Moreover, Roads constructed to support the mining operations will open up the area to shifting agriculturists, permanent farmers, ranchers, land speculators and infrastructure developers. For instance the core of Brazil's Amazon development strategy were infra-structure development projects such as roads providing access to frontier regions, mining area and large hydroelectric reservoirs (Mahar, 1988; Fearnside and Barbosa, 1996; Carvalho *et al.*, 2002, 2004). The construction of roads, railways, bridges, and airports opens up the land to development and brings increasing numbers of peoples to the forest frontier. If wood is used as fuel in mining operations and it is sources from plantations established for the purpose, it can cause serious deforestation in the region. On the other hand, mining can be labour intensive and take labour away from clearing forest.

3.1.7 Urbanization/industrialization and infra-structure

Expanding cities and towns require land to establish the infrastructures necessary to support growing population which is done by clearing the forests (Mather, 1991; Sands, 2005). Tropical forests are a major target of infra-structure developments for oil exploitation, logging concessions or hydropower dam construction which inevitably conveys the expansion of the road network and the construction of roads in pristine areas (Kaimowitz and Angelsen, 1998). The construction of roads, railways, bridges, and airports opens up the land to development and brings increasing numbers of people to the forest frontier. Whether supported or not by the governmental programmes, these settlers have usually colonized the forest by using logging trails or new roads to access the forest for subsistence land (Wilkie *et al.*, 2000; Amor, 2008; Amor and Pfaff, 2008). (Wilkie *et al.*, 2000; Amor, 2008; Amor and Pfaff, 2008). The development of these infrastructure projects are of worldwide concern, since tropical forest clearing accounts for roughly 20 per cent of anthropogenic carbon emissions destroying globally significant carbon sinks (Anon., 2001c) and around 21 per cent of tropical forests have been lost worldwide since 1980 (Bawa *et al.*, 2004).

3.1.8 Air pollution

Air pollution is associated with degradation of some European and North American forests. The syndrome is called "Waldsterben" or forest death. In 1982, eight per cent of all West German trees exhibited damage that rose to about 52 per cent by 1987 (Raloff, 1989) and half of the trees reported dying of Waldsterben in the Alps (Lean, 1990). High elevation forests show the earliest damage including forests in the north-east and central United States.

3.1.9 Wars and role of the military

It is well established that military operations caused deforestation during the Vietnam War and elsewhere (Mather, 1991; Sands, 2005). More recently, linkages have been documented between the civil war in Myanmar and the timber trade between Myanmar and Thailand. Myanmar regime sells timber to the Thais to finance its civil war against the Karen hill tribe. Forest destruction in El Salvador has resulted from war. Apart from military involvements in wars, the role of military in deforestation has been documented in Southeast Asia and South America (Mather, 1991; Sands, 2005). The authors also observed that role of powerful military in Brazilian politics are a major cause of Amazonian forest destruction.

3.1.10 Tourism

National parks and sanctuaries beyond doubt protect the forests, but uncautioned and improper opening of these areas to the public for tourism is damaging. Unfortunately, the national governments of tropical and sub-tropical countries adopt tourism for easy way of making money sacrificing the stringent management strategies. Further, many companies and resorts who advertise themselves as eco-tourist establishments are in fact exploiting the forests for profit. In Cape Tribulation, Australia, for example, the rain forest is being threatened by excessive tourism (Colchester and Lohmann, 1993). Similarly, in the Terai Duars of eastern India foothill Himalaya, eco-tourism is encouraged and we fear this is being done without developing adequate management plans. For instance, the Chilapatta Reserve Forest in this area is opened for eco-tourism for its ancient ruins deep in the forest and a tree species *Myristica longifolia* that exudes a blood like sap when injured. The site has become a popular eco-tourist destination because of the ruins and for this blood exuding tree. In the whole forest only eight individuals were found but two of the trees in the near vicinity of the ruins completely dried away due to repeated injuries caused to the plants by the curious tourists (Shukla, 2010). In fact, in the name of eco-tourism, infra-structure development is taking place mostly be the private players in these wilderness areas which are further detrimental in terms of attracting peoples other than tourists also, causing deforestation especially deep in the forest.

3.2 Indirect causes

The World Rainforest Movement's 'Emergency Call to Action for the forests and their Peoples' asserts that "deforestation is the inevitable result of the current social and economic policies being carried out in the name of development" (Anon., 1990$_d$). It is in the name of development that irrational and unscrupulous logging, cash crops, cattle ranching, large dams, colonisation schemes, the dispossession of peasants and indigenous peoples and promotion of tourism is carried out. Harrison Ngau, an indigenous tribesman from Sarawak, Malaysia and winner of the Goldman Environment Award in 1990 puts the cause of tropical deforestation like this, "the roots of the problem of deforestation and waste of resources are located in the industrialized countries where most of our resources such as tropical timber end up. The rich nations with one quarter of the world's population consume four fifth of the world's resources. It is the throw away culture of the industrialized countries now advertised in and forced on to the Third World countries that is leading to the throwing away of the world. Such so-called progress leads to destruction and despair" (Anon., 1990$_d$)! Such a development leads to overconsumption which is the basic underlying cause of deforestation.

3.2.1 Colonialism

Erstwhile colonies of the colonial powers like Britain, France, Spain or Portugal are now the Third World Countries or the developing nations mostly have the tropical rainforests except Australia and Hawaii were exploited for their natural resources and their indigenous people's rights destroyed by the colonial powers. All these countries have indigenous populations who had their own system of land management and/or ownership in place for thousands of years before the intervention of colonists from rich industrialized nations. Colonialism turned previously self-sufficient economies into zones of agriculture export production. This process continues even today in different form of exploitation and the situation is worsening (Colchester and Lohmann, 1993).

3.2.2 Exploitation by industrialized countries

Wealthy countries or the erstwhile colonial powers having deficit of their own natural resources are mainly sustaining on the resources of the financially poorer countries those are generally natural resource rich. Twenty per cent of the world's population is using 80 per cent of the world's resources. Unfortunately also the governments of these poor resource rich countries had generally adopted the same growth-syndrome as their western neighbours or their erstwhile colonial master giving emphasis on maximizing exports, revenues and exploiting their rich natural resources unsustainably for short-term gains. Moreover, corruption in government, the military and economic powers is well known. The problem is further worsened by the low price of the most Third World exports being realized in the international market (Colchester and Lohmann, 1993).

3.2.3 The debt burden

Pursuing the guided development agenda, the financially poorer countries are on a heavy international debt and now feeling the urgency of repaying these huge debts due to escalating interest rates. Such a situation compels these debt ridden poorer countries to exploit their rich natural resources including their forests partly to earn foreign exchange for servicing their debts. For instance, construction of roads for logging operations in some South-east Asian countries was funded by Japanese aid which allowed the Japanese timber companies to exploit the forests of these countries. Understandably, these timber companies profitably exploited the forests while the South-east Asian countries were left owing Japan money for construction of their roads (Colchester and Lohmann, 1993).

3.2.4 Overpopulation and poverty

The role of population in deforestation is a contentious issue (Mather, 1991; Colchester and Lohmann, 1993; Cropper and Griffiths, 1994; Ehrhardt-Martinez, 1998; Sands, 2005). The impact of population density on deforestation has been a subject of controversy. Poverty and overpopulation are believed to be the main causes of forest loss according to the international agencies such as FAO and intergovernmental bodies. It is generally believed by these organizations that they can solve the problem by encouraging development and trying to reduce population growth. Conversely, the World Rainforest Movement and many other NGOs hold unrestrained development and the excessive consumption habits of rich industrialized countries directly responsible for most forest loss. However there is good evidence that rapid population growth is a major indirect and over-arching cause of deforestation. More people require more food and space which requires more land for agriculture and habitation. This in turn results in more clearing of forests. Arguably increasing population is the biggest challenge of all to achieve sustainable management of human life support systems and controlling population growth is perhaps the best single thing that can be done to promote sustainability. Overpopulation is not a problem exclusive to Third World countries. An individual in an industrialized country is likely to consume in the order of sixty times as much of the world's resources as a person in a poor country. The growing population in rich industrialized nations are therefore responsible for much of the exploitation of the earth and there is a clear link between the overconsumption in rich countries and deforestation in the tropics (Colchester and Lohmann, 1993).

Poverty and overpopulation are inextricably linked. Poverty, while undeniably responsible for much of the damage to rainforests, has to a large extent been brought about by the greed of the rich industrialized nations and the Third World elites who seek to emulate them. Development is often regarded as the solution to world poverty, seldom helps those whose need is greatest. Thus, it is often the cause rather than the cure for poverty. The claim that overpopulation is the cause of deforestation is used by many governments and aid agencies as an excuse for inaction. In tropical countries, pressure from human settlement comes about more from inequitable land distribution than from population pressure. Generally, most of the land is owned by small but powerful elite which displaces poor farmers into rainforest areas. So long as these elites maintain their grip on power, lasting land reform will be difficult to achieve (Colchester and Lohmann, 1993) and deforestation continues unabated. Therefore poverty is well considered to be an important underlying cause of forest conversion by small-scale farmers and naturally forest-dense areas are frequently associated with high levels of poverty (Chomitz *et al.*, 2007). The population also often lacks the finance necessary for investments to maintain the quality of soil or increase yields on the existing cleared land (Purnamasari, 2010). Deforestation is affected mainly by the uneven distribution of wealth. Shifting cultivators at the forest frontier are among the poorest and most marginalized sections of the population. They usually own no land and have little capital. Consequently they have no option but to clear the virgin forest. Deforestation including clearing for agricultural activities is often the only option available for the livelihoods of farmers living in forested areas (Angelsen, 1999).

3.2.5 Transmigration and colonisation schemes

Transmigration of people to the forest frontier whether forced or voluntary due to development policy or dislocation from war is the major indirect cause of deforestation (Mather, 1991; Colchester and Lohmann, 1993; Sands, 2005). Moreover, governments and international aid agencies earlier believed that by encouraging colonisation and transmigration schemes into rainforest areas could alleviate poverty of the areas in the financially poorer countries. Such schemes have miserably failed but hurted the indigenous people and the environment. In Indonesia, the *Transmigrasi* Program of 1974 had caused annual deforestation of two lakh hectares (Colchester and Lohmann, 1993). Dispossessed and landless people bring increased population pressure to the forest frontier. Further, new migrants in the area increase demand for food and other agricultural products which can induce the farmers at the forest frontier to increase their agricultural production by expanding agricultural land by clearing the forests (Levang, 2002). Moreover, the new migrants may not care for conservation of the forests in their new home which further accelerates deforestation of the area.

3.2.6 Land rights, land tenure and inequitable land distribution and resources

Cultivators at the forest frontier often do not hold titles to land (absence of property rights) and are displaced by others who gain tenure over the land they occupy (Mather, 1991; Deacon, 1999; Sands, 2005). This means they have to clear more forest to survive. Poorly defined tenure is generally bad for people and forests (Chomitz *et al.*, 2007). In many countries government have nominal control of forests but are too weak to effectively regulate their use. This can lead to a tragedy of the commons where forest resources are degraded. In frontier areas deforestation is a common practice and legalized way of declaring claim to land and securing tenure (Schneider, 1995).

3.2.7 Economic causes - development/land conversion value, fiscal policies, markets and consumerism

The relationship between development and deforestation is complex and dynamic (Humphreys, 2006; Mather, 1991; Sands, 2005). One point of view is that development will increase land productivity and thereby reduce the need to clear forests to meet food requirements. Another is that development will produce further capital and incentive to expand and clear more forest. The former may be the case when constrained by a fixed food demand. The latter may be the case when food demand may not be satisfied owing to a continuing export market and rising internal population with rising levels of consumption. Profits from deforestation vary from less than a dollar to thousand dollars per hectare depending on location, technologies and land use systems (Chomitz *et al.*, 2007). It is also argued by the workers that richer farmers were better able to finance deforestation while a poor farmer can't afford to clear much forest. Conversely, through transfers, stronger credit markets and better opportunities for off-season employment can increase income as well as deforestation by small land holders. Moreover, land offering higher rents encourage quicker deforestation. Higher prices for crops and lower prices for farm inputs also spur faster deforestation (Chomitz *et al.*, 2007). Wage increase can also stimulate deforestation (Barbier and Cox, 2004). Technological innovations make farming more profitable either prompting the expansion of farms into forest or attract new farmers to forest frontiers (Angelsen and Kaimowitz, 2001; Angelsen, 2006). Even when the increase in commodity price is only temporary, it tends to raise expectation about future prices, increasing the expected probability from land clearance and conversion to agriculture (Angelsen, 1995; Sunderlin *et al.*, 2000). Many development policies have failed because they have supported either wittingly or unwittingly the development of those who already have land, power, influence and political clout. This further alienates the rural poor and puts the pressure back on the forests.

Poor farm households or commercial loggers have little incentive to care about the environmental effects of their actions. Such unaccounted costs give rise to economic failures such as local market failures, policy failures and global appropriation failures (Panayotou, 1990). Market fails due to unregulated market economy which does not produce an optimal outcome. Prices generated by such market does not reflect the true social costs and benefits from resource use and convey misleading information about resource scarcity, providing inadequate incentives for management, efficient utilization and enhancement of natural resources. Policy failures or market distortions are result of misguided intervention or unsuccessful attempts to mitigate failures resulting in worse outcomes (Panayotou, 1990). For instance, lack of respect of traditional land rights make property rights to forest land uncertain and could encourage short-term exploitation of forests rather than long-term sustainable use. Further, global appropriation failures occur as in the case of tropical forests- the benefits of biodiversity conservation and the value of the genetic pool in developing new medicines, crops and pest control agents are poorly reflected in market allocations. For instance, it was argued that improved terms of trade for agricultural and forest product exports and higher real exchange rates make it more profitable to convert forests to other uses (Capistrano, 1994; Southgate, 1994; Kant and Redantz, 1997). The initial empirical analysis done by Scrieciu (2003) appears to confirm that tropical deforestation is caused by the drive for maximizing profits within the agricultural sector. Rampant consumerism by the developed countries frequently has been claimed as a major reason for tropical deforestation. The opening of tropical countries to the world commodity markets accelerated deforestation. The products include coffee, sugar, bananas, cotton and beef in Central America and oil palm, rubber and timber in Southeast Asia.

3.2.8 Undervaluing the forest

Forests gain value only when they are cleared for obtaining legal title through 'improvement' (Mather, 1991; Sands, 2005). The extraction of non-wood forest products has been suggested as a way to add value to the forest but it is not economical when compared to clearing options. If some means could be devised where those who benefit from the environmental values could pay the forest owners or agents of deforestation for them, then the option to not clear would become more competitive. Alternatively, if the national governments value the environmental benefits, it could apply a tax or disincentives to clear. However, even though maintenance of the environmental services is essential for sustained economic development, deforesting nations usually have more immediate goals and are unprepared to take this step.

3.2.9 Corruption and political cause

The FAO identified forest crime and corruption as one of the main causes of deforestation in its 2001 report and warned that immediate attention has to be given to illegal activities and corruption in the world's forests in many countries (Anon., 2001b). Illegal forest practices may include the approval of illegal contracts with private enterprises by forestry officers, illegal sale of harvesting permits, under-declaring volumes cut in public forest, under-pricing of wood in concessions, harvesting of protected trees by commercial corporations, smuggling of forest products across borders and allowing illegal logging, processing forest raw materials without a license (Contreras-Hermosilla, 2000; 2001).

4. Effects of deforestation

4.1 Climate change

It is essential to distinguish between microclimates, regional climate and global climate while assessing the effects of forest on climate (Gupta *et al.*, 2005) especially the effect of tropical deforestation on climate (Dickinson, 1981). Deforestation can change the global change of energy not only through the micrometeorological processes but also by increasing the concentration of carbon dioxide in the atmosphere (Pinker, 1980) because carbon dioxide absorbs thermal infrared radiation in the atmosphere. Moreover deforestation can lead to increase in the albedo of the land surface and hence affects the radiation budget of the region (Charney, 1975; Rowntree, 1988; Gupta *et al.*, 2005). Deforestation affects wind flows, water vapour flows and absorption of solar energy thus clearly influencing local and global climate (Chomitz *et al.*, 2007). Deforestation on lowland plains moves cloud formation and rainfall to higher elevations (Lawton *et al.*, 2001). Deforestation disrupts normal weather patterns creating hotter and drier weather thus increasing drought and desertification, crop failures, melting of the polar ice caps, coastal flooding and displacement of major vegetation regimes. In the dry forest zones, land degradation has become an increasingly serious problem resulting in extreme cases in desertification (Dregne, 1983). Desertification is the consequence of extremes in climatic variation and unsustainable land use practices including overcutting of forest cover (Anon., 1994b).

Global warming or global change includes anthropogenically produced climatic and ecological problems such as recent apparent climatic temperature shifts and precipitation regimes in some areas, sea level rise, stratospheric ozone depletion, atmospheric pollution and forest decline. Tropical forests are shrinking at a rate of about five per cent per decade as forests are logged and cleared to supply local, regional, national and global markets for wood products, cattle, agricultural produce and biofuels (Anon., 2007; 2010). One of the most important ramifications of deforestation is its effect on the global atmosphere. Deforestation contributes to global warming which occurs from increased atmospheric concentrations of greenhouse gases (GHG) leading to net increase in the global mean temperature as the forests are primary terrestrial sink of carbon. Thus deforestation disrupts the global carbon cycle increasing the concentration of atmospheric carbon dioxide. Tropical deforestation is responsible for the emission of roughly two billion tonnes of carbon (as CO_2) to the atmosphere per year (Houghton, 2005). Release of the carbon dioxide due to global deforestation is equivalent to an estimated 25 per cent of emissions from combustion of fossil fuels (Asdrasko, 1990).

4.2 Water and soil resources loss and flooding

Deforestation also disrupts the global water cycle (Bruijnzeel, 2004). With removal of part of the forest, the area cannot hold as much water creating a drier climate. Water resources affected by deforestation include drinking water, fisheries and aquatic habitats, flood/drought control, waterways and dams affected by siltation, less appealing water-related recreation, and damage to crops and irrigation systems from erosion and turbidity (Anon., 1994$_a$; Bruijnzeel *et al.*, 2005). Urban water protection is potentially one of the most important services that forest provides (Chomitz *et al.*, 2007). Filtering and treating water is expensive. Forests can reduce the costs of doing so either actively by filtering runoff or passively by substituting for housing or farms that generate runoff (Dudley and Stolton, 2003). Deforestation can also result into watersheds that are no longer able to sustain and regulate water flows from rivers and streams. Once they are gone, too much water can result into downstream flooding, many of which have caused disasters in many parts of the world. This downstream flow causes soil erosion thus also silting of water courses, lakes and dams. Deforestation increases flooding mainly for two reasons. First, with a smaller 'tree fountain' effect, soils are more likely to be fully saturated with water. The 'sponge' fills up earlier in wet season, causing additional precipitation to run off and increasing flood risk. Second, deforestation often results in soil compaction unable to absorb rain. Locally, this causes a faster response of stream flows to rainfall and thus potential flash flooding (Chomitz *et al.*, 2007). Moreover deforestation also decrease dry season flows.

The long term effect of deforestation on the soil resource can be severe. Clearing the vegetative cover for slash and burn farming exposes the soil to the intensity of the tropical sun and torrential rains. Forest floors with their leaf litter and porous soils easily accommodate intense rainfall. The effects of deforestation on water availability, flash floods and dry season flows depend on what happens to these countervailing influences of infiltration and evapotranspiration- the sponge versus the fountain (Bruijnzeel, 2004). Deforestation and other land use changes have increased the proportion of the basin subject to erosion and so over the long run have contributed to siltation. Heavy siltation has raised the river bed increasing the risk of flooding especially in Yangtze river basin in China, the major river basins of humid tropics in East Asia and the Amazonian basin (Yin and Li, 2001; Bruijnzeel, 2004; Aylward, 2005, Bruijnzeel *et al.*, 2005; van Noordwijk *et al.*, 2006).

4.3 Decreased biodiversity, habitat loss and conflicts

Forests especially those in the tropics serve as storehouses of biodiversity and consequently deforestation, fragmentation and degradation destroys the biodiversity as a whole and habitat for migratory species including the endangered ones, some of which have still to be catalogued. Tropical forests support about two thirds of all known species and contain 65 per cent of the world's 10, 000 endangered species (Myers and Mittermeier, 2000). Retaining the biodiversity of the forested areas is like retaining a form of capital, until more research can establish the relative importance of various plants and animal species (Anon., 1994$_a$). According to the World Health Organization, about 80 per cent of the world's population relies for primary health care at least partially on traditional medicine. The biodiversity loss and associated large changes in forest cover could trigger abrupt, irreversible and harmful changes. These include regional climate change including feedback effects that could theoretically shift rainforests to savannas and the emergence of new pathogens as the growing trade in bushmeat increases contact between humans and animals (Anon., 2005).

Another negative effect of deforestation is increasing incidents of human-animal conflicts hitting hard the success of conservation in a way alienating the people's participation in conservation. Elephant habitat located at northern West Bengal in India is part of the Eastern Himalaya Biodiversity Hotspot which is characterized by a high degree of fragmentation. The heavy fragmentation of this habitat has resulted into an intense human-elephant conflict causing not only in loss of agricultural crops but also human and elephant lives. Mortality of about 50 persons and 20 elephants was reported due to these severe human-elephant conflicts from this hotspot area annually (Sukumar *et al.*, 2003; Mangave, 2004).

4.4 Economic losses

The tropical forests destroyed each year amounts to a loss in forest capital valued at US $ 45 billion (Hansen, 1997). By destroying the forests, all potential future revenues and future employment that could be derived from their sustainable management for timber and non-timber products disappear.

4.5 Social consequences

Deforestation, in other words, is an expression of social injustice (Colchester and Lohmann, 1993). The social consequences of deforestation are many, often with devastating long-term impacts. For indigenous communities, the arrival of civilization usually means the destruction/change of their traditional life-style and the breakdown of their social institutions mostly with their displacement from their ancestral area. The intrusion of outsiders destroys traditional life styles, customs and religious beliefs which intensifies with infra-structure development like construction of roads which results into frontier expansion often with social and land conflicts (Schmink and Wood, 1992).

The most immediate social impact of deforestation occurs at the local level with the loss of ecological services provided by the forests. Forests afford humans valuable services such as erosion prevention, flood control, water treatment, fisheries protection and pollination-functions that are particularly important to the world's poorest people who rely on natural resources for their everyday survival. By destroying the forests we risk our own quality of life, gamble with the stability of climate and local weather, threaten the existence of other species and undermine the valuable services provided by biological diversity.

5. Stretegies to reduce deforestation

Ways to reducing deforestation must go hand in hand with improving the welfare of cultivators at the forest frontier. Any policy that does without the other is unacceptable. There are no general solutions and strategies since these will vary with region and will change over time. All strategies require cooperation and goodwill. Effective implementation is essential including stakeholder participation, development of management plans, monitoring and enforcement. The strategies should be such that on one hand they should recognize the critical roles of national, state and municipal governments and on other hand empower the civil society and the private sector to take a pro-active role in reducing deforestation, often working in conjunction with government.

5.1 Reduce population growth and increase per capita incomes

Reduction of population growth is pivotal in reducing deforestation in the developing countries. Consequent of reduced population, increase in per capita income will occur as a consequence of increased incomes and literacy rates which will reduce pressure on the remaining forests for new human settlement and land use change.

5.2 Reducing emissions from deforestation and forest degradation

Many international organizations including the United Nations and the World Bank have begun to develop programmes to curb deforestation mainly through Reducing Emissions from Deforestation and Forest Degradation (REDD) which use direct monetary or other incentives to encourage developing countries to limit and/or roll back deforestation. Significant work is underway on tools for use in monitoring developing country adherence to their agreed REDDS targets (Chomitz *et al.*, 2007).

5.3 Increase the area and standard of management of protected areas

The provision of protected areas is fundamental in any attempt to conserve biodiversity (Myers, 1994; Myers and Mittermeier, 2000; Nepstad *et al.*, 2006). Protected areas alone, however, are not sufficient to conserve biodiversity. They should be considered alongside, and as part of, a wider strategy to conserve biodiversity. The minimum area of forest to be protected is generally considered to be 10 per cent of total forest area. It is reported that 12.4 per cent of the world's forest are located within protected areas. Tropical and temperate forests have the highest proportions of their forests in protected areas and boreal forests have the least. The Americas have the greatest proportion while Europe the least proportion of protected areas (Anon., 2010).

5.4 Increase the area of forest permanently reserved for timber production

The most serious impediment to sustainable forest management is the lack of dedicated forests specifically set aside for timber production. If the forest does not have a dedicated long-term tenure for timber production then there is no incentive to care for the long-term interests of the forest. FAO (2001) found that 89 per cent of forests in industrialized countries were under some form of management but only about six per cent were in developing countries. If 20 per cent could be set aside, not only could timber demand be sustainably met but buffer zones could be established to consolidate the protected areas. This would form a conservation estate that would be one of the largest and most important in the world (Anon., 2001$_a$).

5.5 Increase the perceived and actual value of forests

There are several ways of achieving increasing the perceived and actual value of forests. Governments can impose realistic prices on stumpage and forest rent and can invest in improving the sustainable productivity of the forest. National and international beneficiaries of the environmental services of forests have to pay for such services (Chomitz *et al.*, 2007). There has been some success in devising schemes to collect payments for environmental services like carbon sequestration, biodiversity conservation, catchment protection and ecotourism. This success can further be more realized by integrating participatory mode of management with these collection schemes to ensure rights and tenure with equity in resource and benefit sharing for improving the livelihood of the rural poors who actually are the primary stakeholders of conservation and management.

5.6 Promote sustainable management

In order to promote sustainable forest management, it must be sustainable ecologically, economically and socially. Achieving ecological sustainability means that the ecological values of the forest must not be degraded and if possible they should be improved. This means that silviculture and management should not reduce biodiversity, soil erosion should be controlled, soil fertility should not be lost, water quality on and off site should be maintained and that forest health and vitality should be safeguarded. However, management for environmental services alone is not economically and socially sustainable. It will not happen until or unless the developing nations have a reached a stage of development and affluence that they can accommodate the costs of doing so. Alternatively, the developed world must be prepared to meet all the costs (Chomitz *et al.*, 2007; Anon., 2010; 2011). There are vast areas of unused land as discussed earlier some of which is degraded and of low fertility. Technological advances are being made to bring this land back into production. This should be a major priority since a significant proportion of cleared tropical forest will eventually end up as degraded land of low fertility.

5.7 Encouraging substitutes

For all purposes where tropical or other timber is used, other woods or materials could be substituted. We can stop using timber and urge others to do the same. As long there is a market for wood products, trees will continue to be cut down. Labelling schemes, aimed at helping consumers to choose environmental friendly timbers are currently being discussed in many countries (Anon., 1990$_d$).

5.8 Increase area of forest plantation

Increasing the area of forest plantations by using vacant or unused lands and waste and marginal lands especially as road side, along railway tracts, on contours, avenues, boundaries and on land not suited for agricultural production should have a net positive benefit. Planting trees outside forest areas will reduce pressure on forests for timber, fodder and fuelwood demands. Moreover the deforested areas need to be reforested.

5.9 Strengthen government and non-government institutions and policies

Strong and stable government is essential to slow down the rate of deforestation. FAO (2010) considered that half of the current tropical deforestation could be stopped if the governments of deforesting countries were determined to do so (Anon., 2010). Environmental NGO's contribution towards conservation management has been enormous. They have the advantage over government organizations and large international organizations because they are not constrained by government to government bureaucracy and inertia. They are better equipped to bypass corruption and they are very effective at getting to the people at the frontier who are in most need.

5.10 Participatory forest management and rights

In frontier areas much of the forest is nominally owned by the state, but the reach of government and the rule of law are weak and property rights insecure. In order for forest management to succeed at the forest frontier, all parties with an interest in the fate of the forest should be communally involved in planning, management and profit sharing. But forest ownership and management rights are almost always restricted and restrictions on ownership and use define alternative tenure systems. The balance of rights can be tilted strongly toward society in the form of publicly owned strictly protected areas. State ownership and management can be retained but with sustainable timber extraction allowed. As of now much of the world's tropical forest are state owned but community participation in forest ownership and management needs to be encouraged with restrictions on extraction and conversion (Chomitz *et al.*, 2007). Land reform is essential in order to address the problem deforestation. However an enduring shift in favour of the peasants is also needed for such reforms to endure (Colchester and Lohmann, 1993). Moreover the rights of indigenous forest dwellers and others who depend on intact forests must be upheld. Therefore, the recognition of traditional laws of the indigenous peoples as indigenous rights will address the conflicts between customary and statutory laws and regulations related to forest ownership and natural resource use while ensuring conservation of forest resources by the indigenous communities. Central to this is the right to 'Prior Informed Consent', ensuring the indigenous communities to know what they are agreeing to. A means must be found to reconcile conservation and development by involving local/indigenous populations more closely in the decision-making process and by taking the interactions between 'societies' and forest resource more fully into account (Chakravarty *et al.*, 2008).

5.11 Support and reforms

Aid organizations like the World Bank have traditionally favoured spectacular large-scale development al projects. In all cases when such projects are proposed there has been a massive opposition from local people. Reducing the demand for southern-produced agribusiness crops and alleviating the pressure from externally-financed development projects and assistance is the essential first/primary step (Colchester and Lohmann, 1993). Campaigns opposing such developments and the campaigns to reform the large aid agencies which fund such schemes should be supported. Local campaigns against specific mining, dams, industrial and tourist developments should be supported. Further reform of the World Bank and other such organizations is largely the demand of time.

5.12 Increase investment in research, education and extension

Training and education of stakeholder's helps people understand how to prevent and reduce adverse environmental effects associated with deforestation and forestry activities and take appropriate action when possible. Research substantiates it and helps to understand the problem, its cause and mitigation. This arena is lagging behind for paucity of funds and investments encourages this arena. There is a lack of knowledge and information in the general community about forests and forestry. Forest managers and those developing forest policies need to be comprehensively educated and need to appreciate the complexity of the interacting ecological, economical, social, cultural and political factors involved.

5.13 Improve the information base and monitoring

Information on the global distribution of biodiversity and forest poverty is inadequate. Knowledge of how much forest, where it is and what it is composed of seems to be straightforward but surprisingly this most basic information is not always available. It is not possible to properly manage a forest ecosystem without first understanding it. New remote sensing technologies make it feasible and affordable to identify hotspots of deforestation. The international community could undertake monitoring efforts that would have immediate payoffs. A priority is to fund and coordinate basic monitoring on the rate, location and causes of global deforestation and forest poverty along with the impacts of project and policy interventions. Without this information, policy makers are flying blind and interest groups lack a solid basis for dialogue (Chomitz *et al.*, 2007).

5.14 Policy, legislative and regulatory measures-enforcement and compliance

A wide variety of policy statements and legislative and regulatory measures have been established to protect forests but need to be effectively enforced. New modifications/adjustments are of course needed for site specific conditions. Laws, policy and legislation should be such that they encourages local people and institutional participation in forestry management and conservation along with safeguarding indigenous people's traditional rights and tenure with rightful sharing of benefits. Many formal and informal enforcement/compliance mechanisms are used to prevent deforestation and environmental problems from forestry activities. These approaches include negotiation, warnings, cancelling work orders, notices of violation, fines, arrests and court action.

6. Conclusion

Economic globalization combined with the looming global land scarcity increases the complexity of future pathways of land use change. In a more interconnected world, agricultural intensification may cause more rather than less cropland expansion. The apparent tradeoff between forest and agriculture can be minimized through spatial management and the use of degraded or low competition lands (Lambin and Meyfroidt, 2011). This can be further addressed by community based forest management which builds on political goodwill and strong community institutions. New challenges from climate change require urgent action to explore and protect the local value of forests for livelihood even more. This is particularly true in the case of emerging activities undertaken as part of REDD+ activities where broad forest governance are aligned with it along with people's participation ensuring livelihood benefits of the people dependant on forests. These renewed activities will safeguard traditional ways of life and the environmentally important forest ecosystems of the world.

7. References

Amor, D. 2008. *Road impact on deforestation and jaguar habitat loss in the Selva Maya*. Ph. D. dissertation. Ecology Department, Nicholas School of the Environment, Duke University.

Amor, D. and Pfaff, A. 2008. *Early history of the impact of road investments on deforestation in the Mayan forest.* Working Paper, Nicholas School of the Environment and Sanford School of Public Policy, Duke University, Durham, NC, USA.

Angelsen, A. 1995. Shifting cultivation and deforestation: a study from Indonesia. *World Development* 23: 1713-1729.

Angelsen, A. 1999. Agricultural expansion and deforestation: modeling the impact of population, market forces and property rights. *Journal of Development Economics* 58: 185-218.

Angelsen, A. 2006. A *stylized model of incentives to convert, maintain or establish forest.* Background Paper for World Bank Policy Research Report entitled "At Loggerheads: Agricultural Expansion, Poverty reduction and Environment in the tropical forests- 2007".

Angelsen, A. and Kaimowitz, D. 2001. *Agricultural technologies and tropical deforestation.* CABI Publishing, Wallingford, United Kingdom.

Angelsen, A.; Shitindi, E. F. K. and Aaarrestad, J. 1999. Why do farmers expand their land into forests? Theories and evidence from Tanzania. *Environment and Development Economics* 4: 313-31.

Anonymous. 1990a. *The Forest Resources of the Temperate Zones, Vol. II.* FAO, Rome.

Anonymous. 1990b. *Situation and outlook of the Forestry Sector in Indonesia, Vol. 1: Issues, findings and opportunities.* Ministry of Forestry, Government of Indonesia; FAO, Jakarta.

Anonymous. 1990c. *Indonesia: sustainable development of forests, land and water.* The World Bank, Washington DC.

Anonymous. 1990d. *Rainforest destruction: causes, effects and false solutions.* World Rainforest Movement, Penang Malaysia.

Anonymous. 1991. *The Forest Sector.* The World Bank, Washington DC.

Anonymous. 1992a. *Forest Products: Yearbook 1991.* FAO, Rome.

Anonymous. 1992b. *Violated Trust: Disregard for the Forests and Forests Laws of Indonesia.* The Indonesian Environmental Forum (WALHI), Jakarta Indonesia.

Anonymous. 1994a. Deforestation Technical Support Package. Third International Conference on Environment Enforcement, Oaxaca Mexico April 25-28, 1994. World Wildlife Fund; U. S. Environmental Protection Agency and U. S. Agency for International Development.

Anonymous. 1994b. Breaking the logjam: obstacles to forestry policy reform in Indonesia and the United States. World Resource Institute, Washington.

Anonymous. 1994c. *Indonesia: Environment and Development.* The World Bank, Washington DC.

Anonymous, 2001a. *Global Forest Resources Assessment 2000-Main Report.* FAO Forestry Paper 140. Rome, Italy.

Anonymous. 2001b. State of the World's Forest 2001. FAO, Rome Italy.

Anonymous 2001c. *Climate Change 2001: Synthesis Report. Contribution of working groups I, II, III to the 3rd assessment report of the IPCC.* IPCC, Cambridge University Press, Cambridge.

Anonymous. 2002. *Forest certification and biodiversity: opposites or complements?* Discussion paper prepared for the GEF, International Tropical Timber Organization, Yokohama Japan.

Anonymous. 2005. *Ecosystems and Human well-being: synthesis*. Millennium ecosystem Assessment. Island Press, Washington DC.

Anonymous. 2007. Three Essential Strategies for Reducing Deforestation. Alianca da Terra, Amigos da Terra, Instituto Centro de Vida, IMAZON, Instituto de Pesquisa da Amazonia, Instituto Socio Ambiental, Nucleo de Estudos e Pratica Juridica Ambiental, Faculdade de Direito- Universidade Federal de Mato Grosso, Woods Hole Research Center and David and Lucile Packard Foundation.

Anonymous. 2008. Mexico: Mining causes ecocide in Coahuayana, Michoacan. *WRM's bulletin N° 136*, November 2008.

Anonymous, 2010. *Global Forest Resources Assessment, 2010-Main Report*. FAO Forestry Paper 163. Rome, Italy. 340p.

Anonymous, 2011$_a$. *State of the World's Forest*. FAO, Rome. 163p.

Anonymous. 2011$_b$. Mining, deforestation cause severe drought. March 5 *The Asian Age*, New Delhi.

Ascher, W. 1993. *Political economy and problematic forestry policies in Indonesia: obstacles to incorporating sound economics and science*. The Center for Tropical Conservation, Duke University.

Asdrasko, K. 1990. *Climate Change and Global Forests: Current Knowledge of Political Effects, Adaptation and Mitigation Options*. FAO, Rome.

Aylward, B. 2005. Land use, hydrological function and economic valuation. In: *Forest, water and people in the humid tropics*, eds. Bonell, M. and Bruijnzeel, L. A. Cambridge University Press, Cambridge United Kingdom.

Barbier, E. B. and Cox, M. 2004. An economic analysis of Shrimp farm expansion and mangrove conversion in Thailand. *Land Economics* 80: 389-407.

Barbier, E. B.; Burgess, J. C. and Folke, C. 1994. *Paradise lost? The ecological economics of biodiversity*. Earthscan.

Barbier, E. B.; Bockstael, N.; Burgess, J. C.; and Strand, I. 1993. *The timber trade and tropical deforestation in Indonesia*. LEEC Paper DP 93-01. Environmental Economics Centre, London.

Barraclough, S. and Ghimire, K. B. 2000. *Agricultural Expansion and Tropical Deforestation*. Earthscan.

Bawa, K. S.; Nadkarni, N.; Lele, S.; Raven, P.; Janzen, A.; Lugo, A.; Ashton, P. and Lovejoy, T. 2004. Tropical ecosystems into the 21st Century. *Science* 306: 227-230.

Bhatnagar, P. 1991. *The Problem of Afforestation in India*. International Book Distributors, Dehra Dun.

Bruijnzeel, L. A. 2004. Hydrological functions of tropical forests: not seeing the soils for the trees? *Agriculture, Ecosystems and Environment* 104: 185-228.

Bruijnzeel, L. A.; Bonell, M.; Gilmour, D. A. and Lamb, D. 2005. Forest, water and people in the humid tropics: an emerging view. In: *Forest, Water and People in the humid tropics*, eds. Bonell, M. and Bruijnzeel, L. A. Cambridge University Pres, Cambridge United Kingdom.

Bryant, D.; Nielsen, D. and Tangley, L. 1997. *The last frontier forests- Ecosystems and Economies on the Edge*. World Resource Institute, Washington DC.

Capistrano, A. D. 1994. Tropical forest depletion and the changing macroeconomy 1967-85. In: *The Causes of Tropical of Tropical Deforestation. The economic and statistical analysis of factors giving rise to the loss of the tropical forest*, eds. Brown, K. and Pearce, D. pp 65-85. UCL Press.

Carvalho, G.; Barros, A. C.; Moutinho, P. and Nepstad, D. 2001. Sensitive development could protect Amazonia instead of destroying it. *Nature* 409: 131.

Carvalho, G.; Moutinho, P.; Nepstad, D.; Mattos, L. and Santilli, M. 2004. An Amazon perspective on the forest-climate connection: opportunity for climate mitigation, conservation and development? Environment, *Development and Sustainability* 6: 163-174.

Carvalho, G.; Nepstad, D.; McGrath, D.; Del Carmen Vera Diaz, M.; Suntilli, M. and Barros, A. C. 2002. Frontier expansion in the Amazon: balancing development and sustainability. *Environment* 44: 34-42.

Caviglia, J. 1999. *Sustainable Agriculture in Brazil. Economic Development and Deforestation.* Edward Elgar.

Chakravarty, S.; Shukla, G.; Malla, S. and Suresh, C. P. 2008. Farmer's rights in conserving plant biodiversity with special reference to North-east India. *Journal of Intellectual Property Rights* 13: 225-233.

Charney, J. G. 1975. Dynamics of deserts and drought in the Sahel. *Quarterly Journal of Royal Meteorological Society* 101: 193-202.

Chomitz, K. M. and Griffiths, C. 1996. *Deforestation, shifting cultivation and tree crops in Indonesia: nationwide patterns of smallholder agriculture at the forest frontier.* Research Project on Social and Environmental Consequences of Growth-Oriented Policies, Working Paper 4. World Bank, Washington DC.

Chomitz, K. M.; Buys, P.; Luca, G. D.; Thomas, T. S. and Wertz-Kanounnikoff, S. 2007. *At loggerheads? Agricultural expansion, poverty reduction and environment in the tropical forests.* World Bank Policy Research Report. World Bank, Washington DC.

Colchester, M. And Lohmann, L. 1993. *The Struggle for land and the fate of forest.* Zed books, London.

Contreras-Hermosilla, A. 2000. *The underlying causes of forest decline.* CIFOR Occasional Paper No. 30. CIFOR, Bogor, Indonesia.

Contreras-Hermosilla, A. 2001. Illegal activities and corruption in the forest sector. In: *State of the World's Forest 2001*, ed. FAO. Pp 76-89. FAO, Rome.

Cossalter, C. and Pye-Smith, C. 2003. *Fast-wood forestry, myths and realities.* CIFOR, Bogor Indonesia.

Cropper, M and Griffiths, C. 1994. The interaction of population growth and environmental qualiy. *American Economic Review* 84: 250-254.

Dauvergne, P. 1994. The politics of deforestation in Indonesia. *Pacific Affairs* 66: 497-518.

Deacon, R. T. 1999. Deforestation and ownership: evidence from historical accounts and contemporary data. *Land Economics* 75: 341-359.

Dickinson, R. E. 1981. *Effects of tropical deforestation on climate. In blowing in the wind, deforestation and language implications. Studies in third world societies*, Publ. No. 14. Dept of Anthropology, College of William and Mary, Williamsburg Virginia. Pp 411-441.

Dick., J. 1991. *Forest land use, forest use zonation and deforestation in Indonesia: a summary and interpretation of existing information.* Background paper to UNCED for the state

Ministry for Population and Environment (KLH) and the Environmental Impact Management Agency (BAPEDAL).

Docena, H. 2010. Philippines: Deforestation through mining subsidized by CDM project. *WRM's bulletin N° 161*, December 2010.

Dove, M. R. 1993. Smallholder rubber and swidden agriculture in Borneo: a sustainable adaptation to the ecology and economy of the tropical forest. *Economic Botany* 47: 136-147.

Dove, M. R. 1996. So far from power, so near to the forest: a structural analysis of gain and blame in tropical development. In: *Borneo in transition: people, forests, conservation and development*, eds. Padoch, C. and Peluso, N. L. Pp 41-58. Oxford University Press, Kuala Lampur.

Dregne, H. E. 1983. *Desertification of Arid lands*. Harwood Academia Publishers, London.

Dudley, N. and Stolton, S. 2003. *Running Pure*. World Bank and WWF, Washington DC.

Ehrhardt-Martinez, K. 1998. Social determinants of deforestation in developing countries: A cross-national study. *Social Forces* 77: 567-586.

Fearnside, P. and Barbosa, R. 1996. The Cotingo dam as a test of Brazil's system for evaluating proposed developments in Amazonia. *Environmental Management* 20: 631-648.

Griffiths, T. and Hirvelä, V. V. 2008. India: Illegal aluminum refinery in Tribal lands in Orissa. *WRM's bulletin N° 126*, January 2008.

Gupta, A.; Thapliyal, P. K.; Pal, P. K. and Joshi, P. C. 2005. Impact of deforestation on Indian monsoon- A GCM sensitivity study. *Journal of Indian Geophysical Union* 9: 97-104.

Hays, J. 2008. Deforestation and desertification in China.
http://factsanddetails.com/china.php?itemid=389&catid=10&s ubcatid=66

Hansen, C. P. 1997. *Making available information on the conservation and utilization of forest genetic resources*. The FAO Worldwide Information System on Forest genetic resources.

Houghton, R. A. 2005. Tropical deforestation as a source of greenhouse gas emissions. In: *Tropical deforestation and Climate change*, eds. Moutinho, P. and Schwartzman, S. Pp 13-20. Amazon Institute for Environmental Research, Belem Brazil.

Humphreys, D. 2006. *Forest Politics*. Earthscan Publications Ltd., London.

Kaimowitz, D. and Angelsen, A. 1998. *Economic models of tropical deforestation. A review.* Center for International Forestry Research, Bogor Indonesia.

Kant, S. and Redantz, A. 1997. An econometric model of tropical deforestation. *Journal of Forestry Economics* 3: 51-86.

Kartodihardjo, H. and Supriono, A. 2000. *The impact of sectoral development on natural forest conversion and degradation: the case of timber and tree crop plantations*. CIFOR Occasional Paper No. 26, Bogor Indonesia.

Lambin, E. F. and Meyfroidt, P. 2011. Global land use change, economic globalization, and the looming land scarcity. *PNAS* 108: 3465-3472.

Lawton, R. O.; Nair, U. S.; Pielke Sr., R. A. and Welch, R. M. 2001. Climatic impact of tropical lowland deforestation on nearby Montane Cloud Forests. *Science* 294: 584-587.

Lean, G. 1990. *World Wildlife Fund Atlas of the Environment*. Prentice Hall, New York.

Levang, P. 2002. People's dependencies on forests. In *Technical Report, Phase I 1997-2001. ITTO Project PD 12/97 Rev. 1 (F) - Forest, Science and Sustainability: the Bulungan model forest*, pp 109-130. CIFOR, Bogor Indonesia.

Mahar, D. 1988. *Government policies and Deforestation in Brazil's Amazon region*. World Bank, Washington DC.

Mangave, H. R. 2004. *A study of Elephant population and its habitats in the northern West Bengal, North East India*. M. Sc. Thesis, Bharathidasan University. Unpubl.

Mather, A. S. 1991. *Global Forest Resources*. International Book Distributors, Dehra Dun.

Mullick, B. and Griffiths, T. 2007. India: Indigenous movement in Jharkhand challenge plans for industrial development that threatens to destroy Adivasi forests, farmlands and way of life. *WRM's bulletin N° 116*, March 2007.

Myers, N. 1992. *The Primary Source: Tropical Forests and Our Future*. Norton, New York.

Myers, N. 1994. Tropical deforestation: rates and patterns. In: *The Causes of Tropical of Tropical Deforestation. The economic and statistical analysis of factors giving rise to the loss of the tropical forest*, eds. Brown, K. and Pearce, D. pp 27-40. UCL Press.

Myers, N. and Mittermeier, R. A. 2000. Biodiversity hotspots for conservation priorities. *Nature* 403: 853-854.

Nepstad, D. C.; Carvalho, G.; Barros, A. C.; Alencar, A.; Capobianco, J. P.; Bishop, J.; Moutinho, P.; Lefebvre, P.; Lopes Silva, Jr. U. and Prins, E. 2001. Road paving, fire regime and the future of Amazon forests. *Forest Ecology and Management* 154: 395-407.

Nepstad, D. C.; Schwartzmann, S.; Bamberger, B.; Santilli, M.; Ray, D.; Schlesinger, P.; Lefebvre, P.; Alencar, A.; Prinz, E.; Fiske, G. and Rolla, A. 2006. Inhibition of Amazon deforestation and fire by parks and indigenous lands. *Conservation Biology* 20: 65-73.

Panayotou, T. 1990. *The economics of environmental degradation: problems, causes and responses*, HIID Development discussion papers 335. Harvard University.

Pearce, D. and Brown, K. 1994. Saving the world's tropical forests. In: *The Causes of Tropical of Tropical Deforestation. The economic and statistical analysis of factors giving rise to the loss of the tropical forest*, eds. Brown, K. and Pearce, D. pp 2-26. UCL Press.

Pinker, R. 1980. The microclimate of a dry tropical forest. *Agricultural Meteorology* 22: 249-265.

Porter, G. 1994. The environmental hazards of Asia Pacific development: the Southeast Asian Rainforests. *Current History* 93: 430-434.

Purnamasari, R. S. 2010. Dynamics of small-scale deforestation in Indonesia: examining the effects of poverty and socio-economic development. *Unasylva* 61: 14-20.

Putz, F. E.; Blate, G. M.; Redford, K. H.; Fimbel, R. and Robinson, J. 2001. Tropical forest management and conservation of biodiversity: An overview. *Conservation Biology* 15: 7-20.

Raloff, J. 1989. Where Acids Reign. *Science News* July 22. Pp 56-58.

Repetto, R. 1988. *The forest for the trees? Government policies and the misuse of forest resources*. World Resource Institute, Washington DC.

Repetto, R. 1990. *Deforestation in the Tropics*. Scientific American April, p. 37.

Ross, 1996. Conditionality and logging reform in the tropics. In: *Institutions for Environmental Aid: Problems and Prospects*, eds. Keohane, R. O. and Leve, M A. Pp 167-197. MIT Press, Cambridge Massachusetts.

Rowe, R.; Sharma, N. P. and Bowder, J. 1992. Deforestation: problems, causes and concern. In: *Managing the world's forest: looking for balance between conservation and development*, ed. Sharma, N. P. Pp 33-46. Kendall/Hunt Publishing Company, Iowa.

Rowntree, P. R. 1988. Review of General Circulation Models as a basis for predicting the effects of vegetation change on climate. In: *Forests, climate and hydrology, regional impacts*, eds. Reynolds, E. R. C. and Thompson, F. B., Pp 162-196. The United Nations University, Tokyo Japan.

Sands, R. 2005. *Forestry in a Global Context*. CABI Publishing.

Schmink, M. and Wood, C. 1992. *Contested Frontiers in Amazonia*. Columbia University Press, New York.

Schneider, R. R. 1995. *Government and economy on the Amazon frontier*. Environment Paper 11. World Bank, Washington DC.

Scrieciu, S. S. 2003. *Economic causes of tropical deforestation- a global empirical application*. Development Economics and Public Policy Working Paper 4. Institute of Development Policy and Management, University f Manchester.

Sheram, K. 1993. *The Environmental Data Book*. The World Bank, Washington DC.

Shukla, G. 2010. *Biomass production and vegetation analysis of Chilapatta Reserve Forest Ecosystem of West Bengal*. Ph. D. Thesis, Uttar Banga Krishi Viswavidyalaya, Pundibari. Unpublished.

Southgate, D. 1994. Tropical deforestation and agricultural development in Latin America. In: In: *The Causes of Tropical of Tropical Deforestation. The economic and statistical analysis of factors giving rise to the loss of the tropical forest*, eds. Brown, K. and Pearce, D. pp 134-43. UCL Press.

Staff, S. 2010. Mining deforestation nearly tripled between 2000-08. *Archives*, Wednesday October 13 2010.

Sukumar, R.; Baskaran, N.; Dharmrajan, G.; Roy, M.; Suresh, H. S. and Narendran, K. 2003. *Study of the elephants in Buxa Tiger Reserve and adjoining areas in northern West Bengal and preparation of Conservation Action Plan. Final Report*. Center for Ecological Sciences, Indian Institute of Science, Bangalore.

Sunderlin, W. D.; Resosudarmo, I. A. P.; Rianto, E. and Angelsen, A. 2000. *The effect of Indonesia's economic crisis on small farmers and natural forest cover in the outer islands*. CIFOR Occasional Paper No. 28 (E). CIFOR, Bogor Indonesia.

Thiele, R. 1994. How to manage tropical forests: the case of Indonesia. *Intereconomics* 29: 184-193.

Van Kooten, G. C. and Bulte, E. H. 2000. *The economics of nature: managing biological assets*. Blackwells.

Van Noordwijk, M.; Agus, F.; Verbist, B.; Hairiah, K. and Tomich, T. P. 2006. Managing watershed services in ecoagriculture land-scapes. In: *The State-of-the-Art of Ecoagriculture*, eds. McNeely, J. A. and Scherr, S. J. Island Press, Washington DC.

Wilkie, D.; Shaw, E.; Rotberg, F.; Morelli, G. and Auzels, P. 2000. Roads, development and conservation in the Congo Basin. *Conservation Biology* 14:1614-1622.

Yin, H. and Li, C. 2001. Human impacts on floods and flood disasters on the Yangtze River. *Geomorphology* 41: 105-109.

2

Depletion of Oak (*Quercus* spp.) Forests in the Western Himalaya: Grazing, Fuelwood and Fodder Collection

Gajendra Singh[1] and G. S. Rawat[2]
[1]Department of Habitat Ecology, Wildlife Institute of India, Dehradun
[2]Deputy Programme Manager / Senior Scientist (ECES), ICIMOD, Kathmandu
[1]India
[2]Nepal

1. Introduction

The Himalaya, youngest mountain range of the world covers about 18% of total geographical area of India. Forests constitute (50% of India's forest cover) an important natural resource base in the Himalaya, most important being the temperate broad leaf forests, which are largely dominated by different species of oak (*Quercus* species). In India Rodgers & Panwar (1988) divided the Indian Himalaya into six provinces *viz.*, Ladakh mountain, Tibetan plateau, North-West Himalaya, Western Himalaya, Central Himalaya and the Eastern Himalaya. The Western Himalaya comprises of the eastern part of Himachal Pradesh and the state of Uttarakhand between the rivers Satluj and Sharada. Oaks (*Quercus* spp.) are the dominant, climax tree species of the moist temperate forests of the Indian Himalayan region (Troup, 1921) where about 35 species of *Quercus* are extensively distributed between 1000-3500 m elevations. Five species of evergreen oak namely *Quercus glauca* (phaliyant/harinj), *Q. leucotrichophora* (banj), *Q. lanuginosa* (rianj), *Q. floribunda* (tilonj/moru) and *Q. semecarpifolia* (brown/kharsu) grow naturally in the western Himalaya.

In the Western Himalaya, oak species assume considerable conservation significance as they are providers of numerous ecosystem services (conservation of soil, water, native flora and fauna) and serve as lifeline for the local communities. Predominantly three oak species (*Quercus leucotrichophora, Q. floribunda and Q. semecarpifolia*) are intricately associated not only with agro-ecosystems but also with the life support system of the inhabitants of the hills in this area. The oak forests are source of fuelwood, fodder and can be correlated with natural springs and wildlife (Singh, 1981).

The Himalaya is the home of many unique and diverse human groups, living in the river valleys and mountain slopes which differ from each other in terms of language, culture, tradition, religion and patterns of resource use. They have been subsisting on the Himalayan natural resources for millennia. In recent few decades, with better access to global market and demand of socio-economic development, local people's dependence on natural resources has increased immensely. Steady increase in human population, tourism, over exploitation, widespread logging, overgrazing, removal of leaf and wood litter from the forests floor has been responsible for the forest degradation in the region (Gorrie, 1937; Chaturvedi, 1985). Other than excessive exploitation by the local communities, replacement of oak by pine (Singh et al., 1984) has become a common and ever increasing phenomenon in the Western Himalayan region. These forests have been burnt from time to time by the local inhabitants in order to encourage growth of grasses and to increase the preponderance of the fire-resistant commercial species (*e.g.*, Chir pine), at the expense of oak species. All the gentle and accessible meadows in the temperate, alpine and sub-alpine regions have undergone extensive habitat degradation, with over 70% of the natural vegetation reported to have been lost (Singh, 1991). Ives and Messerli (1989) called this explanation "overly simplistic" and have named it the "Theory of Himalayan Environmental Degradation".

In this chapter we are mainly concerned with three major forms of activities that have affected the western Himalayan oak, grazing, fuelwood and fodder collection. All these three activities are mainly linked to subsistence living of the local population of the hills. We provide local data on these parameters from Kedarnath Wildlife sanctuary, Western Himalaya.

2. Material and methods

2.1 Study area

The study was conducted in the southern fringe of Kedarnath Wildlife Sanctuary, Chamoli-Rudraprayag district in Western Himalaya. The area is characterized by undulating topography, wide variation in the altitude, rain fall, temperature and soil conditions. The area is an important wintering range of several high altitude animals and is used by a large number of local agro-pastoral and migratory community (gujjars) besides tourists and pilgrims during summer. The intensive study area (~975 km^2) covers a wide altitudinal range from 1500-3680 m asl with the varying mean annual temperature range (-4 to 34^{0}C). About 182 villages are distributed around Kedarnath WS, of which about 50 are close to the wildlife area. The main pressure on the protected area (PA) is in the summer season involving grazers like sheep, goats and buffaloes along with the load of pilgrimage and collection of fuelwood and fodder. About twenty villages situated close to the sanctuary were selected for the present study. Livestock rearing and tourism are the main landuse practices across different altitudinal zones in the region.

2.2 Data collection and analysis

A field survey was carried out from 2006 to 2010, to study the status of plant diversity along the altitudinal and human use gradient across various oak forests. Based on extensive reconnaissance survey and dominance, three types of forests, *Q. leucotrichophora* (1200-2200 m asl), *Q. floribunda* (2201-2700 m asl) and *Q. semecarpifolia* (2701-3300 m asl) were selected for the vegetation study. There were around 16 camping sites present in the study area from where data was collected. These camps are regularly visited by the pastoral communities every year, therefore an attempt was made to understand the impact of grazing and camping on the forest cover. Geometrically corrected Landsat data for the year 1976 (MSS), 1990 (TM) and 1999 (ETM+) were used to focus on the grazing pressure on the forested area over the period. All the three forests had been categorised into different disturbance categories, (i.e. Undisturbed (UD), Moderately disturbed (MD), and Highly disturbed (HD), with the help of the disturbance index (Kanzaki and Kyoji 1986). Canopy cover and grazing by livestock/year is among the most important parameters in the measurement of the disturbance, and were measured by a densitometer (GRS densitometer) and total livestock count respectively. Basal area (area occupied by the base of a tree) is a good indicator of the size and volume or weight of a tree. Girth of cut stumps was measured at ground level and basal area for cut stumps was calculated. Cut stump index was calculated on the basis of the ratio of basal area of cut stumps to the total stand basal area including felled ones.

Stratified random transects with 10 plots at every 200 m interval depending upon the accessibility, were laid in all three types of forests along the altitudinal and anthropogenic pressure gradients. 10 m radius (314 m^2 area) plot for tree species with concentrated plot of 5 m radius (78.5 m^2 area) for shrubs and regenerating individuals within the larger plot was laid. For the ground layer *i.e.,* grasses and herbs 1m × 1m quadrats were laid one each in East, West, North and South directions. In each stipulated plot, name of the species, number, circumference at breast height (CBH, 1.37m) and canopy cover were measured for trees and only number of individuals were recorded for the shrub species. Plants, >30 cm cbh and >3 m straight bole were considered as trees, regenerating individuals of tree species between 10-30 cm cbh were considered as sapling and <10 cm cbh to one or two leaf stage individuals were considered as seedlings. Woody species which had cbh < 20 cm, height < 3 m and those had several branching from base of the stem were considered as shrubs. Transects were laid along pathways and streams in forests in spatially distributed pattern, so as to minimize the autocorrelation in the vegetation. Anthropogenic pressure was assessed in terms of canopy cover (%), cut stumps analysis, livestock availability/year and fuelwood collection. Since fuelwood/fodder collection and livestock grazing are the major cause of disturbance in these forests, total livestock unit available per year and total fuelwood/fodder collection of different forest stands had been used to quantify the intensity of disturbance.

Field data was analyzed both in regress (scientific purposes) and simple (convenience of local people and management authority) methods, where it was quantitatively analysed for frequency, density and basal area following the standard ecological methods. The Importance Value Index (IVI) for tree species was determined as the sum of relative frequency, relative density, and relative basal area. Species richness, the number of species per unit area, Shannon diversity index, and evenness index were also calculated. Chi-square test was performed to test whether the densities of trees in the different categories of forests are (<4 km, > 4 km and sanctuary area) significantly different.

For the estimation of fuelwood and fodder collection, informal interviews were taken in each village. Interviews revealed on an average, two people per day par household were involved in fuelwood and fodder collection. The total number of households in each village were multiplied by two to give an approximation of the total population responsible (TPR) for fuelwood and fodder collection in each village. The identification of major fuelwood and fodder species was mainly based on interviews, informal discussions and personal observations of the authors. The quantity of fuelwood and fodder collection was estimated over a period of 24 hours using a weighted survey method. Traditionally, the woodlot was weighed and left in the kitchen to be burnt and the actual fuelwood consumption was measured following 24 hours. Similarly, the fodder lot was weighed before keeping for the stall feeding and measured on a daily basis.

3. Results and discussion

3.1 Grazing in the oak forested area

Five distinct pastoral practices: (a) nomadic, (b) semi-nomadic, (c) nuclear transhumance, (d) trans-migratory and (e) sedentary have been reported in the area (Rawat, 2007). Few families from every village practice a kind of semi-pastoralism, with animal husbandry (sheep/goats) in higher elevations (2500-4500 m) as their predominant occupation although, over the period of time, several communities have changed their life style from nomadic to semi-nomadic wherein only few members of the family move along with their herds to high altitude areas. Other than that there are nomads, such as "Gujjars" who lead a pastoral life style, traversing almost the entire elevation range that supports oak forests (1200-3300 m asl.) in the western Himalaya to graze their animals (buffaloes, cows and mules).

Livestock mainly feed on small regenerating plants and in this process they gradually eliminate the understory vegetation of the forest. People, who carry livestock with them, meet their needs for fuelwood and timber for making huts and poles for fencing from the surrounding forests. As a result of regular lopping and logging, the canopy density continues to decrease, as younger plants are not available to occupy the open canopies.

3.2 Vegetation structure and regeneration status along grazing gradient

The forests within the 3-4 km distance from villages were frequently used by local communities throughout the year for grazing, fuelwood, and fodder collection. The forests situated away from the villages (>4 km) were occasionally used by the people when they go to higher elevations (sub-alpine and alpine) for grazing in summer and come down before winter. Sanctuary and nearby forests in cool temperate and sub-alpine zones are frequently grazed by livestock of migratory community (gujjar) as well as local people for 6 to 8 months. The forests away from the village (>4 km) significantly showed higher tree density (463 trees ha^{-1}) and regenerating individuals (seedlings 2490 individuals ha^{-1} and saplings 481 individuals ha^{-1}) in comparison with village forests and forests near sanctuary. Chi square test was performed to test whether the densities of trees in the different categories of forests are significant, the null hypothesis was rejected showing marked differences in densities of trees in the different categories of forest at 5% of difference (à=0.05). Shrub density in village forests was higher than away from village and nearer to the sanctuary forests. This is because excessive grazing supports high shrub cover of unpalatable species (Singh and Singh, 1992). High invasive species cover, low canopy and herbaceous cover in village forests indicates the high collection (lopping/logging) of available species (*Q. leucotrichophora*, *Daphniphyllum himalayense*, *Alnus nepalensis* etc.) for fuelwood, fodder and livestock grazing (Table 1).

Among the dominant tree species, *Q. leucotrichophora* (115 trees ha^{-1}) had maximum density in village forests, *Q. floribunda* (76 trees ha^{-1}) had highest density in distant forests (forests away from village) and *Q. semecarpifolia* (178 trees ha^{-1}) had maximum density in forests near the sanctuary (Table 2). *Sarcococca saligna* and *Daphne papyracea* were only shrubs which were frequently found in understory of all the three different forests.

Parameters	< 4 kms from village	>4 km from village	Near Sanctuary area
	n= 40	n=40	n=40
*Tree density ha^{-1} ± SE	408 ± 4.1	463 ± 2.9	386 ± 9.9
*Shrub density ha^{-1} ± SE	5420 ± 63.8	4885 ± 129.3	3936 ± 114.6
*Herb density ha^{-1} ± SE	67375 ± 300	89875 ± 340.4	561188 ± 2052.5
*Seedling density ha^{-1}± SE	917 ± 11.9	2490 ± 28.6	475 ± 17.2
*Sapling density ha^{-1}± SE	455 ± 4.9	481 ± 6.4	217 ± 4.9
% canopy cover (trees)	29.63	54.63	34.25
% Ground cover	4.15	11.63	45.38
% Herb species cover	42.5[a]	15.5[b]	35[c]
Diversity (H′)	3.30	3.1	2.93
Richness (R)	100	86	92

*significant at à=0.05 level, [a]*Eupatorium adenophorum* (invasive species), [b]*Strobilanthes atropurpureus*, [c]*Trachydium roylei*

Table 1. Comparison of vegetation at different distance from villages.

Parameters	< 4 kms from village (Village Forests)	>4 km from village (Distant Forests)	Forests near Sanctuary
	n= 40	n=40	n=40
Tree species			
Q.leucotrichophora	115	10	-
Q. floribunda	-	76	4
Q. semecarpifolia	-	14	178
R. arboreum	72	42	171
Lyonia ovalifolia	57	55	2
Daphniphyllum himalayense	32	3	-
Alnus nepalensis	35	8	-
Pyrus pashia	31	-	-
Shrub species			
Sinarundinaria falcata	1828	229	-
Chimnobambusa jaunasarensis	-	2634	-
Thamnocalamus falconeri	-	-	643
T. spathiflora	-	392	-
Sarcococca saligna	471	1172	2236
Daphne papyracea	16	153	38
Berberis lycium	382	-	-
Randia tetrasperma	672	-	-

Table 2. Density ha^{-1} of important trees and shrubs across the disturbance gradient

The GIS analysis of the study sites shows that the area of grazing land was initially 492.95 ha (1976) and it gradually increased to 601.496 ha (1990), 723.621 ha (1999) and 792.428 ha (2005). This clearly indicates that there has been a steady increase in the area of grazing land at the cost of forest cover around camping sites i.e., ca 108.54 ha during 1976–1990 (14 years), 122.49 ha during 1990–1999 (9 years) and 68.78 ha during 1999–2005 (6 years) (Thakur et al. 2011) (Fig.1).

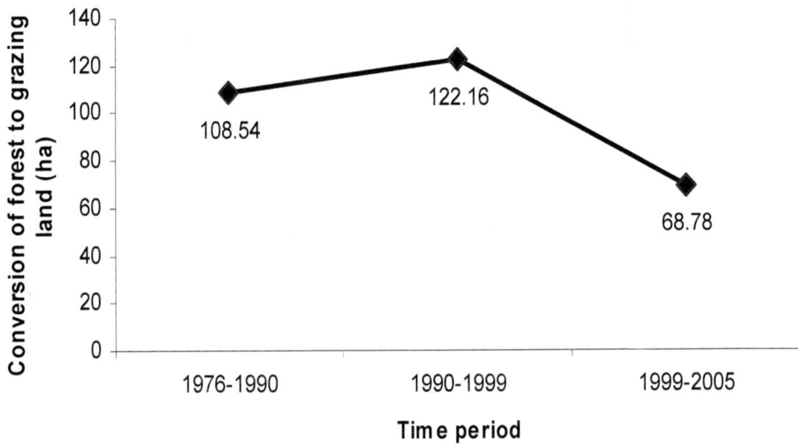

Fig. 1. Conversion of forest to grazing land across various time periods

3.3 Fuelwood consumption by villagers and temporary shops

Fuelwood is an important component of household economies in the Himalaya. In the western Himalaya, about 77.4% of the total human population is rural. Due to low connectivity with the urban areas of the country, the alternative sources of energy resources are not easily accessible hence making the population totally dependent on wood resources (Table 3) of the area. The information on fuelwood resources of the Himalaya is almost scattered. The issue as such has been addressed invariably but there has been no attempt to correlate the fuelwood consumption by villagers and migratory nomads (Gujjars). A total of 865 species of vascular plants (angiosperms, gymnosperms and pteridophytes) belonging to 479 genera and 131 families, were recorded from the region (Singh, 2008). According to growth habits these species were distributed across 501 herbs, 110 shrubs, 74 trees, 43 climbers, 70 graminoides and 61 ferns. Among all, 88 woody species (54 trees and 34 shrubs) are commonly used for fuelwood by the local people (Singh et al. 2010). The diversity of fuelwood use was depended on the species quality, accessibility and availability and also the human population of the adjoining villages (Singh and Singh, 1992).

The density of preferred woody species (for fuelwood and fodder) was highest in the distant forests (463 ± 2.9 ind/ha) followed by village forests (408± 4.1 ind/ha) and least in near sanctuary forests (386 ± 9.9 ind/ha). Seasonal variation of resources extracted from the forests was very much evident from the study. In permanent villages, fuelwood was either cut directly from the forests or dead wood of any kind was collected. Fuelwood consumption at permanent village did not vary across different seasons. The average fuelwood consumption was found to be 20-22 kg/day/household (Table 4a, 4b), where 40-60 species were used by the local communities. Whereas at temporary huts or shops, the fuelwood consumption was restricted between April to October, but due to high tourist season and low temperature at higher altitude, use of fuelwood was quite high compared to permanent residents of villages.

Zone	Important fuelwood species	Important fodder species
Warm temperate (1500-2200m)	*Caesalpinia decapetata, Daphniphyllum himalayense, Alnus nepalensis, Lyonia ovalifolia, Quercus leucotrichophora*	*Grewia optiva, Celtis australis, Ficus roxburghii, Ficus nemoralis, Ficus claveta, Quercus leucotrichophora, Quercus glauca, Sinarundinaria falcata, Chrysopogon gryllus*
Cool temperate (2200-2700m)	*Quercus semecarpifolia, Quercus floribunda, Lyonia ovalifolia, Rhododendron arboreum, Abies pindrow*	*Quercus floribunda, Quercus semecarpifolia,Thamnocalamus spathiflora,Thamnocalamus falconeri,Chimnobambussa jaunsarensis, Chrysopogon gryllus*
Sub-alpine (2700-3300m)	*Quercus semecarpifolia, Rhododendron campanulatum, Rhododendron arboreum*	*Quercus semecarpifolia, Thamnocalamus falconeri*

Table 3. Consumption of major fuelwood and fodder species at different altitudinal zone.

Village	# fuelwood colleting family	# population responsible for collection	# fuelwood collection (kg/day/h.hold[a])	# fuelwood collection (Q.[b]/year)
Siroli	62	124	20-22	2232-2482
Mandal	79	158	20-22	2844-3163
Khalla	53	106	20-22	1908-2122
Koteshwar	46	92	20-22	1656-1841
Bandwara	41	82	20-22	1476-1641
Bairangana	33	66	20-22	1188-1321

a= household, b= Quintal (1 quintal=100 Kilograms)

Table 4a. Fuelwood collection by villagers in the warm temperate region (Singh et al. 2010).

Village	# fuelwood colleting family	# population responsible for collection	# fuelwood collection (kg/day/h.hold[a])	# fuelwood collection (Q.[b]/year)
Dadon	13	26	23-25	538-585
Nail	75	150	23-25	3105-3375
Nauli	40	80	23-25	1656-1800
Kalsir	60	120	23-25	2484-2700

a= household, b= Quintal (1 quintal=100 Kilograms)

Table 4b. Fuelwood collection by villagers in the cool temperate region (Singh et al. 2010).

Fourteen temporary dhabas (hotels) are established every year in the cool temperate region at Chopta (2900m) for 6-8 month during tourist season. Out of 14 dhabas, 12 dhabas consumed fuelwood on an average of 0.9 Q/day/dhaba and 2 dhabas consumed average of 3 Q/day/dhaba of fuelwood. This consumption of fuelwood continued up to 6 months (May to October), during the remaining two months (April and November) 2 dhabas consumed 1.2 Q/day/dhaba fuelwood. The requirement of fuelwood is fulfilled mostly from their respective village forests and partially from the distant forests (Table 5) by only 10-20 species.

In the sub-alpine and alpine region near the famous Hindu shrine "Tungnath" 10 dhabas used fuelwood from nearby forests for six months. Each dhaba owner consumed on average 30 kg/day fuelwood. *Rhododendron campanulatum, Q. semecarpifolia* and *R. arboreum* trees were the main source of fuelwood. The fuelwood consumption by dhaba owners is significantly higher ($p<00.2$) than the fuelwood consumed by a village in the region.

Areas	# Dhabas	# fuelwood collection kg/day/dhaba	# fuelwood collection Q./year	# fuelwood species collected
Cool temperate region	14	90-120	3096.4	20
Sub-alpine region	10	30	540	11

Table 5. Fuelwood collection by dhaba owners in the cool temperate region.

Fodder collection: Over 146 species at different altitudinal zone were used as fodder, among them grasses had the highest contribution (58 species). At permanent villages, most of the livestock remain at home. These were strictly stall-fed throughout the year. Few cattle go into the nearby barren fields or forests for approximately one month every year for grazing. Fodder demands were basically fulfilled through leaf fodder, agricultural by-products and grass fodder (green and dried). There was a seasonal variation in the type of fodder collected. Leaf fodder available in the villages constitutes the major fodder for cattle throughout the year except during the rainy season when green grasses and legumes were used. The leaf fodder (*Q. leucotrichophora* and *S. falcata*) lopped from nearby forests was used from October to May. Agricultural by-products which constituted the staple fodder in winter was collected from May to August. In the permanent village, each family had 3-5 stall fed livestock with fodder requirement of 25-30 kg/day/livestock. Mandal village had maximum fodder requirement per year followed by Siroli and Bandwara according to their village size (Table 6).

Out of 146 species reported as fodder from the study area, 13 species of leaf-fodder were largely used / harvested in the villages. The preferred fodder species in different seasons in the permanent villages are different (Figure 2). From December to May, broadleaf species such as *Grewia optiva, Ficus roxburghii, Q. leucotrichophora* and *Sinarundinaria falcata* constitute major fodder species. In June-September few broad leaf species *(Celtis australis* and *Ficus nemoralis)*, agricultural by-products and grasses fulfil the major requirements of fodder. From October to December, large amount of grasses are collected from the nearby protected grassy slopes and used as a good quality fodder.

Villages	Livestock population	# Availability of fodder in villages (Q/Y)	# Fodder req./year (Q/Y)	# Fodder dependency on forests (Q/Y)
Siroli	146	12204-15224	23268- 27922	11064-12682
Mandal	118	15157-18953	27648- 33178	12491-14225
Khalla	106	9128-11388	16333- 19600	7205-8212
Koteshwar	123	8462-10549	13961- 16753	5499-6204
Bandwara	152	8199-10206	18250- 21900	10051-11694
Bairagana	91	6852-8552	13596- 16315	6744-7766

Table 6. Availability and requirement of fodder (Quintal) per year in the villages.

Few grass species *i.e., Chrysopogon gryllus* (Khor) and *Pennisetum purpureum* (Napier) are cultivated by the villagers in their respective barren fields. Each village has its appointed portion of hillside as its hay preserve. The grass is cut in October- November, dried for sometime and brought home or if trees are very near, hey is hung up on tree branches in wisps to dry. Nowadays, due to increase in population, decrease in forest area and introduction of high bred cattle, people from the valley are facing shortage of fodder at times. It is not an acute problem in the region but improvement of forest blanks near villages and pasture land by using good quality seeds of grasses and trees can help in overcoming problem of fodder scarcity in the area.

Botanical name	Local name	J	F	M	A	M	J	J	A	S	O	N	D
Celtis australis	Kharik						■	■					
Grewia optiva	Bheemal	■	■	■	■								■
Ficus roxburghii	Timla	■	■	■									■
Ficus nemoralis	Thelka						■	■	■				
Ficus clavata	Chanchara	■	■	■									
Q. leucotrichophora	Banj	■	■	■	■	■						■	
S. falcata	Ringal	■	■	■	■	■					■	■	
Chrysopogon gryllus	Khor		■	■	■							■	
Agriculture fodder						■	■	■	■				

Fig. 2. Timing for collection of major fodder species across different months.

3.4 Impact of fuelwood and fodder collection on the village and sanctuary forests

The woody species such as *Q. leucotrichophora, R. arboreum, P. pashia, L. ovalifolia* and *A. nepalensis* fulfil the requirements of fodder and fuelwood in the villages whereas *Q. floribunda, Q. semecarpifolia, R. arboreum* and *R. campanulatum* were used as fuelwood near sanctuary areas. Density of trees, shrubs and saplings were found to be significantly higher in less disturbed village forests as well as sanctuary forests. Whereas, the seedling density was found to be higher in highly disturbed forests with low canopy cover. The ground cover was higher in highly disturbed forests in both the regions, where most of the ground was covered by weedy species. The percentage of lopped trees in village forests (53%) was higher than the sanctuary forests (37%). The village forests species were lopped for fodder as well as fuelwood but in near sanctuary forests, species were mainly lopped for fuelwood requirements (Table 7).

Parameters	Village forests (1700m)		Sanctuary forests (2900m)	
	High collection	Less collection	High collection	Less collection
Tree density/ha	365 ± 8.54	537 ± 8.78	221 ± 12.09	465 ± 32.48
Shrub density/ha	4796 ± 67.52	5195 ± 180.53	3455 ± 130.98	4117 ± 102.01
Species diversity	2.98	2.67	1.89	2.19
Seedling density/ha	2810 ± 86.11	1341± 25.26	696 ± 40.55	263
Sapling density/ha	526 ± 14.41	645 ± 13.87	212 ± 12.99	441 ± 19.16
No. of lopped tree/ha	195	42	83	2
Canopy cover (%)	21.67	47.67	25	41.66
Ground cover (%)	8.8	4.67	64.66	27

Table 7. Comparison of various parameters in village forests and sanctuary forests

4. Impact of grazing

In some pastures, grazing is essential to maintain species diversity (Naithani *et al.* 1992, Negi *et al.* 1993, Saberwal 1996) mainly in the Himalayan alpine region, but some other studies (Ram *et al.* 1989, Sundariyal and Joshi 1990, Singh 1991, Rawat and Uniyal 1993 and Kala 1998) revealed that intermediate level of grazing maintains species diversity. Besides, livestock grazing abiotic factors like snowfall, soil, altitudes and aspects also influence the structure and composition of these pastures (Kala *et al.* 1995).

Presently, it is estimated that more than 15,000 sheep and goats, ~ 2000 buffalos and about ~8000 mules are known to graze in temperate and alpine region of entire Kedarnath WS (Singh 2008). In the present investigation about 900 sheep and goats and ~1100 buffaloes/cows graze temperate to alpine region of the study area. The investigation shows that grazing areas (migratory summer camp = 19) were scattered in almost all altitudes, slopes and aspects. High species density and low diversity in highly grazed areas also indicated the dominance of few grazing indicator species (*Trachydium roylei*). The less disturbed areas were mainly dominated by *Danthonia cachymeriana* and the other herbaceous species had low density and high species richness. Highly grazed areas have shown low above ground biomass (Nautiyal *et al.* 2001), and high density of the species *i.e., Trachydium roylei, Oxygraphis polypetala, Anemone rivularis, Taraxacum officinale and Carex* spp. Camping sites were found to be dominated with opportunistic herbs e.g., *Rumex nepalensis, Polygonum amplexicaule, P. polystachyum* and *Impatiens sulcata* (Singh 1999).

In the present study, it was observed that dependency of villagers for their basic needs in the nearby forests may be sustainable but the use of sub-alpine and alpine forests for livestock grazing and tourism are the major causes of forest degradation. Reserved Forests near villages that possessed small village forests were often badly degraded, while those located at some distance away from villages are in better condition with lower level of disturbance. Maithani (1994) reported a village forest in Gopeshwar (Chamoli) district which has been managed for decades and is "the best protected forest in the area". Such village forests which are closely monitored by the Village Panchayats might have a better regeneration status than the village forests that were observed during our study. Oak regeneration appears to be benefited by moderate levels of disturbance which resulted in the partial opening of the canopy. This is also supported by Rao and Singh (1989), Thadani and Ashton (1995) in the Central Himalayan oak forests *(Quercus leucotrichophora)* and Quintana-Ascencio et al. (1992) in Mexican highlands *(Quercus crispipilis)*. All the studies concluded that these species are unable to regenerate under deep shade, and require open patches relatively free from browsing and trampling by ungulates.

5. Conservation and management approach

Forest resources and agricultural products are the main resources of livelihood for the local people in the study area but these resources seem to be degrading day-by-day. Conservation of such areas cannot be achieved without the involvement of local communities who are directly dependent on these resources for their daily needs. For the long term conservation of forests and other natural resources local people and forest department can help in following ways:

1. Traditional management of grassy slopes: Certain village pastures are closed for grazing during rainy season (July-September) when flowering and fruiting of these species take place. These pastures are opened for fodder collection after peak grazing season when seeds are dispersed for new generation. This practice does not only meet the fodder requirement of local people but also helps in the regeneration of grassy slopes. These tussock forming native grasses *e.g.,* Khor, Napier should be promoted at large scale by the local people and plantation of these species in barren and open pastures need to be undertaken. These species are not only used as fodder but also used for thatching roofs of temporary huts.

2. Protection and sustainable utilization of important fodder tree species (*Ficus roxburghii, Ficus claveta, Ficus nemoralis, Celtis australis* and *Grewia optiva*) in the fringes of agriculture and barren fields help in the conservation of neighbouring village forests and should be promoted. Plantation of these species should be encouraged at large scale by the Forest Department and NGOs. The seedlings and saplings of these species should be distributed to local people at subsidized rates.

3. Fuelwood consumption by 'Dhaba' owners at Chopta (14 numbers) and Tungnath (10 numbers) for six months is higher than the annual consumption of fuelwood by an

average village in the Mandal valley. Therefore, for the proper management of fuelwood at these shops, the Forest Department can provide the fallen and dead trees to these Dhaba owners at low cost. These fallen logs are easily available in Ragsi and Trisula Reserve Forests. Alternative option could be to provide gas stoves and cylinders to these Dhaba owners at subsidized rates so as to decrease their dependency on the forest resources for fuelwood.

Conservation of these valuable species would not be possible without the active participation of the local people. By improving their living standards and by giving benefits of conservation to them, long term conservation goals can be achieved.

6. Acknowledgements

Authors would like to acknowledge Sri. P.R. Sinha (Director), Dr. V.B. Mathur (Dean) Wildlife Institute of India (WII), Dehra Dun and Forest Department of Uttarakhand for logistic support and encouragements. We are thankful to DBT, Govt. of India and UCOST, Govt. of Uttarakhand for funding. Thanks is also due to Ms. Upma Manral, Ms. Preeti (research scholar at WII) and Ms. Maja Bozicevic (Publishing Process Manager, Intech: Deforestation) for consistent encouragement and improving the text and to an anonymous referee for reviewing the manuscript.

7. References

Gorrie, R. M. (1937). Tree lopping on a permanent basis. The Indian Forester. LXIII, 29-31.

Chaturvedi, A. N. (1985). Fuel and Fodder Trees for Man-made Forests in Hills. In: Environmental regeneration in Himalaya: concepts and strategies. Nainital: Central Himalayan Environment Association and Gyanodaya Prakashan.

Ives, J.D. and Messerli, B. (1989). The Himalayan Dilemma: Reconciling Development and Conservation. London and New York: Routledge.

Kala, C.P. (1998). Ecology and conservation of alpine meadows in the Valley of Flowers National Park, Garhwal Himalaya. Ph.D., Thesis. FRI, Dehra Dun.

Kala, C.P., G.S. Rawat and V.K. Uniyal (1995). An interim report on the Valley of Flowers National Park, Mountain Grassland Project. WII, Dehradun.

Kanzaki M, Y. Kyoji (1986). Regeneration in sub-alpine coniferous forests: Mortality and the pattern of death of canopy trees. Bot. Mag. Tokyo, 99 (1053), 37-52.

Mathani, G.P. (1994). Dimensions of fuelwood problems of U.P. hills and solutions thereof. Indian Forester. 120: 202-209.

Naithani, H.B., G.C.S. Negi, R.C. Thaoliyal and T.C. Pokhariyal (1992). Valley of flowers needs for conservation or preservation. Indian Forester. 118 (5): 371-378.

Nautiyal, M.C., B.P. Nautiyal and V. Prakash (2001). Phenology and growth form distribution in an alpine pasture at Tungnath, Garhwal Himalaya. Mountain research and Development. 21 (4): 166-174.

Negi, G.C.S., H.C. Rikhari, J. Ram and S.P. Singh (1993). Foraging niche characterstics of horses, sheep and goats in an alpine meadows of the Indian Central Himalaya. J. Applied ecology, 30: 383-394.

Quintana-Ascencio, P.F., M. Gonzalez-Espinosa and N. Ramirez-Marcial (1992). Acorn removal, seedling survivorship, and seedling growth of Quercus crispipilis in successional forests of the highlands of Chiapas, Mexico. Bull. Torrey Bot. club, 119: 6-8.

Ram, J., J.S. Singh and S.P. Singh (1989). Plant biomass species diversity and net primary production in a Central Himalayan high altitude grassland. J. Ecol.77: 456-468.

Rawat, G.S. (2007). Pastoral practices, wild mammals and conservation status of alpine meadows in Western Himalaya. Journal of Bombay Natural History Society, 104 (1), 251-257.

Rao, P.B. and S.P. Singh (1989). Effect of shade on Central Himalayan species from a successional gradient. Acta Oecologica: 10: 21-33.

Rawat, G.S.,and V.K. Uniyal (1993). Pastoralism and plant conservation: The Valley of flowers dilemma. Environmental Conservation 20:164-167.

Rodgers, W.A. and H.S. Panwar (1988). Planning a Wildlife Protected Area network in India. Vol. I & II. Wildlife Institute of India, Dehra Dun.

Saberwal, V.K. (1996). Pastoral Politics; Gaddi grazing, degradation and biodiversity conservation in Himachal Pradesh, India. Con. Biol., 11(3): 741-749.

Singh, G. (2008). Diversity of vascular plants in some parts of Kedarnath Wildlife Sanctuary (Western Himalaya), Ph.D. Thesis, Kumaun University Nainital.

Singh, G., G.S. Rawat, D. Verma (2010). Comparative study of fuelwood consumption by villagers and seasonal "Dhaba owners" in the tourist affected regions of Western Himalaya, India. Energy Policy, 38, 1895-1899.

Singh, J.S., Y.S. Rawat and O.P. Chaturvedi (1984). Replacement of oak forest with pine in the Himalaya affects the nitrogen cycle. Nature, 311: 54-56.

Singh, J.S. and S.P. Singh (1992), Forests of Himalaya: Structure, Functioning and Impact of Man, Gyanodaya Prakashan, Nainital.

Singh, S.K. (1999). A study on the plant community composition and species diversity in Great Himalayan National Park, western Himalaya. Ph.D. Thesis, Kumaun University Nainital.

Singh, S.P. (1981). Rural ecosystem and development in the Himalaya. In J.S. Singh, S.P. Singh, C. Shastri (eds.) Science and Rural development in Mountains, Gyanodaya Prakashan, Nainital, India.

Singh, S.P. (1991). Structure and function of low and high altitude grazingland ecosystems and the impact of the livestock component in the Central Himalaya. Final Technical Report, Dept. of Envi. Govt. of India, New Delhi.

Sundariyal, R.C. and A.P. Joshi (1990). Effect of grazing on standing crop, productivity efficiency of energy capture in an alpine grassland ecosystem at Tungnath (Central Himalaya) India. Trop. Eco., 31 (2): 84-87.

Thakur, A.K., G. Singh, S. Singh and G.S. Rawat (2011). Impact of pastorl practices on forest cover and regeneration in the outer finges of Kedarnath Wildlife Sanctuary, Western Himalaya. Journal of Indian Soc. Remote Sensing, 39(1), 127-134.

Thadani, R. and P.M.S. Ashton (1995). Regeneration of banj oak (*Quercus leucotrichophora* A. Camus) in the central Himalaya. Forest Ecology and Management: 78: 217-224.

Troup, R.S. (1921). Silviculture of Indian Trees. Vol. I-III. Clarendon Press, Oxford.

3

Nepal's Community Forestry: Need of Better Governance

Ridish K. Pokharel, Santosh Rayamajhi and Krishna R. Tiwari
Institute of Forestry Tribhuvan University, Pokhara
Nepal

1. Introduction

Community forestry promotes the management of forests as Common Pool Resources (CPRs) (Ostrom, 1992; Acharya, 2002). A common pool resource refers to a natural or man-made resource system that is sufficiently large as to make it costly to exclude potential beneficiaries from obtaining benefits from its use (Ostrom, 1990). All CPRs share two attributes: it is costly to exclude individuals from using the goods either through physical barriers or legal instruments, and the benefits consumed by one individual subtract from the benefits available to others (Ostrom and Ostrom, 1977; Ostrom et al., 1994). There are some problems in managing CPRs. The problem of CPR is overuse which is described by Hardin (1968) as "Tragedy of the Commons". Hardin uses meaning of commons as open access i.e., everybody's property and everybody's property is nobody's property (Gordon, 1954). Resources managed as common property are not necessarily open access. They are managed by a community or social group with exclusive rights to use resources. The rights to use resources are limited to the group; not to everybody. One feature of common property is a right to use something in common with others (MacPherson, 1978). As property is in common, the property rights are assigned to a community or social group where the rules of appropriation of resources are assumed to safeguard the community or social group. Members of the group agree to limit their individuals claim on resource by subscribing to rules governing the use of resources. Hardin's notion of the commons was scrutinized under the conceptual differences between resource types and property rights governing their use (Ciriacy-Wantrup and Bishop, 1975). The property rights and governance are closely intertwined and it is one reason that several studies have examined common property institutions to produce different attributes that are conducive for collective action and also for successful governance of resources (Ostrom, 1990; Baland and Platteau, 1996; Hobley and Shah, 1996; Ostrom, 1999; Agrawal, 2001).

Nepal is considered as one of the leading countries in community based forest management as the country has introduced a progressive forest act in favor of community based forest management and also has made progress in rejuvenating forests in denuded hills and naked areas. Forests in Nepal were nationalized and transferred to the control of Department of Forests (DoF) in 1957. However, such transformation created an open access situation due to lack of capacity of DoF to manage the forests (Soussan et al., 1995). During the 1970s, there was a growing recognition that the government could not manage the forests alone and community participation is essential to manage the country forests. The government initiated community based forest management approach in 1978 by enacting legislation that allow transfer of forest management responsibility from the government to local panchayat[1] as Panchayat Forest (PF) and Panchayat Protected Forest (PPF). The regulations specified the provision of transferring a limited area of government – owned, degraded forestland (up to 125 ha) and existing natural forests (up to 500 ha) to the local political unit as PF and PPF, respectively for development and management purposes. His majesty's Government (HMG/N) enacted the rules and regulations by implementing the first national level community forestry project in 1980, covering 29 hill districts with the aim of reducing ecological degradation and increasing the supply of basic forest products for subsistence needs through people's participation (Manandhar, 1981). Since then the community forestry in Nepal has evolved continuously under the supportive forest policies and legislations. The basic institution that implements community forestry is a Community Forest User Group (CFUG[2]). CFUGs are legal entities with autonomy in decision making such as access rules, forest products prices, mechanism for allocation of forest products, user fees, and other important policies are agreed by user members (NORMS 2003, cited by Kanel and Niraula, 2004). This chapter attempts to explore the existing policies and practices of Nepal's community forestry as community forestry in Nepal is now; a well established management form (Pokharel, 2009), successful community based forest governance model (Timsina, 2003; Thoms, 2008) and also plays a dominant role in natural resource management programs.

2. Policy and governance in community forestry

For conservation and management purposes, forests in Nepal are classified into five categories:

- Government managed forests,
- Community forests,
- Leasehold forests,
- Religious forests, and
- Private forests

Community forests are part of national forests transferred to local community, known as CFUGs to conserve, manage and utilize for the basic needs of the community. Community forestry, in most cases, is functioning well in the hills and communities are deriving various benefits.

Nepal's forest policy is considered to be dynamic as there has been a drastic changed in forest management practices transferring management responsibility from state control to local community. The introduction of community forestry in Nepal represents an attempt to decentralize forest resources by allowing local people to control forest resources. The Nepal National Forestry Plan of 1976, developed by Ministry of Forest and Soil Conservation, was the first document indicating the government intentions concerning use and management of forest resources. The First Amendment of the Forest Act of 1961 in 1977 made provision for

---

[1] Lowest level political and administrative unit

[2] A group of people who regularly uses a particular forest for various purposes and organize themselves to protect, manage, and utilize the forest by forming a group

transferring government owned forestland to local communities for protection, development, and utilization purposes (Mahat, 1997). The Master Plan for the Forestry Sector (MPFS) in Nepal was prepared during 1986 – 88 and was approved by the government in 1989. The MPFS, approved by Ministry of Forest and Soil Conservation, was the first long term plan in Nepal's forestry sector which provided a 25-year policy and planning framework. The plan included the following as the long-term objectives of the forestry sector:

- To meet the people's basic needs for forest products on a sustained basis
- To conserve ecosystems and genetic resources
- To protect land against degradation and other effects of ecological imbalance
- To contribute to local and national economic growth

The MPFS guided forestry development within the comprehensive framework of six primary and six supportive programs to achieve its objectives. The main features of the plan lied in an integrated and program oriented approach to forest and watershed management. This program approach was a turning point in the history of Nepal's forestry sector policy (Amatya, 2002). The plan clearly mentioned the following points related to community forestry:

- No ceiling on the area of forests to be handed over
- Handing over of forests to the local users and not to the *panchayats*
- Involvement of women and the poor in the management of community forests
- All accessible forests in the country to be handed over to the user groups to the extent that they are willing and capable to manage them
- A changed role for the forestry staff for advice and extension
- Community forestry to be regarded as the priority program of the forestry sector

The national community forestry workshop series, usually held in every five year interval, has also been key contributing factors in development of Nepal's community forestry. The workshops have helped to define the legal and regulatory framework of community forestry in Nepal and develop consensus on key issues among key players (Ojha and Kanel, 2005). The first national community forestry workshop was organized in 1987 to share the field experiences. Identification of forest users under the political boundary was identified as a major problem to implement the community forestry program since the political boundary did not usually coincide. The workshop recommended a use practice concept to identify the users by traditional use rather than by political boundary. The MPFS also emphasized the CFUGs as the appropriate local institution responsible for the protection, development and sustainable utilization of local forests. The plan facilitated layout of the foundation of the new Forest Act that was introduced in 1993. The act identified a CFUG as a self-governed autonomous entity with authority to independently manage and use the forest according to agreed management plan. The Act also allows the CFUG to fix the price independently, transport and market the forest products from community own forests. The CFUGs can sale forest products to generate income if there is a surplus product. The generated income is not shared with the government rather it adds to CFUGs income. Table 1 shows the timeline of forest policy shift in Nepal focusing on forest governance, particularly in community forestry.

Governance in community forestry addresses the relationships, rights, responsibility and incentives among stakeholders including forest communities, industries and government

Year (AD)	Policy, approach and practices
Before1950	Administered forests as private property (elite class control; authority given by autocratic *Rana* regime)
1950-1956	Period of transition to convert forests as private property to state property
1957-1960	Introduced Private Forest Nationalization Act; declared private forests as state property
1961-1975	Promulgated Forest Act (1961); government took control on forests resources
1976-1986	Emergence of community forestry concept; recognized for the first time the need for community involvement in forest management through national forestry plan; introduced *Panchayat* Forest and *Panchayat* Protected Forest Rules (1978), Decentralization Act (1982), and Decentralization Regulations (1984)
1987-1990	Prepared a 25 years Master Plan for the Forestry Sector and endorsed it by the government in 1989 as a major policy document; recognized community and private forestry as largest program; held 1st community forestry national workshop
1991-1995	Introduced user group concept; introduced Forest Act (1993) and Forest Regulations (1995); held 2nd community forestry national workshop; emergence of **Federation of Community Forestry users in Nepal (FECOFUN)**
1996-2001	Held 3rd community forestry national workshop; set vision for community forestry to contribute to poverty reduction; the ninth five year plan (1997-2002) included poverty alleviation as a primary objective in the forestry sector; prepared community forestry directives (1996); revised forestry sector policy (2000); introduced forest inventory guidelines (2001); made mandatory to invest 25% of generated income from community forestry to forest development and maintenance
2002-2006	Held 4th community forestry national workshop; focused poverty reduction and community empowerment through community forestry; recognized CFUGs as an effective local institution as they survived even in the war and conflict time of the country between the Maoist and Government
2007-onward	Held 1st international community forestry workshop and 5th community forestry national workshop; focused community forestry on governance, poverty reduction and sustainable forest management; linked community forestry with PES, climate change, carbon market mechanism and REDD; made mandatory to invest 35% of generated income from community forest to pro-poor programs; made mandatory to include women in the key posts of the executive committee of CFUGs

Source: Pokahrel et al., 2006; Ojha et al., 2006; MFSC, 2007; Gautam et al., 2004; DoF, 2001

Table 1. Paradigm shift in forest development policy, approach and practice in Nepal

(MFSC, 2007). Similarly, it focuses on pro-poor governance with the aim of benefiting poor and vulnerable people by securing their representation in the executive committee. An executive committee is one forum of CFUGs where management decisions related to community forestry are made through their representatives. CFUGs are required to include 50 per cent of women in the executive committee and are also required to offer the post of either chairperson or secretary to a woman (MFSC, 2009). Similarly, they are required to invest 25 per cent and 35 per cent of their income to forest development and maintenance, and pro-poor programs, respectively (ibid). The remaining income can be used as per the need and interest of the community.

3. Status and achievements of community forestry

Nepal is now considered as a leader in community forestry as the country has had long experience in implementing community forestry programs and also has a new forest act in favor of community forestry. The new act is recognized as innovative and progressive approach in the field of forestry (Pokharel, 1998; Belbase and Regmi, 2002) which recognizes local people as a key partner for managing forest resources. It **has been** observed in the country that the community forestry program has made a remarkable progress in rejuvenating forests in the denuded hills. Several studies indicate that the condition of community forests has been improved substantially (Branney and Yadav, 1998; Gautam et al., 2004; Webb and Gautam, 2001).

The MFSC emphasized community and private forestry as a major program and expected to absorb almost a half (47%) of the total budget allocated to forestry sector in 2010 (MPFS, 1988). The community forest formation process involves identification of users and the creation of a formal forest association i.e., CFUG. According to government policy, all actual users of a given forest should be included in the user group. After formation, a CFUG becomes fully responsible for protecting, managing and utilizing the forest. Community forestry program is based on the policy that emphasizes people's participation in the development and management of forest resources by transferring responsibility from the Department of Forest to the CFUGs, who are willing and able to practice forest management. An area of national forest is transferred as community forest to a particular community for management and utilization purposes. As of December 2010, Nepal had already transferred 1.23 million hectares (almost one-quarter) of national forests to nearly 15,000 CFUGs involving about two-fifth of the country's population (Table 2). This implies that over one-third of potential forest area has already been transferred to community as community forests. Although community forestry program has been implemented throughout the country, there is a great variation in the distribution of community forestry practices between physiographic regions (Chakraborty, 2001). A large portion of community forest areas (about 67%) are located in mid-hills, followed by high mountains (20%) and Tarai (13%) (Sharma, 2009). The forest handing over accelerated fast during mid-1990s, but it is declining now (Kanel, 2004; Sharma, 2009). Annual CFUGs formation rate during mid-1990s was 1,500 (Sharma, 2009) but the trend is declining gradually. The possible reasons for declining CFUGs formation trend are: most of accessible forests of hills and mountains have already been handed over, the government decisions to restrict forest hand over in the Tarai and inner Tarai forests, and also partly donors pulled out from community forestry programs due to Royal take over in 2005.

Table 2 shows the present scenario of Nepal's community forestry in terms of coverage of forest areas, number of group managing forests, number of household involved, and number of women members in the executive committee. Community forestry is one of Nepal's "most successful community based development programs" (Kattel, 2000). The concept and process of community forestry is well appreciated both nationally and internationally (Pokharel, 2008). A Swiss expert on Nepal, Toney Hagen also expressed his view through an interview in Nepal that if the community forestry concept and process were followed in other development sectors, Nepal would soon become a Switzerland in Asia. Although the policy and process of community forestry are good, there are still some challenges in Nepal's community forestry. The success of Nepal's community forestry encouraged the government to initiate some development works as they realized the potential of community forestry to contribute to national development. Such realization made the government to choose community forestry as a tool for poverty reduction rather than limiting to fulfillment of basic forestry needs only. At present, hundred per cent of benefits that come from community forestry goes to CFUGs and contributes to many aspects of local development such as school buildings, temples, road/trail construction, water reservoirs, biogas systems, and children development centers. Similarly, CFUGs are functioning as a small nation delivering services similar to 16 ministries of Nepal Government (Pokharel, 2005).

1	Total land area of Nepal (million hectare)	14.7*
2	Total forest area of Nepal (million hectare)	5.5*
3	Potential community forest area (million hectare)	3.5*
4	Forest area under community forest (million hectare)	1.23
5	Total number of CFUGs	14,572
6	Total number of women CFUGs	778
7	Total number of households involved (million)	1.66
8	Total number of members in the executive committee	163,567
9	Percentage of women members in the executive committee	26

Source: *CFDP, 1991; DoF, 2010

Table 2. Present scenario of Nepal's community forestry

4. Outcomes of community forestry

4.1 Community forest user group income and expenditure

CFUGs in Nepal are not managing forests only but generating products and income for the users as well. Once the forest is handed over as community forest, the CFUG can fix the price of forest products and also sale them to the market if there is surplus. The annual income of the CFUGs is estimated to be over US$10 million (Kanel and Niraula, 2004). Two separate studies conducted by Pokharel (2008a) and Sharma (2009) show that an average annual income of a CFUG is Nrs 63,202 **and Nrs17,887, respectively. Moreover, Pokharel's study has stated that the income** can be increased by nearly five times by removing timber subsidy. Timber subsidy in the CFUGs is considered as an incentive and offer to their members. Timber is generally first sold within the CFUGs and if there is surplus then it is offered for sale to non-CFUG members.

Pokharel conducted a study in 100 CFUGs in three mid-hill districts (Kaski, Tanahu, and Lamjung) to determine income from different sources whereas Sharma carried out a study at macro level using national CFUG data base of year 2004 to determine CFUGs income. The CFUG income of two studies greatly varies by more than three times. One possible reason for variation could be that Pokharel excluded the CFUGs in the samples whose CFUG fund size was below Nrs. 20,000. The other possible reasons for stating low income in Sharma's study could be due to the CFUGs reporting lower income to the government for fear of higher incomes being claimed by the government as tax and may also reflect the peak Maoists insurgency period when extracting forest products through silvicultural operations was limited. Forestry and non-forestry are major source of CFUG income (Table 3). The forestry source includes the sale of timber, fuelwood, poles, NTFPs and fodder/grasses whereas non-forestry are penalty, membership fee, assistance from GOs/NGOs and renting halls and utensils. The forestry sources are further divided into timber and non-timber where non-timber includes fuelwood, small poles, fodder/grasses, and herbs.

SN	Income source		Amount (NRs)	Percentage
1	Forest based	Timber	30,437	68.16
2		Non-timber	3,216	7.20
Sub-total			**33,653**	**75.36**
3	Non-forest based	Membership fee	6,141	13.75
4		Penalty	1,012	2.27
5		Assistance from NGOs/DoF	2,687	6.01
6		Renting halls and utensils	1,165	2.61
Sub-total			**11,005**	**24.64**
Total			**44,658**	100

Source: Pokharel (2008a)

Table 3. Average income of CFUG from different sources

Table 3 clearly shows that three quarters of CFUG income comes from forestry sources. Of the total income, timber and non-timber contribute 68 per cent and seven per cent, respectively. Timber is used for building houses and making furniture. Many CFUGs have a quota system and distribute timber to the members based on needs and availability. The CFUGs charge a price for timber and require advance payment. The price for timber is highly subsidized (Pokharel, 2008a). Fuelwood, grass and leaf-litter are important subsistence products obtained from community forests. Many CFUGs distribute these products freely from the community forests. Some CFUGs do charge the price for them, particularly fuelwood at a subsidized rate. Pokharel (2008a) found in his study areas that some 60 per cent of the CFUGs distribute fuelwood to members freely while the remaining charge for it. They work collectively to gather fuelwood by removing dead, dying, decay and diseased trees and usually distribute it equally to all members. Like timber, fuelwood is also given first to CFUG members and then to non-CFUG members with relatively higher price if there is a surplus. CFUGs obtain income from other sources as well. These sources include membership fees, penalties, assistance from GOs/NGOs and renting halls and utensils (Table 3). All together these sources contribute 25 per cent to the income. Membership fees constitutes a large source of non-forest based income, CFUGs charge relatively high first membership fee to new members. Similarly, they asked high membership fee from someone who builds a new house compared with one who buys existing house in the village. There is a correlation between paying membership fees and timber sale. The CFUGs without timber sales tend to pay higher membership fees than those households of the CFUGs with timber sales (Pokharel, 2010).

CFUGs are legally authorized to sale the forest products and spend the generated income on forest development and various community related development works. There was also a government decision to impose 40 per cent tax on the sale of forest products outside the CFUGs. However, this provision was reviewed and scaled down to 15 per cent only for the sale of two species i.e., *Sal* (*Shorea robusta*) and *Khair* (*Acacia catechu*) after this provision was severely criticize by Federation of Community Forestry Users in Nepal (FECOFUN) and others in the country. The generated funds are being used in different activities including forest development, public infrastructure development, pro-poor activities, and forest administration;   indicating that CFUGs are not limited to forest management but are also involved in different aspects of rural development. The average annual investment of Nepal's CFUGs is estimated to be over US$5 million (Kanel and Niraula, 2004; Kanel, 2004). Similarly, the average annual investment of a CFUG was estimated to be NRs 51,574 (Pokahrel, 2008a, 2009). Number of studies (e.g. Dongol et al., 2002; Acharya, 2002, 2003; Kanel and Niraula, 2004; Pokharel, 2008a, 2009) observed the public infrastructure development constituted a major expenditure of the CFUG funds. It is a matter of debate whether investment made by CFUGs in development activities benefits the poor as more funds are being invested in infrastructure.

Managing CFUG funds in community forestry is becoming a challenging task as the funds have grown in size and become popular with communities. CFUG fund is popular because it has facilitated members to initiate financial transactions in the village by offering loans and investing in other development activities. Borrowing money with an individual who holds cash is a common practice in the village and finding such individuals in the village is difficult now-a-days. There is a tendency of village people moving to urban or sub-urban areas if they can afford it. There is also an increasing trend of moving families to urban or sub-urban areas if household member is **employed in the abroad. As a result, there is less financial transaction in the village**; making CFUG funds more popular as members of CFUGs have an access to loans and also it is simpler in terms of official procedures (no collateral is required and physically nearby).

4.2 Income at household level from community forests

Vedeld et al. (2004) showed that a forest serves as a safety net against crises, prevents someone from falling into deeper poverty, and provides a pathway out of poverty – demonstrating an important income at household level. The first in-depth study of rural livelihood from Nepal Himalaya conducted by Rayamajhi et al. (2010) shows that poorer households are relatively most dependent on forest income. They specify that the forest contributes 22 per cent of the total income account of an average household. More explicitly, households derived as much as 22 per cent of their total income from forest and four per cent from non-forest environmental common goods. When combined, this is higher than income obtained from either crop or livestock.

In Nepal, income inequality increased from 1995/96-2003/04 with the Gini coefficient[3] changing from 34.2 to 41.1% with a net decline in headcount poverty rate from 42% to 31% (**World Bank, 2006**). Rayamajhi et al. (2010) indicated a five per cent improvement in income equality with the inclusion of forest environmental income, indicating that forests play small role in income equalisation. A possible explanation is that all households participate in the extraction of essential forest products. The poor households inclusive of *dalit* (**occupational caste**) have few assets and thus may not immediately be able to use more forest products for improvement of livelihoods and income generation.

4.3 Decision-making in the CFUGs

There are two **levels of decision making body in a** CFUGs: General Assembly (GA) and Executive Committee (EC), also known as Community Forest User Group Committee (CFUGC). The GA represents all members of the CFUGs while the EC is composed of 9 – 15 persons, depending on the size of CFUGs. The EC members are representatives who are either elected or unanimously selected by forest users. Generally, GA meets once a year during mid January to February and EC meets about once a month. GA has a mandate to make any decision related to forest management, such as framing rules on forest use, decision on penalties for rule violators, fixing schedule for silvicultural operations, and managing generated funds with a simple majority. However, there has been an increasing practice of EC decision making, particularly over the use of CFUG funds. The EC makes the decisions on behalf of entire CFUGs and puts forward to GA for endorsement. Generally, GA makes endorsement in the decisions of EC as they believe it might have discussed thoroughly to the best use of resources. In practice, the chairperson and secretary discuss the possible agenda informally before the executive meetings and finalize them accordingly. There is an increasing demand for the funds from various groups such as school management committee, mothers group (*ama samuha*) and water group in the village to invest in their respective areas. Although there is an increasing demand for CFUG funds, the EC makes the decisions according to the interest of the chairperson and secretary.

Year	Representation in the committee		Occupying key positions in the committee	
	Male	Female	Male	Female
2004	76	24		-
2008	73	27	93	7
2010	64	36	76	24

Source: Kanel, 2004; Pokahrel, 2008a; Pokharel et al., 2010

Table 4. Representation and occupying key positions in the executive committee by gender

Table 4 shows that women representation in the executive committee and also occupying the key positions has increased significantly. Adhikari et al. (2004) conducted a study in two mid-hill districts of CFUGs and found 15.7 per cent women in the EC. Similarly, Kanel (2004) **stated that women** representation the EC is 24 per cent. Two different studies conducted by Pokharel (2008a) and Pokharel et al. (2010) in three mid-hill districts

---

[3] Gini coefficient is good measure of income equality and has been applied in analyzing the role of forest income in rural income equalization

and five different districts of CFUGs, respectively and found 27 per cent and 36 per cent, respectively women in the EC. These studies also reported that women occupying the key positions in the executive committee were seven per cent and 24 per cent, respectively. This shows that there is an increasing trend of women's representation in the EC including occupying the key positions. Finding women in the village to serve in the EC is difficult due to gender disparity. Gender disparity in Nepali society begins right after the birth (Lamichhane, 2006; Pokharel, 2008). As a male dominated society, women are encouraged not to play a role in the EC since it is considered as public sphere and it is the role of men rather than women. Such trend in the EC shows that these perceptions are changing gradually. Similarly, the recent **community forest** policy also facilitated to bringing about the change in women representation in the EC. About 778 women run CFUGs, indicating that women are coming forward and taking the leadership. There is a representation not only from women but also from poor and marginalized groups in the EC. Pokharel et al. (2010) reported that the EC is more or less inclusive in terms of gender, poor and marginalized groups. The recent community forestry policy encourages CFUGs to form an inclusive EC by representing different classes including poor, women and marginalized groups.

In the EC, the position of chairperson, vice-chairperson, secretary and treasurer are considered to be important as these positions hold some kind of authority and their individual involvement in the respective field is necessary. For instance, the decisions are not considered as final unless the chairperson endorses them. Similarly, the secretary keeps the record by maintaining minutes, schedules executive meetings and general assembly with the consent of chairperson and determines the agenda for the meeting and general assembly as well. The treasurer looks after the financial activities and maintains its record accordingly. Among these posts, chairperson and secretary are considered more powerful as individuals occupying these positions have authority to invite meetings as well as make final decisions. Similarly, the secretary has the authority to invite meetings and also put forward the agenda for discussion in the meetings.

4.4 Networking and institutions

One of the major successes of the community forestry is institutionalization of CFUGs for the management of community forests. It has established a strong institution at local level. In fact, CFUGs are the only institution at the local level that survived during the period of Maoist insurgency in Nepal. They were effective in conducting development work, and holding meetings and elections regularly. There is no elected body at the local level for the last 10 years because the government has not been able to hold the election yet. But the CFUGs have been holding elections regularly to choose their representatives in the EC – proving them as an effective local institution. CFUGs have also served as a good model for development and attracted the planners and policy makers to follow the model in other sectors as well. There is an increasing tendency for different development providers to use CFUGs as entry point for the development work in the rural areas. A network to build national level federation of community forestry has also been established in the name of **Federation of Community Forestry Users in Nepal** (FECOFUN) which has emerged as a strong civil society. It raises the voices for users' rights over forests at policy levels and works in favor of forest users.

5. Issues and challenges in community forestry

Despite having the most innovative policies to promote community forestry and forest governance in place, CFUGs are unable to provide a significant contribution to livelihoods of poor and marginalized groups such as women and *dalit*. Traditionally, fuelwood collection is linked closely with livelihood. For instance, making local liquor is one of the livelihood strategies for some poor ethnic groups such as Gurung and Tamang. **Similarly, making charcoal is one livelihood strategy of blacksmith** (occupational caste). Making charcoal and local liquor requires large quantities of fuelwood which is collected from the forest. The trade of these groups is affected with the introduction of community forestry (Soussan, 1998). Although community forestry is a user focused program rather than absolute poverty focused, its aim is to contribute to achieving national goals of poverty reduction and this directed its activities accordingly. It is argued that domination of local elite in decision making process and also passive management are some of the reasons for not contributing significantly to livelihoods of poor and marginalized groups. So, one of the major challenges in community forestry is to ensure the poor's meaningful involvement in the decision making process. Several studies (e.g. Baral and Subedi, 2000; Adhikari, 2002; Malla et al., 2003) have noted that the poor and marginalized groups have not received the benefits from community forestry as expected. A study conducted by Pokharel (2008a) has clearly shown that the non-poor are getting more benefits from Nepal's community forestry.

Several studies describe that leadership of the CFUGs in Nepal is in the hand of local elites who often influence the decisions (Banjade et al., 2006; Baral and Subedi, 2000; Malla et al., 2003). A study conducted by Hills and Shields (1998) in India also made similar observations about Forest Protection Committees (FPC) of Joint Forest Management programs. They observed that leadership of the FPC tends to be in the hands of better educated local elites who tend to be less dependent on the forests. Leadership is one of the factors that made the community forestry program successful (Pokharel et al., 1999) and the succession of leadership is seen as a potential problem in Nepal's community forestry (Agrawal and Ostrom, 2001). Currently, there is a discussion in Nepal whether local elites are unwilling to include poor and marginalized groups in the EC including key positions or they themselves are not willing to serve in the EC as well as in the key positions. It is observed in some cases that the same individuals from the local elites have been serving as chairperson for many years and also have shown an interest to transform their role to younger generation or marginalized group but could not do it due to an increasing trend of youth migration from the village and the socio-economic condition of the marginalized groups. The work in community forestry such as attending meetings, and patrolling the forests is voluntary and this has become costly for the poor. Mr. Badri Prasad Jangam in *Gaukureshwor* community forest has been serving as chairperson for the last 19 years and he is now willing to pass on his duties to younger, more innovative hands (Shahi, 2011).

Community forestry is a major program in forestry sector in Nepal which is being implemented throughout the country as a blanket approach. Such approach has created imbalance in demand-supply situation of forest products within the CFUGs and has also led to ineffective forest management. For instance, some CFUGs with higher population have smaller forest areas – raising conflicts among users because of not fulfilling required forest product. Raising conflict among the users is likely to affect their participation in forest management and in turn lead to ineffective forest management practices. Similarly, CFUGs with lower population have larger forest size whose demands are often met from the community forests. As a lower population, they may not have capacity to manage larger forest size effectively. Therefore, a blanket approach is not appropriate for community forestry programs.

Most of the CFUGs have adopted protection oriented management strategies to manage their forests by harvesting dead, diseases, dying and deform trees only. The focus of protection-oriented management approach has been to allow regeneration. So, the focus of CFUGs is to regenerate rather than management of the existing forest resources. As a result, CFUGs have not been **able to harvest the products from community forest with its potentiality**. Protection oriented management practice has made the limit of the amount of take home forest products from the community forest; forcing people to walk further to find forest products.

After the United Nation Framework on Climate Change Convention (UNFCCC) meeting in Bali in 2007, forests in developing countries were identified as an important source of carbon sink under the concept of Reducing emissions from deforestation and forest degradation (REDD). In this context, Nepal's community forestry can be a potential source for extra benefits. In recent years **there is considerable discussion on community forest** about payment for environmental services (PES) and climate change. These contributions **of community forests** have not been captured yet due to the lack of policy and methodological framework.

6. Way forward

The concept and process of community forestry is well appreciated both nationally and internationally. Although the policy and process of community forestry are well appreciated, there are still some challenges in involving all users, particularly women and the poor in forest management. Finding a way to reach all the households of forest users is important in increasing people's involvement in community forest management. One way of reaching a greater number of women and other marginalized members is by organizing meetings at the *tole* level. Organizing small meetings before GA would encourage women and marginalized groups to participate in forest management actively. Similarly, increasing the number of people in the EC, especially for women, poor and marginalized groups would also encourage them to voice their interests and concerns regarding resource management.

Focusing the activities on capacity building and technical training may enhance the participation of women and marginalized groups in the decision making process. There is also a need to improve the socio-economic condition of the marginalized groups and women to promote them in the community forestry leadership. Transferring forest management responsibility to CFUGs should not follow blanket approach. It should be location specific as forest management and forest use practice varies from one place to another. A judgment for handing over of forest resources to CFUGs should be made based on demand, household and forest ratio, and capability.

Forests are important source of timber, fuelwood and fodder and they also provide crucial ecosystem services. There is a great role of forests in addressing climate change. Promoting active forest management could be one way of addressing the needs and concern of local people and also capturing the benefits from climate change mitigation. There is a need to develop the methodological framework for REDD which should focus on benefiting the poor as they are more depended on forests for their livelihoods.

7. References

Acharya, K. P. 2003. Sustainability of support for community forestry in Nepal, *Forests, Trees and Livelihoods*, 13: 247 – 260.

Achayra, K. P. 2002. Twenty four years of community forestry in Nepal, *International Forestry Review*, 4 (2): 149 – 156.

Adhikari, B., Falco, S. D. and Lovett, J. C. 2004. Households characteristics and forest dependency: Evidence from common property forest management in Nepal, *Ecological Economics*, 48: 245 -257.

Agrawal, A. 2001. Common property institutions and sustainable governance of resources, *World Development*, 29 (10): 1649 – 1672.

Agrawal, A. and Ostrom, E. 2001. Collective action, property rights, and decentralization in resource use in India and Nepal, *Politics and Society*, 29 (4): 485 – 514.

Amatya, S. M. 2002. A review of forest policy in Nepal, In T. Enters and R. N. Leslie (eds.), *Information and Analysis for Sustainable Forest Management: Linking National and International Efforts in South and Southeast Asia, Proceedings of the Forest Policy Workshop*, EC-FAO Partnership Programme, Thailand: 79 – 83.

Baland, Jean-Marie, Platteau, 1996. *Halting Degradation of Natural Resources: Is There a Role of Rural Communities*, Oxford: Clarendon Press.

Baral, J. C. and Subedi, B. 2000. Some community forestry issues in the Tarai Nepal: Where do we go from here, *Forests, Trees and People Newsletter*, No. 42: 20 – 25.

Banjade, M. R., Schanz, H. and Leeuwis, C. 2006. Discourses of information in community forest user groups in Nepal, *International Forestry Review*, 8 (2): 229 – 240.

Belbase, N. and Regmi, D. C. 2002. *Potential for Conflicts: Community Forestry and Decentralization Legislation in Nepal*, Kathmandu/Nepal: International Center for Integrated Mountain Development

Branney, P. and Yadav, N. P. 1998.Changes in Community Forest Condition and Management (1994 – 1998), Report G/NUKCFP/32, Kathmandu: Nepal-UK Community Forestry Project.

Chakraborty, R. N. 2001. Stability and outcomes of common property institutions in forestry: Evidence from the Traia of Nepal, *Ecological Economics*, 36 (2): 341 – 353.

Ciriacy-Wantrup, S. V., and Bishop, R. C. 1975. Common property as a concept in natural resource policy, *Natural resource Journal*, 15: 713 – 727.

CPFD, 1991. The Community and Private Forestry Program in Nepal, Kathmandu/Nepal: Community and Private Forestry Division, Ministry of Forests and Soil Conservation.

DoF (Department of Forests) 2010. Management information system, Kathmandu, Nepal: Department of Forests.

Dongol, C. M., Hughey, K. F. D. and Bigsby, H. R. 2002. Capital formation and sustainable community forestry in Nepal, *Mountain Research and Development*, 22 (1): 70 – 77.

Gautam, A. P., Shivakoti, G. P. and Webb, E. L. 2004. A review of forest policies, institutions, and changes in the resource condition in Nepal, *International Forestry Review*, 6 (2): 136 – 148.

Gordon, H. R. 1954. The economic theory of a common property resource: The fishery, *Journal of Political Economy*, 62: 124 – 142.

Hardin, G. 1968. The tragedy of commons, *Science*, 162 (3859): 1243 – 1248.

Hobley, M. and Shah, K. 1996. What makes a local organization robust? Evidence from India and Nepal, Natural Resource Perspective No. 11, London: Overseas Development Institute

Hill, I. and Shields, D. 1998. Incentives for joint forest management in India: Analytical methods and case studies, World Bank Technical Paper No. 394, Washington D. C.: The World Bank.

Kanel, K. R. and Niraula, D. R. 2004. Can rural livelihood be improved in Nepal through community forestry? *Banko Janakari*, 14 (1): 19 – 26.

Kanel, K. R. 2004. Twenty five years' of community forestry: Contribution to millennium development goals, In K. R. Kanel, P. Mathema, B. R. Kanel, D. R. Niraula and M. Gautam (eds.), *Twenty five years of community forestry, Proceedings of the fourth national workshop on community forestry*, Community Forest Division, Department of Forests, Kathmandu: 4 – 18.

Kattel, G. 2000. Community based development, *Sunday Post* (March 20), Retrieved March 20, 2001, http://www.nepalnews.com.np/contents/englishweekly/sundaypost/2000/sep/sep17/recollection.htm

Lamichhane, R. C. 2006. Social conflict, inclusion, peace and development in Nepal: An analysis with reference to appreciative inquiry approach, Unpublished Master Thesis, Tribhuvan University, Kathmandu, Nepal.

MacPherson, C. B. 1978. The meaning of property, In: C. B. MacPherson (ed.), *Property: Mainstream and Critical Positions* (pp. 1 – 13), Toronto: University of Toronto Press.

Mahat, T. B. S. 1997. Development of Community Forestry in Nepal: An Overview, *Draft Working Paper*, Pokhara/Nepal: Institute of Forestry/International Tropical Timber Organizational Project.

Malla, Y. B., Neupane, H. R. and Branney, P. 2003. Why aren't poor people benefiting more from community forestry? *Journal of Forest and Livelihood*, 3 (1): 78 – 93.

Manandhar, P. K. 1981. Introduction to Policy, Legislation and Program of Community Forestry Development in Nepal (NEP/80/03), Field Document # 1a, Kathmandu/Nepal: HMG/UNDP/FAO Community Forestry Development Project.

Master Plan for Forestry Sector Nepal (MPFS) 1988. Master Plan for Forestry Sector Nepal, Main Report, Kathmandu, Nepal: Ministry of Forests and Soil Conservation.

Ministry of Forests and Soil Conservation (MFSC) 2009. Directives for Community Forest Development Program, Kathmandu/Nepal: MFSC.

Ojha, H. and Kanel, K. R. 2005. Twenty-five years of community forestry in Nepal: A review of 4th national workshop proceedings, *Journal of Forest and Livelihood*, 4 (2): 56 – 60.

Ostrom, E. 1999. Coping with tragedy of commons, *Annual Review of Political Science*, 2 (1): 493 – 535.

Ostrom, E., Gardner, R., Walker, J. M. 1994. *Rules, Games, and Common Pool Resources*, Ann Arbor, MI: University of Michigan Press.

Ostrom, E. 1992. The Rudiments of a Theory of the Origins, Survival, and Performance of Common Property Institutions, In: D. Bromley (ed.), *Making the Commons Work: Theory, practice and Policy*, San Francisco: ICS Press.

Ostrom, E. 1990. *Governing the Commons: The Evolution of Institutions for Collective Action*, Cambridge, United Kingdom: Cambridge University Press.

Ostrom, V. and Ostrom, E. 1977. A theory for institutional analysis of common pool problems, In: G. Hardin and J. Baden (eds.), *Managing the Commons* (157 – 172), San Francisco: W. H. Freeman

Pokharel, R. K., Gyawali, A. R., Yadav, R. L. and Acharya, K. P. 2010. Benefiting People through Pro-Poor Investment of Nepal's Community Forestry Funds, A Research Report, Pokhara: ComForM/IOF (Community Based Natural Forest and Tree Management in Himalaya/Institute of Forestry)

Pokharel, R. K. 2010. Generating income from Nepal's community forestry: Does timber matter? *Journal of Forest and Livelihood*, 9 (1):16-20.

Pokharel, R. K. 2009. Pro-poor programs financed through Nepal's community forestry funds: Does income matter? *Mountain Research and* Development, 29 (1): 67 – 74.

Pokharel, R. K. 2008. Good forest governance: A key aspect in achieving the Nepal's community forestry vision, *Nepalese Journal of Development and Rural Studies*, 5 (2): 62 – 68.

Pokharel, R. K. 2008a.Nepal's Community Forestry Funds: Do They Benefit the Poor? Working Paper No. 31-08, Kathmandu, Nepal: SANDEE.

Pokharel, S. 2008. Gender discrimination: Women perspectives, *Nepalese Journal of Development and Rural Studies*, 5 (2): 80 – 87.

Pokharel, B. K. 2005. Community forest user groups: Institution to protect democracy and vehicle for local development, *Journal of Forest and Livelihood*, 4 (2): 64.

Pokharel, R. K., Adhikari, S. N. and Thapa, Y. B. 1999. Sankarnagar forest user group: Learning from a successful FCU in the Tarai, *Forests, Trees and People Newsletter*, No. 38: 25 – 32.

Rayamajhi, S., Smith-Hall, C. and Helles, F. 2010. Empirical evidence of the economic importance of central Himalayan forests to rural households, Submitted to *Forest Policy and Economics*, in review process since 30.04.2010.

Sharma, A. R. 2009. Impact of community forestry on income distribution, unpublished doctoral thesis, Tribhuvan University, Kathmandu, Nepal.

Shahi, Pragati 2011. In search of a successor, *The Kathmandu Post*, XIX (99) (May 28): 10.

Soussan, J. G., Shrestha, B. K. and Uprety, L. P. 1995. *Social Dynamics of Deforestation: A Case Study from Nepal*, New York and London: UN Research Institute for Social Development and Parthenon Publishing Group.

Thoms, C. A. 2008. Community control of resources and the challenges of improving local livelihoods: A critical examination of community forestry in Nepal, *Geoforum*, 29 (3): 1452 – 1465.

Timsina, N. P. Promoting social justice and conserving montane forest environments: A case study of Nepal's community forestry program, *Geographical Journal*, 169 (3): 236 – 242.

Vedeld, P., Angelsen, A., Sajaastad, E. and Berg, G. K. 2004. Counting on the environment – forest incomes and the rural poor, Environmental Economics Series, Paper no. 98, The World Bank Environment Department, Washington DC.

Webb, E. L. and Gautam, A. P. 2001. Effects of community forest management on the structure and diversity of a successional broadleaf forest in Nepal, *International Forestry Review*, 3 (2): 146 – 157.

World Bank, 2006. Poverty in Nepal, Retrieved August 2011,
        http://web.worldbank.org/WBSITE/EXTERNAL/COUNTRIES/SOUTHASIAEX
        T/EXTSAREGTOPPOVRED/0,,contentMDK:20574069~menuPK:493447~pagePK:3
        4004173~piPK:34003707~theSitePK:493441,00.html

4

The Influence of Decisional Autonomy on Performance and Strategic Choices – The Case of Subcontracting SMEs in Logging Operations

Etienne St-Jean and Luc LeBel

University of Quebec at Trois-Rivieres & Université Laval
Canada

1. Introduction

The emergence of new actors in the globalized economy has given rise to increased competition in a number of economic sectors, including the forest industry. The logging industry in Canada has seen challenging times in recent years, as it faces problems at both the structural and cyclical levels. Increases in the cost of fibre, currency exchange rates and compensatory fines on timber that are detrimental to exports to the United States, higher energy costs, a decrease in the price of timber and some paper products, the increasing scarcity of resources: the industry faces many serious problems.

As of the late 1970s, industrial activity in the forests of Eastern Canada quickly evolved from a structure that was almost entirely controlled by the large pulp and paper companies to a more flexible and decentralized organization, characterized by the generalized use of subcontractors (Mercure, 1996). Once salaried workers employed by a company, subcontracting forest entrepreneurs have become business managers. As the orchestrators of the operations with which they have been entrusted, entrepreneurs must possess a diversified skill set composed of ever more complex abilities. Pressure from competition within the industry pushes the large companies to demand even more from their subcontractors, who must improve their performance to survive.

Since those operations had previously been conducted by the large companies, who may tend to view their subcontractors as an extension of their own business, such a situation could reduce their decisional autonomy and entrepreneurial behaviours, two elements likely to limit their performance. In a highly competitive context, where the performance of one player in the value chain affects all of the businesses involved, this could prove to be dramatic. The objective of this study was to see whether the strong commercial dependency of subcontracting SMEs in logging operations, as well as their potential loss of decisional autonomy, has a negative influence on their performance and strategic choices.

2. The forest industry of eastern Canada

In Quebec, logging is still predominantly conducted by the large corporations (Blais & Chiasson, 2005). Accordingly, forest entrepreneurs essentially provide these companies with logging, forest road construction or log transport services. As a consequence, they find themselves in a subcontractor business relationship, as indicated by Legendre (2005). Studying the evolution of subcontracted contracts in the forest industry, the author notes that risks and responsibilities have been relegated to the small logging businesses, which are "[...] completely dependent economically and financially on the [large corporations] and have almost lost all of their organizational independence" [translation] (Legendre, 2005). This fact becomes more tangible when we consider statements by Canadian economic analysts, who view small logging companies as dependent on large organizations, with the main goal of providing these organizations with the flexibility they need to restructure in a post-Fordist economy, as highlighted by Bronson (1999).

Public forests account for a little over 90% of Quebec's woodlands. The vast majority of Quebec's forest harvesting, transport and road construction entrepreneurs are commercially dependent, working for only a few main contractors which determine the felling areas and, notably, the rates paid out for wood supplies,. Commercial dependency is defined as a situation where a limited number of relatively undiversified customers (or suppliers) generate more than three-quarters of a company's turnover (Rinfret *et al.*, 2000). This situation is prevalent in the logging industry, where 49.4% of Québec forest entrepreneurs have a single customer which accounts for their entire turnover, and 81.1% have only three or fewer customers (PREFoRT, 2007). This commercial dependency could significantly influence their performance and reduce their strategic leeway. Since the emergence of forest entrepreneurs is the consequence of a strategic choice on the part of large forest product companies to focus on their core competencies and to contract out logging, transport and logging road construction activities, it is reasonable to assume that not all owner-managers of logging companies exhibit entrepreneurial behaviours such as innovation and the quest for growth and profits (Carland *et al.*, 1984; Filion, 2005; Gartner, 1989). From this perspective, the dependency situation could influence entrepreneurial behaviour and, as a result, logging SME performance.

3. Commercial dependency and performance of the SME

The literature pertaining to commercial dependency suggests that SME dependency on one or a limited number of clients leads to a decrease in entrepreneurial behaviours by the manager, who may be tempted to settle for his subcontracting role (Raymond & St-Pierre, 2002; Wilson & Gorb, 1983). In the forestry industry, the contract provider issues a number of set specifications governing a forest entrepreneur's operations, for example, by setting restrictions on log length, the quantity of wood cut, and even on certain work methods and tools to be used (Legendre, 2005). From this perspective, it becomes increasingly difficult for an entrepreneur to innovate, even though innovation is often considered as fundamental in evidencing entrepreneurial action (for example, Risker (1998)).

In many cases, as noted by Holmlund and Kock (1996), subcontractors have no choice but to follow the contractor's orders, even though they may result in unprofitable operations. In fact, it has been noted that the profit margins of commercially dependent SMEs are lower than those who work for a broader range of clients, even taking into account their lower sales and administrative costs (Rinfret *et al.*, 2000). Explanations could lie in their reduced ability to negotiate pricing with the few big companies subcontracting them.

Furthermore, commercial dependency also involves the notion of power (Emerson, 1962), which is essentially concentrated in the hands of the contractor. It is therefore not surprising to discover that relationships characterized by strong commercial dependency, and therefore unequal power, are often dysfunctional, unstable, or devoid of trust (Corsten & Felde, 2005; Kumar *et al.*, 1995). Relationships where contractors have considerable power over their subcontractors result in lower profits for the latter. (Cox *et al.*, 2004). The difference in size of the businesses could explain the difficulties SMEs experience in negotiating as equals with large contractors (Ramsay, 1990). Moreover, significant changes in the volume of business they receive from a large contractor could lead to radical fluctuations in the growth rate of SMEs (St-Jean *et al.*, 2008). As a result, it is recommended that small businesses which are dependent on one major client diversify their operations to reduce risk (Henricks, 1993; Kalwani & Narayandas, 1995). Reducing commercial dependency thus enables SMEs to better negotiate pricing for their products and services with large organizations (Wilson & Gorb, 1983).

Thus, decisional autonomy is the more appropriate term in the context of interaction between a subcontractor and his contractor. For example, an entrepreneur could be commercially dependent, given that his order book is essentially filled by a single client, but nevertheless be left with decisional autonomy with regard to the manner in which the contract is fulfilled and prices set, remaining free to do business with the contractor's competitors, etc. The concept of decisional autonomy has in fact been raised by Lyons *et al.* (1990), who indicated that losing this decisional autonomy could even result in reduced autonomy with regard to strategic choices. The contractor's power can go so far as to influence strategic choices such as decisions related to innovation or product range (Inderst & Shaffer, 2007; Inderst & Wey, 2007). Although client concentration is a risky strategy for subcontracting SMEs (Kalwani & Narayandas, 1995), the negative effects could be less pronounced for SMEs which do not experience some form of "strategic dominance" on the part of the contractor, which underscores the significance of considering the decisional autonomy of commercially dependent subcontracting SMEs.

However, the relationship between commercial dependency and performance is not very clear, in particular with regard to certain specific industries. For example, Mäkinen (1993) reported that Finnish forest entrepreneurs achieved satisfactory financial results if they worked for a single "good" contractor, which provided enough suitable work (therefore more profitable) to the entrepreneur. This is also the case with businesses operating in highly competitive global markets such as, electronics, tooling/machinery and the automobile industry, where commercially dependent subcontractors enjoy greater growth (Kalwani & Narayandas, 1995). In the textile industry, on the other hand, strong commercial dependency on the part of subcontractors and pressure exerted by the work-providing manufacturers force these subcontractors to lower their production costs to avoid having to close down their business for lack of contracts (Kilduff, 2005; Remili & Carrier, 2006). Within the aerospace industry, for example, three subcontractor categories can be identified, each characterized by a typical contractor/subcontractor relationship: subcontracting of economy (strong dependency), subcontracting of specialization (complementary relationship) and subcontracting of supply (power equilibrium) (Amesse *et al.*, 2001). Thus, commercial dependency comes with certain advantages and inconveniences for subcontracting SMEs, which can negatively but also positively influence performance (Barringer, 1997).

These elements suggest that the notion of decisional autonomy should be examined alongside commercial dependency in a parallel manner. The notion of commercial dependency, based on the high concentration of sales in the hands of a single or limited number of clients, does not make it possible to consider the level of decisional autonomy, which may be significantly, albeit not systematically, reduced by commercial dependency. As mentioned above, SMEs may maintain decisional autonomy, even in a situation of commercial dependency, where they hold a competitive advantage that renders them essentially indispensable to the contractor, as is the case with subcontracting of intelligence (Julien, 2000).

4. Research hypotheses

This study aimed at examining the effect of commercial dependency and decisional autonomy on subcontracting logging SME performance. Although certain nuances may be noted according to the studies under review, commercial dependency negatively affects SME performance. In particular, subcontractors achieve lower profits where contractors hold considerable power over them (Cox *et al.*, 2004). Such was the case in the automobile industry in the 1970s, for example, where the large manufacturers used their power to constantly negotiate lower prices, to the detriment of the subcontractors (Perrow, 1974). Inversely, client diversification into foreign markets, which reduces dependency on national clients, allows for higher profits (Daniels & Bracker, 1989). This suggests the following hypothesis:

H_1:   Commercial dependency negatively influences logging SME performance.

As with commercial dependency, contractor interference with a subcontracting SME's strategic choices could hinder their performance. Inversely, greater decisional autonomy should positively influence performance, which leads to the following hypothesis:

H_2:   Decisional autonomy positively influences performance.

In addition, commercial dependency could considerably reduce the decisional autonomy of the SME's managers. Aware of their subcontractors' dependency, contractors could end up viewing them as an extension of their own business and use their power to impose constraints with regard to their strategic choices. This suggests the following hypothesis:

H_3:   Commercial dependency reduces decisional autonomy.

Lastly, in a situation where a subcontractor's decisional autonomy is limited by the power exerted by a contractor, an SME manager's strategic choices may be limited. This situation is commonly found among commercially dependent subcontractors, in low-technology contexts in particular. These subcontractors eventually develop specific strategies to reduce their dependency and, as a result, reduce risk (Jansson & Hilmersson, 2009). In addition, it is noted that the contractor's strategic choices must be aligned with those of the subcontractor, for example, when they choose a strategy of innovation (Isaksen & Kalsaas, 2009). As a result, greater decisional autonomy can allow for a broader range of strategic choices, which suggests the following hypothesis:

H_4:   Decisional autonomy influences strategic choices.

5. Research methodology

5.1 Population and sample

To answer our research questions, we used the data collected during a follow-up survey of a sample of 717 Québec subcontracting logging SMEs, which represents 28% of the total SME population in this sector, and who had responded to a prior investigation in 2006. The follow-up with this SME sample was conducted in 2009, where 265 SMEs responded to our questionnaire, for a response rate of 37%. The questionnaire was sent to, and answered by, the SME owner.

5.2 Measures

Commercial dependency is defined as a situation where a small number of clients accounts for more than 75% of the turnover (Remili & Carrier, 2006; Rinfret *et al.*, 2000). However, rather than providing a statement where respondents must determine whether or not their business is in a situation of commercial dependency, which would have created a dichotomous variable, respondents were asked to determine the approximate percentage of their turnover that was attributable to their main client. This provided us with a metric variable (from 0 to 100).

We also measured the *decisional autonomy* of subcontracting logging SMEs with respect to six (6) components: the price of services, selection of employees, their working conditions, the nature of the contracts to be fulfilled, the manner in which the work is to be carried out and the tools or technology to be used. These dimensions were selected by researchers following comments collected during round-table discussions on the subject conducted during a symposium that drew over one hundred forest entrepreneurs. For each dimension, respondents were asked to select the most appropriate situation, from a graduated seven (7)-point scale from 1-My clients make all the decisions to 7-My business makes all the decisions.

SME *performance* is obviously a multidimensional concept (Wolff & Pett, 2006). This suggests that measures of performance should be based on multiple indicators related to a business's strategic goals (Kaplan & Norton, 1992). We therefore selected eleven (11) performance indicators: increases in revenue, number of employees and profit margin (Le Roy, 2001; McMahon, 2001), improvements in production techniques and use of new technology (Beamon, 1999), quality and variety of services and client satisfaction (Perera *et al.*, 1997), personnel recruitment and retention (Ulrich, 1999), investment in the community (Graves & Waddock, 1994), and respect for the environment and sustainable development (Gondran & Brodhag, 2003).

These measures are subjectively based on the respondent's perception of his performance compared to his two main competitors. The measurement scale varies from 1-*Very inferior to the others* to 5-*Very superior to the others*. Although this manner of measuring performance has its limitations, due in particular to increased measurement error, previous research has demonstrated that perceptual performance measures correlate significantly with objective measures (Murphy & Callaway, 2004; Murphy *et al.*, 1996). Given the fact that the businesses under study were all independent and private, that managers are rarely prepared to divulge financial information, and that it is unlikely that objective data is available for certain dimensions, we selected a subjective approach, which is the suggested course of action for these types of situations (Dess & Robinson, 1984).

Lastly, with respect to the strategic choices, we asked the managers to indicate their strategies for the next five (5) years from the following options: *no significant change, growth, diversification within the forest, diversification outside the forest* and *selling/closing the business*. Multiple choices were allowed, and each variable was coded "1" (selected) or "0" (not selected).

5.3 Data analysis

Before testing our hypotheses, we first verified the factor pattern for decisional autonomy, since this is a new measure and this concept may conceal more than one dimension. Correlations were subsequently calculated to test hypotheses H_1, H_2 and H_3. Lastly, binary logistic regressions were calculated to verify the effect of decisional autonomy on the probability of making certain strategic choices (H_4).

5.4 Results

5.4.1 Descriptive data

Subcontracting logging SMEs have an average of 5.54 employees, excluding the owner (median of 3) and 84.8% of the sample had fewer than ten (10) employees. The average turnover is $2.05 million CDN (median of $500,000). With regard to commercial dependency, our sample can be considered to be strongly dependent on contractors, since the main client represents an average of 85% of the turnover (median of 100%). In fact, for 73.8% of logging SMEs, the main client represents 75% or more of the turnover and 55.7% of SMEs only have one client.

With regard to decisional autonomy, we noted certain differences in distribution among each of the components (see Table 1). For example, where the price of services is quite well distributed, such is not the case for selection of employees, where a strong majority of SMEs have low to complete autonomy. It should be specified that respondents were asked to indicate the SME's decisional autonomy, from a unilateral decision on the part of the client (1), to a decision entirely made by the SME (7), where the neutral response (4) is the equivalent of a negotiation between equal parties.

With respect to performance, all measures had a median of 3.00 (the measure varied from 1 to 5) and averages varied from 2.91 to 3.57, which corresponds fairly well to normal distributions and which is to be expected for this type of measure.

	Client (1-3)	Equals (4)	SME (5 to 7)	Average	Median
Price of services	49.4%	16.0%	34.6%	3.54	4.00
Selection of employees	19.9%	5.8%	74.3%	5.51	7.00
Working conditions	19.6%	13.7%	66.7%	5.24	6.00
Nature of the contract to be fulfilled	51.6%	11.2%	37.2%	3.47	3.00
Manner in which work is to be carried out	37.2%	13.4%	50.6%	4.19	4.00
Tools or technology to be used	20.9%	13.5%	65.6%	4.97	6.00

Table 1. Distribution of Decisional Autonomy Components

5.4.2 Factor analysis of decisional autonomy

An exploratory factor analysis was conducted with the decisional autonomy components. Only one item with a communality of less than 0.5 was removed, namely *autonomy with respect to tools or technology to be used*. Table 2 presents the two factors with an Eigenvalue greater than 1. These factors explain 80.62% of the total variance. The first factor, which we have entitled *autonomy of human resources (HR)*, includes autonomy with regard to selecting employees and working conditions. Cronbach's Alpha Cronbach (1951) for autonomy of human resources is 0.844.

	Autonomy of HR	Managerial autonomy
Selecting my employees	0.936	
Working conditions	0.899	
Price of our services		0.847
Nature of contracts to be fulfilled		0.893
Manner in which work is to be carried out		0.847
Cronbach's Alpha	**0.844**	**0.837**

Table 2. Factor Analysis of Decisional Autonomy

The second factor, entitled *managerial autonomy*, includes autonomy with respect to the price of services, the nature of the contracts to be fulfilled and the manner in which the work is to be carried out. Cronbach's alpha for managerial autonomy is 0.837. Autonomy of HR has an average of 5.38 out of 7.00 (median 6), whereas commercial autonomy has an average of 3.75 out of 7.00 (median 3.8).

At first glance, it may seem rather odd that the autonomy of human resources emerges as a distinct factor. These results must be replaced in their historic context to understand their relevance. As mentioned above, during the late 1970s, the large forestry companies moved away from logging operations and began hiring subcontractors, which led to the birth of a number of forest entrepreneurs. Many of these employees were unionized and did not want to lose their benefits as a result of the logging operations being subcontracted out. In Quebec, the *Labour Code* governs union rights. Section 2 of the Code provides that the logging operator (e.g. the big contracting company) shall be deemed to be the employer of all the employees engaged in his logging operations, including the subcontractor's (e.g. SME) employees. This statutory exception, which is specific to the forest industry, means that subcontractors working for a unionized contractor must abide by the agreements entered into between the union and the contractor, even if they were not consulted during the collective bargaining process. Thus, some subcontractors have little leeway with regard to the management of human resources, which explains why this factor emerges as distinct.

5.4.3 Testing the hypotheses

Since the commercial dependency variable does not follow a normal distribution (several SMEs in the sample are in a strong dependency situation), Spearman's Rho correlation test was used. As illustrated in Table 3, three (3) measures of performance are negatively influenced by commercial dependency, that is, the quality and variety of services, as well as customer satisfaction. This partially confirms H_1. With respect to HR and managerial autonomy, only respect for the environment and sustainable development are positively influenced by managerial autonomy, which partly confirms H_2.

	Commercial dependency	Autonomy of HR	Managerial autonomy
Increase in revenues	-0.144	0.076	0.013
Increase in the number of employees	0.130	-0.113	-0.155
Profit margin	-0.017	0.112	-0.139
Improvements in production techniques	-0.135	0.098	0.003
Use of new technology	-0.133	0.105	0.078
Quality of client service	**-0.269****	0.121	-0.075
Variety of services	**-0.346*****	0.157	0.061
Client satisfaction	**-0.198***	0.127	0.010
Personnel recruitment and retention	-0.042	0.022	-0.108
Investment in the community	0.004	0.091	-0.138
Respect for the environment and sus. dev.	-0.160	0.094	**0.164***
*$p \leq 0.05$ ** $p \leq 0.01$ *** $p \leq 0.001$*			

Table 3. Correlations Among Measures of Dependency and Autonomy and Performance

Our analyses also reveal an inverse relationship between commercial dependency and the two measures of decisional autonomy (Table 4). This confirms H_3.

	Commercial dependency	Autonomy of HR	Managerial autonomy
Commercial dependency	1.00		
Autonomy of HR	-0.272***	1.00	
Managerial autonomy	-0.315***	0.241**	1.00
*$p \leq 0.05$ ** $p \leq 0.01$ *** $p \leq 0.001$*			

Table 4. Correlations Among measures of Dependency and Autonomy

To verify the effect of decisional autonomy on strategic choices, a binary logistic regression was calculated for each strategic choice. The goal was to determine whether decisional autonomy influenced the probability of choosing (or not) this strategy for the coming years. The "step by step" method was used, where business size (number of employees) and the number of sectors were included as control variables. Business size could influence choices of growth and diversification. The number of sectors is a variable that computes the various services offered by the SME, such as harvesting, logging-road construction and wood transport. The more services an SME offers, the less likely it is to want to diversify within the forest, since it has already done so.

As illustrated in Table 5, only the strategies of growth and diversification of logging activities can be influenced by decisional autonomy. Specifically, managerial autonomy positively influences growth. The model enables us to correctly predict the strategic choice of growth in 88.1% of cases with these variables. With respect to diversification of logging activities, autonomy of human resources is the influential factor. The results therefore partly validate H_4.

6. Discussion

As we have noted, commercial dependency negatively affects performance with regard to the client, namely, the quality and variety of services offered and client satisfaction. This result is interesting because it highlights the fact that commercially dependent subcontractors provide poorer client service. This result suggests that over the course of a long-term relationship, subcontractors could deploy fewer efforts to satisfy the client, and limit themselves to initial contractor demands. However, where subcontractors have several clients, they must stay tuned to their specific needs, which helps improve client satisfaction as well as the efficiency of the overall value chain of their partners (Heikkilä, 2002). Nonetheless, any difficulty on the part of subcontracting SMEs to meet client requirements could reduce their performance (Bourgault, 1998). In a situation where the larger contractors are reducing their involvement in logging operations, SMEs less capable of meeting their requirements could disappear for lack of contracts.

	No change 67 cases		Growth 23 cases		Diversification within the forest 42 cases		Diversification out of forest 38 cases		Close/ Sell 20 cases	
	β	Exp (β)	β	Exp(β)	B	Exp (β)	B	Exp(β)	β	Exp(β)
Size	-0.02	0.98	0.00	1.00	-0.01	0.99	0.25	1.03	0,02	1.02
No. sectors	0.21	1.23	-0.16	0.85	0.31	1.36	0.18	1.20	-0.26	0.77
Comm. Autonomy.	-0.04	0.96	**0.36***	**1.44**	0.02	1.02	0.02	1.02	0.19	1.21
Autonomy HR	-0.14	0.87	0.143	1.15	**0.32***	**1.38**	0.16	1.18	-0.14	0.87
Fit index										
Number of cases[1]	135		135		135		135		135	
Nagelkerke R^2	0.038		0.117		0.117		0.080		0.062	
χ^2 (d.l. = 4)	3.866		8.408		10.796		7.216		6.329	
% Correct pred.	63%		88.1%		77.0%		79.3%		63.0%	

$* p \leq 0.05$ $** p \leq 0.01$ $*** p \leq 0.001$

Table 5. Influence of Decisional Autonomy on Strategic Choices

We were somewhat surprised by these results, since no other measure of performance is influenced by commercial dependency. These results could lend credence to the proposal by Julien (2000), who argues that the subcontractor's dependency is strongly dependent on the nature of the relationship itself, which varies from "subcontracting of capability" to

---

[1] The cases observed with missing data were removed from the analysis.

"subcontracting of specialty" and "subcontracting of intelligence". According to this author, contractors are more dependent on their subcontractors in a "subcontracting of intelligence" relationship than in a "subcontracting of capability" relationship where they can more freely impose their directives and demands. According to Legendre (2005), the large logging companies haven't had in-forest expertise for some thirty years, since this work has been sub-contracted out. They must ensure that their subcontractors benefit from their relationship, or the SMEs will shut down their operations, and the contractors will suddenly face supply shortages. Unlike major contract providers in the manufacturing sector, which can import offshore parts through delocalized production (Lecler, 1991), the larger logging companies must deal with local subcontractors, since their services are not as easily delocalized.

Moreover, as was demonstrated in a study that highlighted the relationship between two producers and a distributor (Bergès & Chambolle, 2009), we note that the distributor is prepared to concede a greater portion of the added value to the producers (*i.e.* pay a higher price for their products) if he is concerned about one of them going out of business, thereby reducing his number of potential suppliers, which he feels may negatively affect his own future performance. In this context, commercial dependency is not a determining factor to explain performance. Other dimensions, such as the development of relationships geared towards the long-term between a contractor and a supplier, could be more useful in influencing subcontractor performance (Paulraj *et al.*, 2008).

Hence, in a relationship based on trust geared towards the long-term, we note that commercially dependent subcontractors are more inclined to innovate, even if they are unable to correctly assess the benefits they will reap, as long as they have identified a need to be fulfilled on the part of the contractor (Kamath & Liker, 1990). In other words, if the subcontractor strongly believes that he will reap benefits by following the contractor's directives, he will take more risks since he expects better performance. The subcontracting relationship, in fact, shares similar mechanisms with an employment relationship, where the contract (employment or subcontracting) is in fact an exchange of promises where both parties take a gamble (Baudry, 1992). In this case, as the author suggests, the relationship of authority only ensures partial coordination and requires a certain amount of trust and confidence to validate this gamble.

This study has enabled us to highlight a concept that is specific to subcontracting SMEs, that is, decisional autonomy. In a situation of commercial dependency, some contractors use their power over the SME to interfere in the management of their business, thereby curtailing their decisional autonomy. Our results suggest that these concepts are inter-related, but remain nevertheless distinct. Decisional autonomy does not influence performance in quite the same manner as commercial dependency. Furthermore, this concept is much more specific than having the order book being concentrated in the hands of a few contractors, and takes into consideration not only the number of clients involved, but rather the nature of the relationship between contractors and their subcontractors.

Our analyses demonstrate the existence of two distinct factors with regard to decisional autonomy, namely, managerial and human resources autonomy. The former corresponds to interference on the part of the contractor with regard to the price of services (or products, in another context), the nature of the contracts and the manner in which they are fulfilled. The nature of contracts refers to the decision on the part of the contractor to determine which felling areas[2] are to be awarded to a particular subcontractor. In this type of situation where there's little negotiation over price, which may even be imposed by the contractor, selecting the felling area becomes a strategic issue. Some areas are more easily accessible, trees are bigger, etc. As a result, dollar for dollar, the better felling areas allow for greater profits since operating costs are lower.

The human resources autonomy factor, as mentioned above, is rooted in a historical situation where some subcontractors are required to abide by union agreements with the contractor to ensure that their employees continue to enjoy the benefits they had gained while employed by the contractor. However, this factor is not necessarily unique to the forestry sector. In the automobile industry, Laval (1998) highlighted the influence of the contract provider's human resources policies on subcontractors through their purchasing policies. In other industries, contract providers may suggest that subcontractor employees undergo training to meet industry standards (Esposito & Raffa, 1996). Although there may well be a sectoral and historic effect on our data, it is reasonable to suggest that this factor could be equally significant in other industries where SMEs are deeply involved as subcontractors.

It was surprising to note that autonomy of HR did not positively influence SME performance. As suggested by Naro (1990), human resources must be aligned with the strategic directions of the SME and its management in order to gain a competitive advantage. As a result, it is not surprising to note that HRM positively influences SME performance (Lacoursière et al., 2005). Thus, in a situation where SMEs lose their autonomy over decisions about their employees and working conditions, an effect on performance would be expected. However, it is possible that SME managers believe that the contractor imposes certain restrictions with respect to human resources, but that even without these restrictions, they would have to hire the people who are available and offer them the same benefit packages as their competitors. In other words, despite the perceived loss of autonomy, the sectoral constraints that affect human resources, such as the shortage of qualified labour and salary standards, impose, in themselves, restrictions that reach beyond any potential interference on the part of the contractor. As a result, all SMEs are equally affected, since our study is uni-sectoral, which could cancel the effect of autonomy of HR on SME performance. Studies within other industries would be required before drawing any conclusions in this area.

Managerial autonomy of subcontracting SMEs positively influences performance with regard to respect for the environment and sustainable development. This suggests that the more leeway SME managers have to make decisions about their commercial strategy, the more involved they become in achieving excellence with regard to the environment. It would appear that once managers lose their managerial autonomy, they simply do what the contractor asks them to do. Inversely, when managers maintain their autonomy with regard to their business decisions, their personal goals are likely to influence their performance objectives and the performance of their SME (St-Pierre & Cadieux, 2009).

---

[2] A felling area represents the surface are for which a sub-contractor has been given a harvesting contract. Its boundaries are determined according to the plans provided by the contractor.

Autonomy of human resources influences the choice to diversify in the future. These results suggests that such autonomy enables entrepreneurs to have greater leeway to assign staff to other operations, for example, to train logging staff to be able to repair equipment, use machinery to build logging roads or to carry out sylviculture duties. Entrepreneurs feel more confident if they feel they can get their employees to contribute to the development of new activities.

Also, managerial autonomy influences the desire to grow in the future. When subcontracting SMEs have little autonomy with regard to their commercial choices, they are not driven to find other clients or to increase sales. It is likely that the little control they have over their commercial choices increases their perceived risk and, as a result, reduces their intention to further develop their activities. As has been noted with other subcontracting SMEs in the defence sector, contractors influence the strategic choices of SMEs/SMIs, in particular so that they can diversify their order books toward other industries (Frigant & Moura, 2004). Thus, they can avoid slowdowns caused by a temporary drop in demand within the industry and remain available for the contractor when business picks up again. Other contractors indirectly impose external growth on subcontracting SMEs, by requiring constraints that only average-sized businesses can implement (for example, a computerized "just-in-time" system, quality assurance practices, etc.) (Tréhan, 2004). In the forest industry, it would appear that the reverse occurs. Subcontractors who do not have choices imposed on them by contractors are more likely to want to grow and diversify to offer turnkey services. Thus, decisional autonomy enables forest entrepreneurs who show significant entrepreneurship to exercise their strategic choices (St-Jean *et al.*, 2010).

7. Limitations

Although this study enables us to discuss a rarely studied concept with regard to the relationship between a contractor and his subcontractors, namely, decisional autonomy, the results should be interpreted within the limitations of the methodology used. First, it should be emphasized that this study used data from a single sector, that is, the forest industry. This may influence the results, for example, with regard to autonomy of human resources, although this aspect has been discussed above. Research on this concept in other industries would be necessary to confirm its accuracy and relevance. The data used were taken from a follow-up survey with initial respondents of a survey conducted in 2006. Given the widespread restructuring that occurred in this industry during this period (factory closures, corporate mergers, etc.), the sample of respondents could be less representative than in the initial probe in 2006, where the entire SME population was targeted for the investigation. However, given the relatively high response rate and number of respondents, the sample can nevertheless be considered reliable.

It should be underscored that the performance measures used in this study are subjective. This is a valid course of action where objective data is unavailable (Dess & Robinson, 1984), but subjective data may be less reliable. With regard to business strategy, we did not require entrepreneurs to choose a predominant strategy, but rather make a selection from possible applicable options. In addition, this did not reflect their observed strategy, but rather their strategic intentions, which could be less stable over time. Lastly, only the subcontractor's vision was considered. No measure of decisional autonomy on the part of the contractor was taken into consideration. This would be an interesting area to investigate in the future.

8. Conclusion

In a context of SME commercial dependency on contractors, the concept of SME decisional autonomy provides an opportunity to consider a dimension that has received little or no attention in past research, the significance of which can be felt in the strategic choices of managers and its influence, if only partial, on performance. Additional research is necessary, however, to validate its effect with regard to other areas that could influence performance, such as the development of SME-specific skills or innovation. Businesses which find themselves in a situation where they have to obey the orders of a contractor may have limited resources to innovate or develop new and promising niche markets. Further studies in this area, and many others, would help contribute to a better understanding of contractor/subcontractor relationships.

This study adds nuance to prior research that did not take into consideration the effect of subcontractor type as a moderator in the relationship between commercial dependency and subcontractor performance (for example, Rinfret *et al.* (2000), Cox *et al.* (2004) or Kalwani et Narayandas (1995)). Since this study was conducted exclusively within the forest industry, our results invite further comparative studies where the nature of the relationship between the subcontractor and the contract provider would be controlled (subcontractor type), which would help better explain the effect of commercial dependency, or even decisional autonomy, on SME performance.

9. References

Amesse, F., Dragoste, L., Nollet, J. and Ponce, S. (2001), "Issues on partnering: evidences from subcontracting in aeronautics", *Technovation*, Vol. 21, No. 9, pp 559-569.

Barringer, B. (1997), "The Effects of Relational Channel Exchange on the Small Firm: A Conceptual Framework", *Journal of Small Business Management*, Vol. 35, No. 2, pp 65-79.

Baudry, B. (1992), "Contrat, autorité et confiance: La relation de sous-traitance est-elle assimilable à la relation d'emploi?", *Revue économique*, Vol. 43, No. 5, pp 871-893.

Beamon, B. (1999), "Measuring supply chain performance", *International journal of operations and production management*, Vol. 19, No., pp 275-292.

Bergès, F. and Chambolle, C. (2009), "Threat of Exit as a Source of Bargaining Power", *Recherches économiques de Louvain*, Vol. 75, No. 3, pp 353-368.

Blais, R. and Chiasson, G. (2005), "L'écoumène forestier canadien : État, techniques et communautés - l'appropriation difficile du territoire", *Revue canadienne des sciences régionales*, Vol. 28, No. 3, pp 487-512.

Bourgault, M. (1998), "Performance industrielle et contribution des sous-traitants nationaux : analyse du secteur aérospatial canadien dans le contexte nord-américain", *Revue internationale PME*, Vol. 11, No. 1, pp 41-63.

Bronson, E.A. (1999), *Openings in the forest economy: A case study of small forest operators in the Bulkley Valley, B.C., Canada*, Unpublished Thesis, University of British Columbia, Vancouver, B.C.

Carland, J.W., Hoy, F., Boulton, W.R. and Carland, J.A.C. (1984), "Differientiating Entrepreneurs from Small Business Owners: A Conceptualization", *Academy of management review*, Vol. 9, No. 2, pp 354-359.

Corsten, D. and Felde, J. (2005), "Exploring the performance effects of key-supplier collaboration", *International Journal of Physical Distribution & Logistics Management*, Vol. 35, No. 6, pp 445-461.

Cox, A., Watson, G., Lonsdale, C. and Sanderson, J. (2004), "Managing appropriately in power regimes: relationship and performance management in 12 supply chain cases", *Supply Chain Management: An International Journal*, Vol. 9, No. 5, pp 357-371.

Cronbach, L. (1951), "Coefficient alpha and the internal structure of tests", *Psychometrika*, Vol. 16, No. 3, pp 297-334.

Daniels, J. and Bracker, J. (1989), "Profit performance: do foreign operations make a difference?", *Management International Review*, Vol. 29, No. 1, pp 46-56.

Dess, G.G. and Robinson, R.B. (1984), "Measuring organizational performance in the absence of objective measures: the case of the privately-held firm and conglomerate business unit", *Strategic Management Journal*, Vol. 5, No. 3, pp 265-273.

Emerson, R. (1962), "Power-dependence relations", *American sociological review*, Vol. 27, No. 1, pp 31-41.

Esposito, E. and Raffa, M. (1996), "L'évolution de la gestion de la qualité totale dans les petites entreprises sous-traitantes du secteur de l'aéronautique", *Revue internationale PME*, Vol. 9, No. 2, pp 57-80.

Filion, L.J. (2005), "Entrepreneurs et propriétaires-dirigeants de PME", in P.-A. Julien (Ed). *Les PME - Bilan et perspectives*, Presses InterUniversitaires, Cap-Rouge (Québec), pp 143-182.

Frigant, V. and Moura, S. (2004), "Les déterminants des stratégies réactives des sous-traitants de la défense. Le cas des PME aquitaines, lombardes et du North-West dans la décennie 1990", *Revue internationale PME*, Vol. 17, No. 3-4.

Gartner, W.B. (1989), ""Who Is an Entrepreneur ?" Is the Wrong Question", *Entrepreneurship Theory and Practice*, Vol. 13, No. 4, pp 47-68.

Gondran, N. and Brodhag, C. (2003), "Rôle des partenaires des PME/PMI dans l'amélioration de leurs performances environnementales", *Revue internationale PME*, Vol. 16, No. 2, pp 35-59.

Graves, S. and Waddock, S. (1994), "Institutional owners and corporate social performance", *Academy of Management Journal*, Vol. 37, No. 4, pp 1034-1046.

Heikkilä, J. (2002), "From supply to demand chain management: efficiency and customer satisfaction", *Journal of operations management*, Vol. 20, No. 6, pp 747-767.

Henricks, M. (1993), "Too big, too few, too risky?", *Small Business Reports*, Vol. 18, No. 10, pp 49-57.

Holmlund, M. and Kock, S. (1996), "Buyer Dominated Relationships in a Supply Chain - A Case Study of Four Small-Sized Suppliers", *International Small Business Journal*, Vol. 15, No. 1, pp 26-40.

Inderst, R. and Shaffer, G. (2007), "Retail mergers, buyer power and product variety", *Economic Journal*, Vol. 117, No. 516, pp 45-67.

Inderst, R. and Wey, C. (2007), "Buyer power and supplier incentives", *European Economic Review*, Vol. 51, No. 3, pp 647-667.

Isaksen, A. and Kalsaas, B. (2009), "Suppliers and strategies for upgrading in global production networks: the case of a supplier to the global automotive industry in a high-cost location", *European Planning Studies*, Vol. 17, No. 4, pp 569-585.

Jansson, H. and Hilmersson, M. (2009), "Escaping the trap of low-cost production and high dependency: a case study of the internationalization networks of small subcontractors from the Baltic States", in EIBA (Ed). *Progress In International Business Research*, Emerald Group Publishing Limited, pp 225-247.

Julien, P.-A. (2000), *L'entrepreneuriat au Québec: Pour une révolution tranquille entrepreneuriale 1980-2005*, Les Éditions Transcontinental inc., Montréal, Québec.

Kalwani, M. and Narayandas, N. (1995), "Long-term manufacturer-supplier relationships: do they pay off for supplier firms?", *The Journal of Marketing*, Vol. 59, No. 1, pp 1-16.

Kamath, R., R. and Liker, J.K. (1990), "Supplier dependence and innovation: a contingency model of suppliers' innovative activities", *Journal of Engineering and Technology Management*, Vol. 7, No. 2, pp 111-127.

Kaplan, R.S. and Norton, D.P. (1992), "The Balanced Scorecard - Measures That Drive Performance", *Harvard Business Review*, Vol. 70, No. 1, pp 71-79.

Kilduff, P. (2005), "Patterns of strategic adjustment in the US textile and apparel industries since 1979", *Journal of Fashion Marketing and Management*, Vol. 9, No. 2, pp 180-194.

Kumar, N., Scheer, L. and Steenkamp, J. (1995), "The effects of perceived interdependence on dealer attitudes", *Journal of Marketing Research*, Vol. 32, No. 3, pp 348-356.

Lacoursière, R., Fabi, B., St-Pierre, J. and Arcand, M. (2005), "«Effets de certaines pratiques de GRH sur la performance de PME manufacturiers: vérification de l'approche universaliste»", *Revue internationale PME*, Vol. 18, No. 2, pp 43–73.

Laval, F. (1998), "La gestion des ressources humaines des entreprises fournisseurs-partenaires : l'impact de la politique d'achat d'un grand groupe donneur d'ordres", *Revue internationale PME*, Vol. 11, No. 2-3, pp 75-94.

Le Roy, F. (2001), "Agressivité concurrentielle, taille de l'entreprise et performance", *Revue internationale PME*, Vol. 14, No. 2, pp 67-84.

Lecler, Y. (1991), "Les fournisseurs/sous-traitants japonais : quasi-ateliers ou partenaires de leurs donneurs d'ordres ?", *Revue internationale PME*, Vol. 4, No. 2, pp 137-162.

Legendre, C. (2005), *Le travailleur forestier québécois : Transformations technologiques, socioéconomiques et organisationnelles*, Presses de l'Université du Québec, Sainte-Foy (Québec).

Lyons, T., Krachenberg, A. and Henke Jr, J. (1990), "Mixed motive marriages: what's next for buyer-supplier relations?", *Sloan Management Review*, Vol. 31, No. 3, pp 29-36.

Mäkinen, P. (1993), "Metsäkoneyrit-tämisen menestystekijät" (Report No. 818), Folia Forestalia - The Finnish Forest Research Institute, Helsinki, Finland.

McMahon, R. (2001), "Business growth and performance and the financial reporting practices of Australian manufacturing SMEs", *Journal of Small Business Management*, Vol. 39, No. 2, pp 152-164.

Mercure, D. (1996), *Le travail déraciné : l'impartition flexible dans la dynamique sociale des entreprises forestières au Québec*, Boréal, Montreal, Canada.

Murphy, G. and Callaway, S. (2004), "Doing well and happy about it? Explaining variance in entrepreneurs' stated satisfaction with performance'", *New England Journal of Entrepreneurship*, Vol. 7, No. 2, pp 15-27.

Murphy, G.B., Trailer, J., W and Hill, R.C. (1996), "Measuring performance in entrepreneurship research", *Journal of Business Research*, Vol. 36, No. 1, pp 15-23.

Naro, G. (1990), "Les P.M.E. face à la gestion de leurs effectifs : comment adapter les ressources humaines aux impératifs stratégiques ?", *Revue internationale PME*, Vol. 3, No. 1, pp 57-74.

Paulraj, A., Lado, A. and Chen, I. (2008), "Inter-organizational communication as a relational competency: Antecedents and performance outcomes in collaborative buyer-supplier relationships", *Journal of operations management*, Vol. 26, No. 1, pp 45-64.

Perera, S., Harrison, G. and Poole, M. (1997), "Customer-focused manufacturing strategy and the use of operations-based non-financial performance measures: a research note", *Accounting, Organizations and Society*, Vol. 22, No. 6, pp 557-572.

Perrow, C. (1974), *Organizational analysis: A sociological view*, Tavistock London.

PREFoRT (2007), "Résultats préliminaires - Sondage sur les entrepreneurs forestiers", Université Laval, Québec, Canada.

Ramsay, J. (1990), "The myth of the cooperative single source", *Journal of Purchasing and Materials Management*, Vol. 10, No. 8, pp 19-28.

Raymond, L. and St-Pierre, J. (2002), "Performance effects of commercial dependency for manufacturing SMEs", National Meeting of the Academy of Business Administration, Key West, Florida.

Remili, N. and Carrier, S. (2006), "Exploration du concept de dépendance commerciale dans un contexte de turbulence : le cas de l'industrie québécoise du vêtement", Conférence de l'Association des Sciences Administratives du Canada (ASAC), Banff, Alberta.

Rinfret, L., St-Pierre, J. and Raymond, L. (2000), "«L'impact de la dépendance commerciale sur les résultats financiers des PME manufacturières»", *V ème Congrès International Francophone de la PME, Lille, France*, Vol., No., pp 1-15.

Risker, C.D. (1998), "Toward an Innovation Typology of Entrepreneurs", *Journal of Small Business and Entrepreneurship*, Vol. 15, No. 2, pp 27-41.

St-Jean, E., Julien, P.-A. and Audet, J. (2008), "Factors associated with growth changes in "gazelles"", *Journal of Enterprising Culture*, Vol. 16, No. 2, pp 161-188.

St-Jean, E., LeBel, L. and Audet, J. (2010), "Entrepreneurial Orientation in the Forestry Industry: A Population Ecology Perspective", *Journal of Small Business and Enterprise Development*, Vol. 17, No. 2, pp 204-217.

St-Pierre, J. and Cadieux, L. (2009), "La conception de la performance : quel lien avec le profil entrepreneurial des propriétaires dirigeants de PME", 6e Congrès de l'Académie de l'entrepreneuriat, Sophia-Antipolis, France.

Tréhan, N. (2004), "Stratégies de croissance externe des moyennes entreprises patrimoniales sous-traitantes", *Revue internationale PME*, Vol. 17, No. 1, pp 9-35.

Ulrich, D. (1999), "Measuring human resources: an overview of practice and a prescription for results", in R. S. Schuler and S. E. Jackson (Eds). *Strategic human resource management*, Wiley-Blackwell, Malden, MA, pp 462–482.

Wilson, P. and Gorb, P. (1983), "How large and small firms can grow together", *Long Range Planning*, Vol. 16, No. 2, pp 19-27.

Wolff, J.A. and Pett, T.L. (2006), "Small-Firm Performance: Modeling the Role of Product and Process Improvements", *Journal of Small Business Management*, Vol. 44, No. 2, pp 268-284.

5

Member, Owner, Customer, Supplier? – The Question of Perspective on Membership and Ownership in a Private Forest Owner Cooperative

Gun Lidestav and Ann-Mari Arvidsson
Department of Forest Resource Management, SLU, Umeå
Sweden

1. Introduction

1.1 Forest owner cooperatives – Association and corporation at a market in change

In response to their exposed position on the timber market in the beginning of the last century, Swedish private forest owners started to organize themselves in forest owner cooperatives. Initially the cooperatives' only business was collecting timber from the members in order to bring larger volumes to the timber market (Andersson *et al*, 1980). Through these joint deliveries, the forest owners (members) gained an improved bargaining position and could get better pay for their timber deliveries (Glete, 1987). In the early 1940, when the cooperatives could not reach their economic goals only by trading their members' timber, some of the cooperatives bought or established new sawmills and other wood processing industries. From the board of the cooperatives the main motive put forward was that, by owning their own industry, members could achieve surplus values (Gummesson 1993). Thus, the Swedish forest owner cooperatives follows the general characteristics of a cooperative summarized by Skår (1981) such that the cooperative constitutes of an economic business with joint action between members and consists of a democratic association and an enterprise (corporation). Further, individuals are assumed to become members for social and other reasons, but their interests lie in their individual activities and benefits. However, for members who join a cooperative, dilemmas arise when members' decisions are made as joint decisions that can be very different from the individual's own decision (ibid). This could, according to Nilsson & Björklund (2003), cause organizational problems when the association and the enterprise are two different sides of the same coin. The analytical implications of this organizational duality and complexity of the cooperative will be developed further in the next section. In practice, the Swedish private forest owner associations has, as one way of dealing with the duality and the multiple needs of members, introduced other services to their members such as management planning, providing tax advice, undertaking silviculture on the forest owner's request, arranging forest-days and evenings for the members. Additionally, employees at the cooperative represent the private forest owners in dialogue with authorities and advocate for good policies concerning business in the timber market and in various forest policy issues. The lobbying to the government and other authorities, is however mostly handled from The Federation of Swedish Family Forest Owners, an umbrella organization for the Swedish private forest owners´ cooperatives (http://www.lrf.se/In-English/Forestry/).

Norra skogsägarna
16,000 members
4 processing industries
400 employees
(http://www.norra.se/
templates/Page.aspx?id=496)

Skogsägarna Norrskog
13,000 members
6 processing industries
300 employees
(http://www.norrskog.se/Upload/
web_Norrskogs_årsberättelse%)

Mellanskog
32,000 members
Part owner in Setra Group AB
200 employees
http://www.mellanskog.se/t
emplates/MS_InfoPage.aspx?id=528

Södra skogsägarna
51,000 members
29 processing industries
4000 employees
(http://www.sodra.com/sv
/Om-Sodra)

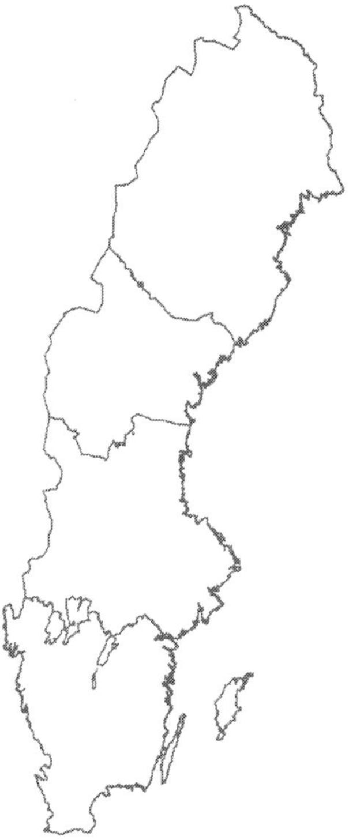

Fig. 1. The four major private forest owner cooperatives in Sweden.

With an increasing industrial demand for timber and forest fuel, there are, different to earlier situations, other actors in the forest sector who are eager to serve and start business with the private forest owners and offer comparable services as cooperatives (Törnqvist 1995). Further, due to the Swedish Competition Act, the cooperatives are not allowed to restrict or complicate member´s mobility on the market. For example, a cooperative member can sell to any buyer, while the forest owner cooperative cannot refuse a delivery from one of its members, if nothing else is said (Swedish Government, 1992/93; Swedish Government, 1999/2000; Swedish Codes of Statues 1993).

Similar to the structural changes in other parts of the society, the cooperatives have gradually merged and today there are four major cooperatives, namely Norra Skogsägarna, Norrskog, Mellanskog and Södra skogsägarna that cover the entire Sweden (Figure 1). All together they organize 112 000 members (management units) corresponding to 53 % of private forest ownership in Sweden (Swedish Forest Agency, 2010).

1.2 A conceptual model of cooperative

Built on the basic assumptions of a cooperative as constructed by a democratic association of members and an economic activity (enterprise) outlined by Georges Fauquet, Skår (1981) has developed a conceptual model for the analysis of cooperatives. The economic activity is presumed to be directed by collective decisions of the members, which give rise to a number of cooperative coordinating decisions and cooperative plant(s). Between the two sides of the model; I) the Individuals/Members and II) the Economic environment, there exist three forms of relationships: A) an organizational relationship made up by the participation, information, communication and control aspects, B) the stationary or structural relationship, which refers to the activities executed in plants of members and the cooperative, and C) the functional relationship, which consists of the flow of economic activity and tangible assets from the members and the plant (Figure 1). Due to the integrative nature, members become mutually interdependent in their efforts to pursue their own individual objectives, which give rise to the cooperative dilemma mentioned previously.

According to Skår (1983) a well-functioning cooperative must have a well-developed organizational relationship. It is through this relationship the collective decision-making will be developed and processed and the operation be controlled. This requires most likely a sense of belonging or fellowship to the cooperative, such as those expressed by a common language, knowledge and education, norms etc. As the members have different ideas, needs and capacity to act, the organizational relationship may be regarded as a negotiating body. Thus, the form of decision making is of great importance in a cooperative, as those who master the forms can guide the decisions (Skår, 1981).

1.3 Values and benefits of forest owners

With respect to "the sense of belonging and fellowship" it is well-known that, private forest owners in Sweden as well as in Finland, Norway and USA, have become more heterogeneous, and in addition less dependent on forestry as their only income (for an overview see Fischer et al 2010). The same can be claimed for cooperative members (Berlin 2006). Further there are now more non-resident forest owners (Lidestav & Nordfjell 2005), which presumably undermine the feeling of fellowship (solidarity) to some extent. The increase of joint ownerships may on the other hand have the opposite effect. From earlier studies it is also known that resident forest owners in Sweden regard forestry income, residence, availability of firewood and timber for own use, outdoor life and recreation highly important, while non-residence owners regard outdoor life and recreation, maintaining contact with native locality, availability of firewood and timber for own use and keeping up a tradition in forestry as highly important. In general, residence owners regard benefits that could be considered as monetary more important whereas non-residence owners regard non-monetary benefits more important (Lidestav & Nordfjell, 2005). Looking at cooperative members specifically, Berlin et al (2006) has found that resident members regard housing, and timber and firewood for own use more valuable than non-resident member do. It also seems that resident members tend to place higher values on forest income, keeping in contact with native locality and preserving a forestry tradition than non-resident members do. In accordance with Skår (1981) Berlin (2006) argues that member´s perceptions, motives and values are likely to play a major role in the future of cooperative organizations. It is therefore urgent to increase knowledge about the relation (agreement or lack theeof) between the cooperative and its members. In particular, the mismatch between the defined goal of the cooperative (to optimize the economic result for its members) and the members´ multiple goals is something that the forest owner cooperatives need to consider (Berlin 2006).

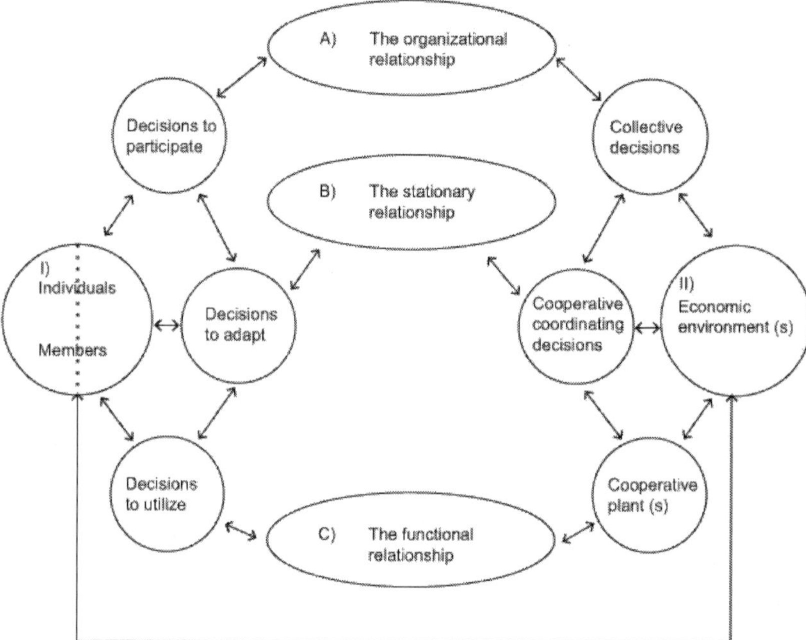

Fig. 2. The cooperative model suggested by Skår. (Figure adapted from Skår 1981:240).

The **aim** of the paper is to more thoroughly identify fields and aspects of agreement or lack thereof between how members themselves, and inspectors and managers understand the forest owner cooperative as a provider of benefits, including economic results for the members. By that, ideas on how to increase the members' comprehensive picture, the entirety, of the organization are developed. The organizational relationship, marked with A in the analytical model presented in Figure 2, will therefore be the scope of this study.

1.4 Theoretical and methodological approach

To explore and understand the complexity of motives and interactions, or lack thereof between members and the cooperative (i.e. the organizational relationship) a set of theories were used and applied on empirical data gathered by qualitative methods. Focus has been set on the concept of *identity, benefit* and *agreement*. The forest owners' identities and benefits have been analyzed through mode of life theory, while agreement is analyzed through the theory of new institutional organizing in organizations and the theory of meaningful communication.

It was considered necessary to investigate the identities of the private forest owners from two angles. First, according to the forest owners themselves based on their own subjectively experienced identity. Second, based on the cooperatives employees (inspectors' and the managers') perceptions of the members´ identities as forest owners. This qualitative data was then analyzed through the theory of mode of life. This theory provides a way of placing people into structured categories based on the way they live their lives in the society. The theoretical structures will work as a tool for understanding how people choose their way in life, which will impact their choices and solutions in different issues (Jacobsen, 1999). It has previously been applied by Törnqvist (1995) in his comprehensive sociological study on private, non-industrial forest ownership in Sweden. According to Törnqvist, owner of private forest land, even if several categories can be distinguished, have its main structure in the independent mode of life, although there are also those with wage earner life mode and mixes in between (Törnqvist, 1995). The theory makes it possible to talk about "everyday life" in a systematic, categorized way, and it is from this individual level as the identity of the members "everyday life" can be categorized.

An organization with many decision levels tends to be organized in a classical, hierarchical way to achieve its goal. According to Nilsson & Björklund (2003) the contemporary Swedish forest owner cooperatives are organized in this way, with a superior leadership and with one goal - the best economic results for the members. However, their present members have different goals and desired benefits from the ownership of a forest property (Berlin, 2006; Lidestav & Nordfjell, 2005). In order to keep and develop the engagement of the members, Nilsson & Björklund (2003) therefore advocate a renewed membership, which better corresponds to the goals and desired benefits of the members. Equally important is a renewed perception of the cooperative from the members´ viewpoint, since the way of gaining and distributing the profit has changed. Previously the business activity was purely based on trading timber. Nowadays, much of the profit appears in the cooperatives' processing industry and will be delivered to the members by three different financial instruments based on the investment capital of the individual member. To deal with this increased complexity, the members must look upon the cooperative as a whole (Nilsson & Björklund, 2003). If the organization is looked upon in a new institutional way, the wholeness will become more obvious. This theoretical perspective allows a more inclusive way of understanding a variety of benefits and goals, and also provides a better understanding of the delivery of profits to the members (Nilsson & Björklund, 2003).

The organization shall also be defined, conceptualized, and what is very important, even act in that way throughout the levels (Högvall Nordin, 2006 p. 51). Also Abrahamsson & Andersen (2000) argue through the theory of the new institutional organization, that rationality does not have the biggest impact, while the people in the organization go outside these rolls and also work for other important benefits and goals. In this organization the system is open, influenced by things around, which are allowed to be considered important.

It must also be considered important that the communication between the different levels in the organization should work satisfactory; it should result in a clear agreement between the levels. The theory of meaningful communication can give an insight and understanding in that field (Weick et al, 2005 p. 413). To be effective, the communication must be allowed to grasp complex, creative processes where the issues can be constructed and understood through interaction between individuals in the whole organization, through all levels. Evidently, individuals have to share a common understanding of norms, symbols, concepts and "words" (c.f Skår 1981). Here the theory of structuralism, with emphasis on the impact of language, can be useful e.g. when investigating how members are addressed; what words are used to referrer to the members, and what definition corresponds to the specific words. It should also be asked what the words imply, as meanings, in the mind, and ultimately the social world, in which people as well as other aspects more or less are shaped by the structure of the language (Ritzer, 2000). When people shall shape their surroundings, this is important because out of the definition, steps are taken to treat that person in a certain way where the language more or less gives the perspective of individuals (Ritzer, 2000). These different perspectives are used as communication for interaction between the actors in the different levels in the organization.

2. Design of the study

The host company for the present case study was the forest owner cooperative Norra Skogsägarna economic association (from here on referred to as Norra skogsägarna), situated in the north of Sweden, with head office in the city of Umeå. The 16 thousand memberships in Norra Skogsägarna, are divided into 4 regions and 26 management district (abbreviated SBO hereafter). The democratic principle, one member one vote, is important (www.norra.se 20110911). It is in the members' custody to vote for the major board and also for the board in each SBO, which both are handled by election committees. To bring their voice, the members can make proposals that will first be discussed by the board of the SBO, and further at the superior board and finally brought to the representatives (assembly) meeting. At the annual meeting in each SBO, members are elected to represent the SBO members at the assembly which is held every year. Any member can raise a proposal, either to be treated at the SBO annual meeting or can request that it shall be treated at the Assembly meeting. The proposal can also be treated directly in the major board and is then called an official letter. In the regulations of the cooperative the main and first paragraph is: Work for the economic interests of the members through the association or through anybody else: 1) run business with and processing of forest and forest products from principally the members and according to this run financial business 2) work for a safe and well adapted marketing of the members' timber deliveries to satisfactory prices (www.norra.se 20110910). As a member you are also owner and thus have to build up an owner capital by a 2% payment from every timber delivery to the forest owner cooperative up to a value corresponding to 10 % of the property´s ratable value. The minimum investment capital (capital share) is 2000 SEK and the maximal is 50 000 SEK. (10 SEK =1.1 Euro)

Norra Skogsägarna has 27 offices for inspectors, each of them having 1 to 4 inspectors at work, with a total of 56 inspectors. The cooperative also has 4 wood-processing industries and 3 department stores for building materials, and 400 people are employed at Norra Skogsägarna. In 2008 the cooperative had 1. 8 billion SEK in turnover (www.norra.se 20110910). The profit generated from the business of Norra Skogsägarna goes back to the members. Before 2002 there was only one single instrument for this: a post-payment on the timber delivery done during the actual financial year. However, since 2002 there are also dividend and interests on invested capital. As an example, between 2004-2008, 93 million

SEK (~ 10 million Euro) has been distributed to the members in these ways. The superior board gives suggestions about the amount of the profit to be distributed to the members and which financial instrument shall be used, while the final decision is taken by the representatives (assembly) meeting.

The design of the study was elaborated with the assistance of an advisory group, consisting of one of the regional managers, the previous long-lasting member principal at Norra Skogsägarna, three elected members of the board of SBOs and the membership principal at the forest owner cooperative Mellanskog. Key questions were elaborated, pre-tested on a test panel, and then addressed through focus group discussions following the guidelines by Wibeck (2000). The main questions were:

- What are the benefits for you being a member of Norra skogsägarna forest owner cooperative?
- What are the benefits for you being part-owner of Norra skogsägarna forest owner cooperative?
- In what ways can Norra Skogsägarna forest cooperative assist you better in those issues that are important to you?
- What activities are poorly informed about to the members?
- What do you think about the interaction between the cooperative and the members?
- Are there new activities that both you and the cooperative could benefit from?

For each of the 12 SBOs a list of 15 memberships, randomly selected, was provided by the membership section, and from that list the local chairman was asked to recruit 4 – 6 persons for a focus group discussion. In order to capture the opinion of woman and non-resident members, who are often overlooked, it was decided that at least two of the SBO should have "women-only" focus groups, and further that one focus group discussion should be conducted with nonresident forest owners in Stockholm. One group with four managers from head office and another with four inspectors, who are active in day-to-day operation in the SBOs, were recruited. These discussions had a different course, as the inspectors and the managers were asked to tell how they thought the members' answers to these questions were. They were also asked to reflect on the agreement or lack thereof between members and the cooperative from their own perspective as inspectors or managers. These group discussions were made late in the study, which was an advantage to the study, as many organizational questions had appeared during the discussions, and thus could be followed up with the inspectors and managers. In addition, webpages and printed materials, such as membership paper and an anniversary book, were surveyed in order to examine the discourse of the word 'member'.

Prior to the first discussion, the recruited members got an invitation with a welcome letter together with the two stimulus material to get them familiar with the questions before coming to focus group discussion. The purpose of the first stimulus material was to initiate the discussion in the focus group and to explore their forest owner identity. This was done by asking the participants to select personal "key words" reflecting their personal opinion from a list of words related to the categories, experiencing the forest/nature, forestry as an occupation, management, family relations, emotions, work and tools. The second stimulus material dealt with benefits and values, and the participants were asked to indicate how they considered the importance (on a five graded scale) of the values and the benefits; forest incomes, hunting/fishing, berries/mushrooms, firewood/timber for household use, residence/housing, outdoor life/recreation, opportunity to keep in touch with native locality and relatives/friends, opportunity to keep a forestry tradition, and feeling of ownership and decision-making belonging to a forest ownership.

Management district (SBO)	Sequence and place for FG discussion	No. of men	No. of woman
Sorsele – Storuman	1) Storuman, 20060418	5	1
Jörn- - Arvidsjaur	2) Arvidsjaur, 20060419	4	0
Lycksele	3) Lycksele, 20060421	4	1
Bjurholm - Fredrika	4) Agnäs, 20060423	3	0
Skellefteå - Södra	5) Burträsk, 20020424	0	4
Örnskjöldsvik - Norra	6) Gideå, 20060426	0	6
Vännäs – Umeå södra	7) Vännäs, 20060427	4	0
Sävar – Umeå norra	8) Umeå, 20060502	2	1
Bygdeå - Nysätra	9) Bygdeå, 20060503	4	0
Norsjö – Malå	10) Norsjö, 20060504	4	0
Nordmaling	11) Nordmaling, 20060515	4	0
Nonresident	12) Stockholm, 20060611	2	3
Inspectors	13) Umeå 20061110	4	1
Managers	14) Umeå 20061110	4	0

Table 1. Composition of focus group (FG) discussions with members, inspectors and managers Norra Skogsägarna economic association.

The FG discussions were recorded (total recorded time 32 hours), and transcribed (180 pages). The first statement related to the identity of the participants and it provided a list of identity" words to choose from. The selected "identity" words from each focus group were counted and put into a table for further analysis. The scores for benefits and values were processed with an mean value for each SBO and put into a table (Table 3). The results from the questions targeted directly to Norra Skogsägarna, were processed in a written form through existing theories, as a summary of the member's, inspector's and the manager's thoughts according to the private forest cooperative.

3. Analysing the organizational relationship

For each of the focus issues; identity, benefits and agreements, results are organized and presented according to three levels; members, inspectors and managers.

3.1 Identity

From the discussions, initiated by the first stimulus material, two general views were expressed by the members, regarding their identity as forest owner (Table 2). Those that practice forestry work themselves identify as forest managers ("skogsbrukare"), while those who do not, identify themselves as forest owners ("skogsägare"). *"To own a property is not the most important, that is to manage a forest property, to achieve something. The ownership is of a secondary importance, it is the right to manage that is important. You can own a lot but, but not decide anyhow"* (FG discussion 9), *"A forest manager is one that do the forestry work by themselves, a forest owner is one that pays to get the work done"* (FG discussion 11). From the inspectors' point of view, the identity of the members as forest owners is not connected to the forestry work but to the membership. *"They are forest owners; it is like a trademark for them, Norra Skogsägarna. They do not think like that if they sell to Holmen* (a forest company, authors comment), *with Norra Skogsägarna they are organized, have like a red thread in their business"* (FG discussion 13). According to the managers, the members identify themselves neither as a forest manager nor forest owner, but with their professions, i.e. forestry work or membership have not a major impact on their identity. *"I do not think that forest owners think of themselves in the first place as forest owners, they think of themselves as their profession, doctor, combat pilot, electrician or...* (FG discussion 14). Altogether there is a significant discrepancy concerning the view of the identity of the members. This is further expressed by the use of different words (definitions) when referring to the members. *"That we, as managers, shall have an understanding in what our customers are, the members......"* *"The members, in the first place, do not identify themselves as owners of the cooperative, but as suppliers and partner" "They are absolutely the best customers we have....."*. (FG discussion 14). The different wording or way of addressing members to Norra Skogsägarna are; member (all levels), owner (mostly in printed materials), customer (inspectors and managers) and supplier (managers).

In conclusion, the results show that some members consider themselves as forest managers others as forest owners. According to the inspectors the membership constitutes the forest owner identity, while the managers do not consider the forest owner identity as an issue.

The majority of the members live in a mix of independent-, wage-earner- or career mode of life, however their aspiration for independence by means of the forest property is often expressed. Thus, independence itself is the main function or goal of forest ownership. By owning and managing a private forest property, whether they do the work themselves or hire someone to get it done, this business will be the security for the member to stay with the independent mode of life. In line with Törnqvist (1995) our results indicates that members business is both the goal and the means to secure independence. Yet, many of the members live a waged-earned- or career mode of life for many reasons; make a career, have to move and so forth. However, even among the non-resident members, living in a mixed mode of life, their forest-land ownership is motivated by their desire for independence. One of them expressed: *"Whatever happens I have my forest-land to go to. It is a feeling of safety and independency"* (FG discussion 12). To make a living only on the forestland is a reality only for some of the members (often in combination with farming) and they live in the genuine independent mode of life.

The inspectors and managers view of the members' identity is different from the members themselves. The inspectors understand that the members consider the independency important but not that the members themselves have the means to achieve this independency. Instead it is the cooperative that provide them with this. That may also explain why the inspectors name the members as customers. According to the managers, all members only belong to the waged–earner- or career mode of life, and they express that the members are their off-farm profession, which could explain why the managers call the members "suppliers". In the waged-earned mode of life the work time and leisure time are distinctly separated part of life, and it is the revenue from work (wage earnings or forest income) that serves as means for up keeping this mode of life to which leisure time activities are very important. The managers understanding is that the timber sales that serves members to improve their leisure time; by that they can pay the holyday abroad or the boat, as the managers pointed out.

Experience the forest/nature	The forest as an occupation	Managing the forest	The forest in family relations	Emotions on the forest	Work and tools in the forest
Outdoor recreation (10)	Forest owner (11)	Silviculture (11)	Family estate (11)	Confidence (11)	Cleaning saw (11)
Nature experience (10)	Forest manager (11)	Cleaning (10)	Inheritance (9)	Relaxing (10)	Chain saw (10)
Hunting (9)	Manager (11)	Fire wood preparation (9)	Single owner (9)	Satisfaction (10)	Plants (9)
Berry & mushroom picking (8)	To have many occupations (6)	Pension insurance (9)	How will take over? (9)	Responsibility (9)	Risky work (8)
Fishing (7)	Woman (6)	Thinning (7)	Alternation of generations (7)	Sense of belonging (5)	Workload (7)
Wildlife (6)	Farmer (6)	Harvesting (6)	Part-owner (6)	Peace, tranquility (5)	Planting pipe (6)
Seasonal changes (6)	To run my own business (5)	Planting (5)	Ownership history (6)	Preservation (4)	Tractor (5)
"Every man's right" (5)	Man (5)	Repairs (5)	"My place on earth" (6)	Lifetime achievement (4)	Injuries (3)
Show respect and considerations (5)	To have two jobs (5)	Forest owner movement (4)	Property history (5)	Re-creation (3)	Forwarder (3)
Ancient remains (4)	Self-employed (4)	"Bankbook" (4)	Forest farmer tradition (4)	Happiness (3)	Grapple loader (3)
Spruce forest (4)	Decision-maker (2)	"Gilt edge" (4)	Family forestry (3)	Make duty for the family (2)	Harvester (1)
Plants (2)	Investor (2)	Elected representative (3)	"Have to" (2)	Insecurity (1)	Trowel (-)
Birds (2)	Non-resident (1)	Nature conservation (3)	Disagreement (-)	Make a duty for the authorities (-)	
Broad leaved forest (1)	Problem solver (-)	Game management (2)	Forced inheritance (-)	Economic stress (-)	
Primeval forest (-)		Protection of cultural heritage (2)		Emotional stress (-)	
Clear felled area (-)		Costs (2)			
		Club activities (1)			

Table 2. Selected "key words" reflecting FG participants personal opinion on their identity as owner of a forest property (Number of times selected)

3.2 Benefits

Several forest property benefits and values of ownership are looked upon in a similar way by the members, the inspectors and the managers. Yet there is variation between individuals as well as between SBO-districts. To the owners of foothill forest land (Storuman and Arvidsjaur), hunting and fishing are more important than forest income. This is likely reflecting the generally low profitability in foothill forestry due to poor growing conditions and long distance to processing industries. Also, hunting and fishing are traditionally central activities in an independent mode of life in these areas. Housing is important in those focus groups where participants live at the property, while it is unimportant to the non-resident forest owners living in Stockholm. They, on the other hand, value the property for its function of keeping the relations to native locality, family and friends. Firewood and timber for own consumption are generally considered as rather important. In four of the focus groups, participants indicate that owning and ruling of a property is most important to them. In Nordmaling and Vännäs/Umeå södra, this opinion is likely to have been influenced by an ongoing court process between private forest owners and reindear hearders, where the latter claim their immemorial right to graze their reindears during winter time on any forest land from the mountain to the Baltic Bay. In Bjurholm, the regional authority's (Länsstyrelsen) actions in a specific nature conservation case have upset the local opinion.

Focus group	Forest property benefits								
	Forest income	Hunting & Fishing	Wild berries & Mushrooms	Fire-wood &timber	Recidence & Housing	Outdoor life & recreation	Native locality	Tradition in forestry	Owning & ruling
Storuman	3.2	**3.6**	2.5	3.0	3.0	3.3	2.0	1.6	2.8
Arvidsjaur	3.0	**4.5**	3.2	3.0	3.2	4.2	2.2	2.2	3.2
Lycksele	**4.2**	3.0	2.6	4.0	3.4	3.2	2.8	2.6	4.0
Bjurholm/ Fredrika	**4.6**	3.3	2.0	3.6	2.6	3.3	2.3	3.6	**4.6**
Norsjö/Malå	4.0	3.0	2.0	3.6	**4.6**	3.6	4.0	4.0	**4.6**
Ö-vik Norra	**4.6**	2.3	2.6	3.5	2.3	3.1	3.5	3.0	3.8
Vännäs/Umeå södra	3.7	2.5	2.2	3.5	3.5	3.2	1.7	2.7	**4.2**
Sävar/Umeå norra	**5.0**	1.6	3.0	4.0	0.3	3.6	*0.6*	3.0	2.6
Skelleftå södra	**5.0**	2.5	3.7	4.2	4.5	3.7	2.7	3.5	4.2
Bygdeå/Nysätra/ Lövånger	**5.0**	2.5	2.0	3.5	4.2	3.2	3.0	4.0	3.7
Nordmaling	4.5	3.2	3.0	4.0	3.5	2.5	3.0	3.5	**4.7**
Non-resident owners, Stockholm	2.0	1.2	1.8	2.0	1.4	1.8	**3.4**	1.0	*0.6*
Inspectors	**4.4**	3.6	2.4	3.6	3.6	2.8	2.6	3.6	4.0
Managers	4.3	3.5	2.5	2.3	3.3	3.3	2.5	3.0	**4.5**

Table 3. Mean values of forest property benefits by each focus group. 1=Not important, 5=Very important Table 4. Mean values of forest property benefits by each focus group. 1=Not important, 5=Very important. Figures in Bold indicate the value with highest mean value, while figures in Italics indicate that only one or two of the participants have given a response on that particular issue.

The most frequently discussed and valued benefit was the forest incomes, mainly in terms of timber price (Table 3). Some members who expressed that forest income was not important, had another income, but for others the forest income supports their living and provides them with "a gilt edge" (FG discussion 6). Also it was frequently referred to as a kind of pension insurance. Many of the members said that they used to compare the pricelists of various forest enterprises before deciding to harvest and chose the one that offered the "best price". Only a few expressed a responsibility of selling to the cooperative due to their membership *"I have not had any contact with Norra Skogsägarna, so I cannot speak in this issue. It was a long time ago I delivered to them. We have delivered to Svea Skog while they offered the best price"* (FG discussion 5). In addition, many members ask for transportation if timber suitable for firewood to their farmyard in connection to the harvesting operation *"Firewood and timber would surely be high ranked for members, higher than keeping in touch with relatives and friends. But I also think it is important, the firewood"* (FG discussion 6).

The inspectors have the same opinion; members chose the enterprise that offers the" best price" to contact for harvesting. However, they point out that this attitude and behavior is based on an illusion, as the pricelists are constructed as to make comparisons impossible. *"The price is built up in such a complicated way in order to not be comparable between different actors"* (FG discussion 13). In case the members don´t strictly go for what they assume is the "best price"; the personal relationship to the inspector is important. The inspectors have the crucial position, to establish a good relationship with the members in the SBO. *"We fail in explaining about member benefits, some of them have an understanding but many act due to old habits, to sell to that certain person"* *"They sell to you, not to the cooperative"* (FG discussions 13). Further, the inspectors understand that household firewood is important for the members and has to be dealt with seriously. *"You can see, if you are in charge of the harvesting and they can keep the firewood, it is very important. They say do not transport the last because I use that for firewood. It is almost more important than the felling"* *"Sometimes the firewood initiates the harvesting"* (FG discussions 13).

The managers share the inspectors´ opinion regarding the incomparable price lists, and the members unconsciously accept "the best price", argument. *"All actors are working for the prices not to be comparable at all. They are making the pricelists not to be comparable. It is impossible to compare, it is like that, it just does not work"* (FG discussion 14). Further, they argue that forestry income and to own a forest property is the most important benefit. When reflecting upon the latter benefit, they realize that the forest owner views that particular property as being important, otherwise more forest properties should be put on the market for sale. According to the managers, household firewood is not considered as important to the members.

Participating in a cooperativ, can also give (intangible) abstract benefits as the sense of belonging and a common ground and common interest (c.f. Skår, 1981). According to the member opinion, they get good service in practical forestry operations, but whish for a more sensitive organizational approach when it comes to a more diversified refinement of their own timber, more resistant plant material, more social activities like forest-days and evenings. In political matters, they ask for more help representing the members in conflict issues on hunting, nature protected areas and reindeer husbandry. Another important aspect is that there must be a distinction between members and non-member. *"It shall make a difference if you are a member or not, if you can have a discount. In that way you shall feel your membership"* (FG discussion 3). The inspectors pointed out the importance to support the members in political matters, which distinguish Norra Skogsägarna from the other forest enterprises. *".....It is not the other enterprises that stands on the barricades for the forest owners, in terms of nature protected areas, dealing with reindeer husbandry, they do not care about those issues, and I think this is a great advantage with us that we really must show the forest owners"* (FG discussion 13). However they admit that many members are not aware of these benefits as they have not been explaining these services well enough. *"We have not been plain enough in showing advantages and special values you get as a member, unfortunately I think the members are not aware of this values"* and further *"It has been a too little difference if a member makes a timber delivery versus a non-member"* (FG discussions 13) (c.f. Skår 1983 and Nilsson & Björklund 2003 regarding the cooperative dilemma). Regarding the political argument the managers fully agree with the inspectors but not regarding the importance of making differences between members and non-members. Beside the political issue, the managers do not consider intangible benefits as important. They stress that being offered a good timber price, being at presence on a short notice, and the quality of the work, is what matters. *"The tools we have are price, presence at time and quality"* (FG discussion 14). They also mark the importance of take active part in political matters. *"We work for the forest owners, it is exclusive for us and we must indicate that for the members. No one of the other enterprises put time and money on that. It is easy to argue for, like service in time, it is concrete (tangible) and objective.* (FG discussion 14).

With regards to the stationary relationship and functional relationship (c.f. Skår 1983, Figure 1), another benefit, should be the part-ownership of the cooperative Norra Skogsägarna and its industry. However, a majority of the discussioned members expressed a skeptical attitude both to the ownership and the price on their timber deliverance. *"You cannot look at the saw mill and think it is something I own, it is impossible"* (FG discussion 3). *"I do not feel like an owner to a cooperative nor to an industry"* (FG discussion 12). *"But the forest owner cooperatives, they have like two feet, they act like timber brokers and they have their own saw mills and do not want to pay too much for the timber to the own saw mill. Is that for cutting down expenses to make as big profit as possible? I think it is easier to have business with the others"* (FG discussion 2). Yet, they think of benefit in other terms; the cooperative cannot refuse a delivery from members if nothing else is said, the existence of the cooperative helps to keep the prices of the raw material on an acceptable level even if they do not know how it works. With an industry, that cannot be moved abroad, they can gain knowledge of processing costs.

The inspectors claim, that only few members understand the member- and part-ownership concept, and therefore something has to do be done to enlighten them on the entire workings of Norra skogsägarna cooperative. Until now this issue has not been communicated, i.e. the organizational relationship aspect (c.f. Figure 1) does not work well enough. *"We have not been successful to implement the advantages of being a member; I do not think the members have it clear" "One got to do something radical to really point out the benefit of being a member and an owner to the cooperative" "It is less than 5 % of the members that understand the connection between the industry and the membership"* (FG discussions 13).Yet, it should be stressed that the members do not show such trust to the cooperative, as the inspectors thought because they regard the cooperative as any other forest enterprise, questioning their motives and actions. In contrast to the inspectors, the managers don´t think it´s necessary for the members to identify as part-owners, it´s enough if they identify as suppliers and partners. *"I think they look upon themselves as partners, because the owner capital they have is, it is a small amount, in most cases it is a small capital"*... *"They do not identify as an owner to the cooperative, I do not think so, but as a supplier and a partner"* (FG discussions 14). However, one manager had a different opinion, a wish that the members should show more interest in being a part-owner, while it´s the intention from the superior board. *"The superior board has decided to mark the ownership through sharing the profit, it is important, they want to strengthen the connection with the ownership, in any case that is their intension, that this is important that the members also have an obvious engagement in the ownership"* (FG discussion 14). Information about the part ownership is provided from the cooperative at every annual SBO-meeting, where the members get informed of what they own and why they are owners of a cooperative. However, *"It is a drawback when only 7 to 8 % come to the annual meetings, the account of the business. If we have done a good presentation with accessible information it had been good if 20 % would come, it would have been a strenght, so one can say, it is very few that show owner engagement when you shall tell what the cooperative really works for, what you as a member own and why"* (FG discussion 14). Another manager argues in opposite direction; *"I think this member/owner benefit is exaggerated, I do not think the expectations is that big, as long as you are pleased with price, service, quality and accessibility it is good enough. That is what is valuable for the members.........and if there are something else that makes it important to be a member I think is ideological perspective, that it has a value, I think they are members more for that reason......*(FG discussion 14).

3.3 Agreement

The members do not feel as cooperative part-owners, and express a lack of information beside the activity for the members. They have noticed that they are mostly regarded as suppliers or customers. *"They talk so much that we shall act as owners, we shall make demands on the economic association, but it is hard to get by this and not to look upon this issue as we versus them. After all, what should be done should be initiated from us, the members, but it is always coming from the top, from the managers. We should say what we want but it is hard to get by, to have the knowledge and most of all understand what the managers are saying"* (FG discussion 9). Predominant among the participating members is a *"we versus them"* feeling. We are the forest owners/members/customers/suppliers and *they* are the managers/the cooperative/the industry. The members contact the cooperative as suppliers, mostly in questions about the "best price". Some of them also state the importance of a good relation to the inspector of their SBO, if business shall be done. *"What forest enterprise I shall chose, is up to the personal contact I have with the buyer"* (FG discussion 10). The inspectors are aware of the members *"we versus them"* feeling and admit that the cooperative must improve to make the "we-feeling" stronger. *"To strengthen the we-feeling is our opportunity"* but at the same time they call them customers. *"They are our customers, you have had the feeling of saying this and that....but you cannot do anything "*. The only measure they can think of is more information about the entire working. *"I think they would like to have more information about the cooperative, I guess it is like that. From the cooperative out to the members, and on the annual meeting they can have this information"* (FG discussions 13). One of the managers agrees with the inspectors that *"We have failed in informing about advantages and values the members have access to, just because they are members, so unfortunately I think they are not aware of this"* (FG discussion 14). Other manager thought that there is no need for improved agreement within the cooperative. *"How the members apprehend the cooperative in the local sphere together with the inspectors of hers/his SBO, that is what Norra Skogsägarna is for many members. Other contacts are not so important for them, I believe". "If they got the service, confidence and are listened to, that is good enough, they do not ask about anything else"* (FG discussions 14).

Going over the 75 years anniversary book (Jonsson, 2008) and the member magazine No.1 Norra Skogsmagasinet, searching for how members are addressed, it is found that the term member is used frequently while the term owner only is used as being an owner to a forest property, but never in relation to the economic enterprise Norra Skogsägarna.

3.4 Towards a new organizational approach

The cooperative Norra Skogsägarna is organized towards the achievement of one single goal; to optimize the economic results for the members. To reach this goal, the cooperative is hierarchically organized in levels and therefore it becomes hard for the members to realize agreement while communicating with the cooperative. If the cooperative was organized in line with the theory of the new institutional organizing (c.f. Nilsson & Björklund, 2003) which leaves "open doors" between the levels and where different benefits and goals, even those beyond the cooperative business are given importance, better communication and trust could be reached. The cooperative ought to incorporate a wider range of benefits and goals, and define, express and also act in accordance to this approach through all organizational levels.

According to our results, to make a new institutional organizing successful like for the different benefits and goals of the members, these issues needs to be communicated and the communication has to be meaningful. In this context, the communication is not only a technical matter as e. g. printed materials for information, as pointed out by Skår (1983) and Weick et al (2005). The communication has to be looked upon as a complex social teamwork between different people in different levels within the cooperative. Further, the communication has to create meaning and importance in this teamwork. In all the different categories of employees and members in the cooperative there are different interpretations of the "real world", also what they consider as benefits and goals differ; and these differences ought to be communicated.

In this study, we found a mismatch between the desired benefits of the members and those on offer from the cooperative, which could be seen as a reflection of the lack of meaningful communication. Further the members could not comprehend the entire workings of Norra Skogsägarna. This was also was expressed as desirable by the superior board. Both managers and inspectors argue that there are occasions when the members could ask about the entirety, such as forest days and evenings and at the annual meeting in every SBO, but less than 10% of the members do come to these meetings. However, to create agreement, meaningfulness and a sense of entirety when you address the members as suppliers and customers may be impossible. Rather, this is part of the organizational relationship problem, as it emphasizes the "we versus them" feeling. Still the circumstances can become different as one of the managers finally said, *"This is a request from the board; we must communicate the organization"* (FG discussion 14*)*. To try to understand the agreement problem, one can look at it in a further direction, as the title of this paper also points out, the perspective of the members are so different. Who and what are the members: owners, members, customers or suppliers? And will it be possible to organize in a new institutional way or create a meaningful communication for people that are looked upon in four different perspectives?

Sociologically it should be mentioned that the words the employees of Norra Skogsägarna use to referrer to the members, who also are the owners of the cooperative, is an urgent issue to analyze and explain. The inspectors say customer, the managers say supplier (contractor) and customer, and no one ever mentioned the word owner, not even the members themselves. But they are mentioned as owners in printed materials from the cooperative e.g. in the annual report: "Our economic surplus goes back to our owners – the member" (Norra Skogsägarna, 2006, p 3). The theory of structuralism is useful to explain this phenomenon. The usage of the different definitions, of what the members really are, and what the members are called, must be considered important, since out of that definition steps are taken, in developing the suitable language, the way one speaks to, and as a result of that, the way one treats that certain person (member). Language usage counts, while through the words that are used, the receiver, and also other persons within the cooperative and outside, gets an impression, what that word really imply. Within that frame of definition, a perspective emerges in which way those people belonging to the definition are treated, it states the way one deals with the persons in question (members). The language, the choice of word, can make the basis how the individual is going to be looked at, treated, spoken to verbally and in writing (Ritzer 2000, Lash 1991). On the other hand it also states the way the individuals are going to think, talk and act towards the cooperative where the language are used as a communication for interaction between actors. To put it into the Thomas theorem; "If men define situations as real, they are real in their consequences" (Thomas and Thomas 1928: 572), which underlines the importance of what people think and how this affects what they do.

4. Conclusion

This study confirms the need for better agreement between the members, the inspectors and the managers as pointed out by Berlin (2006), and it becomes very obvious when looking at the forest owner identity. Upon exploring the identities of the members, it was found they belong to the independent mode of life or mixed with wage-earned mode of life or carrier mode of life, where the independence itself is still very important. It is essential to consider this, in order to create an agreement between the forest owner member and hers or his cooperative and also to other institutions in the society as the authorities, forest-business and the science.

The forest owning members has to be attended with respect as an independent forest owners, where the forest property guarantee her/his independence. When the member's expectations on the cooperative to support their life mode of independency are not satisfied, and also there is a "we versus them" feeling, they expressed their disappointment. I. e., this example of the "cooperative dilemma" becomes very apparent when superior levels in the cooperative organization look upon them as dependent owners and members, respectively customers and suppliers. As the members were addressed differently, this phenomenon adds to the scarcity of agreement and this makes the perspective of the members different. It really makes a difference if you are addressed as a member, an owner, a customer or a supplier. This is a vital dilemma to resolve in order to get an agreement followed by a meaningful communication between the levels and parts of the cooperative.

The members are part owners and obtain a share of the cooperative's profit, and regularly receive information about the ownership from printed materials and at the annual SBO-meetings. However, this information seems to be insufficient as they do not internalize the information and perceive themselves as an owner. It is obvious that the members do not internalize the entirety of the cooperative and this, as well as the lack of agreement is important to resolve.

To talk, act and think of the member as the independent forest owners they are, and with organizing the cooperative with "open doors", in a new institutional way, will facilitate the communication to become meaningful and improve the situation. This could create an environment where members can mediate values and benefits which are important to them, and also become more receptive to the information from the organization. By this, better agreement between the levels can be achieved, and thus it will be easier for members to internalize the entirety, which must be considered crucial for the long-time survival of the cooperative. In other words, the *"we versus them"* feeling will fade away.

Finally, this study stressed the illogical way of defining and addressing the members in four different ways - as member, owner, customer or supplier. According to what is found so far, they are members and nothing else. It is not clear if it is juridically correct to call them owners, because this is not a personal ownership, it is a collective ownership. This latter issue could be a topic of another study.

5. Acknowledgment

We thank Professor Ljusk Ola Eriksson (SLU) for valuable comments on earlier draft of the manuscript, and PhD Mahesh Poudyal (SLU) for suggested improvements of language and style of writing. The study has been supported by the Brattås foundation, The Swedish Farmers Foundation for Agricultural Research, and not at least Norra Skogsägarna economic association.

6. References

Abrahamsson, B. &, Andersen J. A. (2000). Organisation – att beskriva och förstå organisationer. Liber ekonomi, ISBN 91-47-06103-0 Malmö, Sweden (Available in Swedish only).

Andersson, B., Häckner, J., & Lönnstedt, L. (1980). Skogsägareföreningarna i ett historiskt perspektiv. [History of The Forest Owner Associations in Sweden] SSR. ISBN 91-7446-017-X Borås, Sweden.

Berlin, C, (2006) Forest Owner Characteristics and Implications for the Forest Owner Cooperative. Dept. of Forest Resource Management and Geomatics. Swedish University of Agricultural Sciences. Licentiate Thesis. Rapport/Report 17 ISBN 91-567-6862-0. Umeå, Sweden.

Berlin C.; Lidestav, G. & Holm S. (2006). Values Placed on Forest Property Benefits by Swedish NIPF Owners: Differences between Members in a Forest Owner

Association and Non-members. Small-scale Forest Economics, Management and Policy, 5(1): 83 – 96. ISSN 1447-1825.

Bjuggren, P-O. (1994). Vertikal integration i Företaget – en kontraktsekonomisk analys [Vertical integration in The Company – a contractual economic analysis] pp 104 – 111, SNS Förlag ISBN 91-7150-300-5 ; Stockholm. Available in Swedish only.

Eriksson M. (1990). Ägostrukturens Förändring inom Privatskogsbruket, ett Framtidsperspektiv. [The Ownership in Alteration, in a Future Perspective]. Dept. of Forest Products and Markets. Swedish Universityof Agricultural Sciences. Report nr. 13. ISBN 91-576-4150-1 Uppsala, Sweden. Available in Swedish only.

Fischer,P., Bliss, J., Ingemarson, F., Lidestav, G., and Lönnstedt, L. 2010. From the small woodland problem to ecosocial systems: The evolution of social research on small-scale forestry in Sweden and the USA. Scan. J. For. Res. 2010: (25) 390-398. ISSN 0282-7581.

Glete, J. (1987). Ägande och Industriell omvandling [Ownership and Industrial Development]. SNS Förlag, Stockholm, Sweden. ISBN 9789171503008.

Gummesson O. (1983)., Utan kamp ingen seger - om Gösta Edström och Södra Skogsägarna , Trydells Tryckeri AB, Laholm, Sweden. ISBN 91-971414-9-6.

Högvall Nordin M. (2006), "Dom brukar jämföra det med en stridspilot" ["They usally compare it to a combat pilot"]A Study in organizational communication . Science of Media and Communication, Umeå. Dissertation. Print & Media. ISBN 91-7264-186-X. Umeå, Sweden

Ingemarson F. (2004), Small-scale forestry in Sweden –owners´objectives, silvicultural practices and management plans. Acta Universitatis agriculturae Sueciae. Silvestria, 1401-6230 ; 318 Swedish University of Agricultural Sciences Uppsala. Sweden ISBN 91-576-6702-0.

Jacobsen, L. (1999), Livsform, kön och risk. TA-tryck Bjärnum, Sweden. ISBN 91-7924-130-1.

Jonsson, E. (editor) (2008)"En samlad kraft, Norra Skogsägarna 75 år", Umeå Sweden. ISBN 978-91-633-3338-5.

Kittredge, D. (2005).The Cooperation of Private Forest Owners on Scales Larger than one Individual Property: International examples and Potential Application in the United States. Forest Policy and Economics, 7, pp 671-688 ISSN 1389-9341.

Lidestav, G. (1997). Kvinnliga skogsägare och skogsägande kvinnor. [Female Forest Owners and Female Forest Holdings- a structure analysis]. Dept. of Operational Efficiency. Research Notes no. 298. Swedish University of Agricultural Sciences, Umeå, Sweden.

Lidestav, G. & Nordfjell, T. (2005). A Conceptual Model for understanding Social Practices in Family Forestry. Small-scale Forest Economics Management and Policy, 4:391 – 408. ISSN 1447-1825.

Nilsson, J. & Björklund, T. (2003). Kan kooperationen klara konkurrensen? Report / Swedish University of Agricultural Sciences (SLU), Department of Economics (149). Uppsala.

Norra Skogsmagasinet (2009). No 1 Norra Skogsägarna ekonomisk förening. ISSN 1653-5154.

Reynolds, B. J., Decision-making in Cooperatives with Diverse Member Interests. United states Dept. of Agriculture. Rural Buisness Cooperative Service. Research report no. 155 April 1997.

Ritzer. G. (2000) Sociologisk teori [Sociological Theory]. McGraw-Hill Book Co – Singapore. International Edition.

Skår, J. (1981) Kooperativa företag. En studie av det kooperativa företaget och dess särart utförd på uppdrag av kooperationsutredningen. SOU 1981: 54 Stockholm, AB ISBN 91-38-06309-3.

Swedish Code of Statues (1993), Konkurrenslag, SFS 1993:29 (Competition Act), Government of Sweden, Stockholm ISSN 0346-5845.

Swedish Government (1992/93) Government Bill, 1992/93:56, Ny Konkurrenslagstiftning (A New Competition Act), Prepartory work for Swedish Competition Act, Stockholm ISSN 99-1211253-9.

Swedish Government (1999/2000), Government Bill, 1999/2000:140, Konkurrenspolitik för förnyelse och mångfald (Competition Policy for renewal and multitude), Prepatory work for Swedish Competition Act, Stockholm. ISSN 99-1211253-9.

Swedish Forest Agency (2010) Skogsstatistisk årsbok 2010, [Statistical Yearbook of Forestry 2010], Jönköping 2010, Sweden. ISBN 978-91-88462-93-0.

Törnqvist Tommy, Skogsrikets arvingar, En sociologisk studie av skogsägarskapet inom privat, enskilt skogsbruk. [A Sociological Study of the Ownership in Private Forestry]. Dept. of Forest Products and Markets. Swedish University of Agricultural Sciences. Dissertation. Available in Swedish only with English abstract. Report nr. 41, Repro HSC Uppsala 1995. ISSN 0284-379X.

Weick E. K., Sutcliffe K. M., Obstfield, D. (2005).Organizing and the Process of Sensemaking Organization Science. Vol 16. No 4. Pp 409-421. ISSN 1047-7039.

Thomas, W. I. & Thomas, D.(1928) *The child in America: Behavior problems and programs*. New York: Knopf.

Lash, S. 1991:ix, "Introduction" In Poststructuralist and Postmodernist Sociology, Aldershot, Eng.:Edward Elgar ix – xv ISBN 1-85278-183-1.

Wibeck, V. (2000) Fokusgrupper, om fokuserade gruppintervjuer som undersökningsmetod. Studentlitteratur Lund. ISBN 91-44-01060-5.

Federation of Swedish Family Forest Owners
    http://www.lrf.se/In-English/Forestry/, accessed 20110908.

Mellanskog
    http://www.mellanskog.se/templates/MS_InfoPage.aspx?id=528
    accessed 20110910

Norra skogsägarna
    http://www.norra.se/templates/Page.aspx?id=496, accessed 20110910

Skogsägarna Norrskog
    http://www.norrskog.se/Upload/web_Norrskogs_årsberättelse%
    accessed, 20110910
Södraskogsägarna
    http://www.sodra.com/sv/Om-Sodra. accessed 20110910

6

Economic Feasibility of an Eucalyptus Agroforestry System in Brazil

Álvaro Nogueira de Souza, Humberto Ângelo, Maísa Santos Joaquim,
Sandro Nogueira de Souza and John Edward Belknap
Universidade de Brasília
Brasil

1. Introduction

Agroforestry systems consist of cultivation practices, which combine arboreal species with either annual or perennial crops in a manner which seeks the optimal use of land together with the maximum return per unit area. Macelo (1992) considered that agroforestry systems could be used for the recovery of degraded soils and for fragmented pastures under agropastoral activities, in both cases the factors of production are insufficient for the natural recovery of the land's productive potential.

Hoeflich (1995) believed that these agroforestry systems are forms of land use, which offers a sustainable return in which management activities combine with the cultivation patterns of the local population, thus creating greater diversity and sustainability. For Franco (2000), these agroforestry systems help to control soil erosion and replace nutrients lost to the soil by intensive use. Such overuse, often not viable from an economic point of view, given the price of fertilizer, can be substituted by agroforestry, with the decomposition of plant material, which remains in the soil.

a.   The use of legumes in agroforestry systems

A principal reason for the use of legumes in an agroforestry system is for the benefits offered by green fertilizer. Atmospheric nitrogen is captured by bacteria present in the legume roots and incorporated in the soil composition (Ferreira, 1994). According to Malavolta (1981), the guandu bean is a legume capable of fixing 50 to 150 kg of nitrogen per ha per year. This same observation was made by Mielnickzuk (1988). Evaluating various cultivation systems, Mielnickzuk (1988) observed that when the systems included legumes, the quantities of nitrogen in the soil increased significantly with emphasis on guandu bean, which after three years presented an increase in total nitrogen of 900 kg per ha.

b.   Presence of animals in agroforestry systems

The silvopastoral system is an association of species of trees and grazing, with or without the presence of crops. In this type of system, one can encounter commercial timber stands with the presence of cattle or pasture as a complementary factor to subsistence agriculture.

Tree varieties can include recent reforestation pastures, lumber grade trees, fruit trees, forage trees, windbreaks, pastureland, shade trees, and trees to provide conservation and soil improvement in pastures (Santos, 1990).

The problem of soil compaction in silvopastoral systems cannot be neglected. According to Silva (1999), the physical conditions of the soil are affected principally in the upper surface layers by animal trampling.

1.1 The use of eucalyptus in agroforestry systems

When one considers agroforestry cultivation with eucalyptus, the protection of the soil is especially important because, the wet period is when the soil is exposed, as a result of the soil preparation for planting annual crops.

The *Eucalyptus grandis* has been one of the most utilized varieties in agroforestry systems. The most common combination is with legumes, cited in the works of Passos (1990), Passos et al. (1993) and Ferreira Neto (1994). The association of reforestation with crops and pasture reduces substantially the costs of planting and maintaining of stands of eucalyptus, it augments the productivity of the site besides minimizing erosion and other negative environmental impacts (Dube et al., 2002).

Silva (1999) evaluated the possibility of combinations of *Eucalyptus grandis* with the grasses *Braquiária decumbens* and *Melinis minutiflora*, in steep land. Four spacings were tested for Eucalyptus: 3x2m, 4x2m, 5x2m, and 6x2m. The control was Eucalyptus alone with a spacing of 3x2m. The results showed a smaller diameter tree, with a reduction of the useful area per plant. The best treatment was at 3x2m combined with *Braquiaria*. This test plot reached a volume 55.2% greater than the control. With respect to lumber quality and quantity of grass biomass, the best spacing was 6x2m, which produced less volume, however, the timber had commercial use as lumber. The *Braquiária* produced 2.45 times with this spacing in relation to the 3x2m spacing. *Milinis minutiflora* disappeared after two years, demonstrating its lack of suitability for the conditions of this study.

Since the start of agricultural and pastoral activities on Brazilian savanna soils, which presented a principal barrier to cultivation in the high acidity level, varied cultivation practices have been tried. Among the crops are rice, soy, corn, coffee, sugar cane, eucalyptus and pastoral activities, in which cattle were raised exclusively.

In the beginning of the decade of 1990, agro-silvopastroral systems were started in commercial scale in regions of savanna soils in the South East Region of the State of Minas Gerais, Brazil. In the studied case, the system developed according to these stages: In December 1999, the project began with the planting of rice and eucalyptus (Figure 1).

The Eucalyptus seedlings were clones adapted to the North East Region of Minas Gerais. The area was prepared by one deep tilling and two grader levelings, as well as the application of calcium zincal MMA 85% with PRNT, at a rate of 2.5 tons per hectare. The eucalyptus seedlings were planted in lines east/west, in order to permit greater solar exposure to the combined crops. The space between eucalyptus rows was planted with rice, (*Orizza sativa*, cultivar Guarany) with a spacing of 0.45 meters.

Soy (*Glycine max (L), Merril*, cultivar Conquista), at a spacing of 0.45m was the crop introduced at the end of the first year. At this time, the eucalyptus had one year of growth (Figure 2).

Planting of pastures began at the end of the second year / start of the third year, with planting of *Brachiaria brizantha* (Figure 3).

Fig. 1. Cultivation of rice (*Orizza sativa*, cultivar Guarany) with eucalyptus (natural hybrid of *Eucalyptus urophylla x Eucalyptus camaldulensis*) in a spacing of 10 x 4m in the first year.

Fig. 2. Soy crop (*Glycine max (L), Merril*, cultivar Conquista) in place of rice in year 2 of an associated system of eucalyptus (natural hybrid of *Eucalyptus urophylla x Eucalyptus camaldulensis*) before the first pruning.

Fig. 3. Formation of pasture (*Brachiaria brizantha*) in year 3 of agroforestry association with eucalyptus (natural hybrid of *Eucalyptus urophylla x Eucalyptus camaldulensis*) after the second pruning of trees.

In the fourth year of eucalyptus growth, nitrogen fertilizer was applied (300 kg/ha) to the pasture grass. At this time, yearling steers were placed in the pasture to fatten and the cattle used the eucalyptus for shade and rest areas (Figure 4).

The forest eucalyptus stands had three prunings. At 2.5 years after planting, the first pruning was done with loppers and saw cutting of all the branches to a height of 2 meters. At 3.5 years of growth, pruning with saws was done to cut all branches to the height of 4 meters. At 4.5 years, pruning was done to a height of 6 meters, leaving a tree trunk free of all knots (Figure 5).

The trees were sold standing in the market for pole logs, used as utility posts in rural electrification. The demand for this product was stimulated by a Federal Government program, called "Light for All" which sought to bring electricity to all those living in rural communities.

Prior to this use for utility poles, the wood from agroforestry plantings was used to make pallets, or, for unusable wood, converted to charcoal for use in steel mills. These uses did not have a high market value, but were profitable for the landowning companies. This study is reported in Souza et al. (2007).

Given the new alternative use of wood in agroforest systems, a new economic analysis was completed and the results are compared with those found in Souza et al.

Fig. 4. Yearling cattle in the shade of eucalyptus trees at the start of year 4 of tree growth.

Fig. 5. Trees in the system with six meters of knot free trunk, allowing for a higher use of lumber.

Fig. 6. Logs stacked on ground already cut to dimension for utility poles.

2. Characteristics of the area of cultivation in the agroforestry study

The area of study belongs to a large forestry enterprise, active in the savannas of the State of Minas Gerais, located in the Municipality of Vazante, in the Northeast Region of the State. The latitude is 17° 36′ 09″ S and the longitude is 46° 42′ 02″ West of Greenwich. It is at an altitude of 550 meters, with a climate Aw, that is, tropical humid savanna, dry winters, rainy summers, according to the classification of Köppen (Antunes, 1986). The average temperature is 24° C, with an average precipitation of 1,450 mm.

The items of study included seedlings of clones of a natural hybrid of *Eucalyptus urophylla x Eucalyptus camaldulensis*, planted in association with rice, soy, and pasturage with a spacing of 10 x 4m (Figures 1-6). The initial objective was to produce wood for a lumber mill and for energy and after a few years, the forestry component could provide timber for sale as utility poles for the rural electrification program.

3. Structure of costs

Table 1 shows the costs of diverse activities related to the agroforestry system. For eucalyptus, the costs considered were planting, annual maintenance, and timber sold standing. In the case of soy and rice, the costs summed all of the expenses from panting until harvest, since these activities have a production cycle of less than one year. For cattle ranching, the costs were itemized according to the time period of occurrence, since these are spread over several years. Cattle density was 1.5 steer per hectare, and the yearlings are acquired at 8.25@ (1 @ is equal to 15kg). Each animal should gain 5.5@ per year, which results in a weight gain of 8.25@ per hectare per year. The cost of land was calculated as the interest on the land value.

Itemized costs	Year	Value[1]
- Creation and installation of plan (R$/ha)	0	1,956.78
- Rice cultivation (R$/ha)	0	690.40
- Eucalyptus maintenance (R$/ha)	1	299.24
- Soy cultivation (R$/ha)	1	856.91
- Eucalyptus maintenance (R$/ha)	2	263.81
- Pasture seeding (R$/ha)	2	323.42
- Eucalyptus maintenance (R$/ha)	3	237.08
- Cattle infra-structure (R$/ha)	3	171.31
- Eucalyptus maintenance (R$/ha)	4 a n-1	144.17
- Eucalyptus maintenance (R$/ha)	n[2]	188.31
- Cattle inputs (R$/ha)	3 a n	64.03
- Cattlemen (R$/ha)	3 a n	17.69
- Depreciation of goods related to cattle raising (R$/ha)[1]	3 a n	2.49
- Acquisition of yearling cattle for fattening (R$/ha)	3 a n	519.75
- Administration (R$/ha)	1 a n	99.24
- Land (R$/ha)	1 a n	90.00
- Harvest (R$/m³)	n	11.54

[1] The goods related to cattle ranching are: housing for ranch hands, barns and sheds, corral, electric fence, water and salt supply, riding tack, and service horses. The value of depreciation of those goods related to other activities has been added to their costs.
[2] n is the age of eucalyptus, to be defined based on economic criteria.

Table 1. Costs of the diverse activities of the agroforestry system.

4. Structure of income

Table 2 shows the prices, quantities produced, and income received from the sale of agricultural products and fattened cattle. In the case of forestry products, the wood is sold standing. Only forestry price is reported since the quantity produced and income vary with the age of felling to be determined by economic analysis.

Itemization	Year	Unit	Price (R$un)	Quantity/ha	Income (R$/ha)
- Rice	0	Sc	26.00	20.16	524.16
- Soy	1	Sc	29.00	21.60	626.40
- Fatten Steer	3 a n	@	57.00	16.50	940.50
Timber standing	n	m³	190.00		

n: rotation of trees defined by economic analysis.

Table 2. Prices, quantities and income from products of the Agroforestry System.

5. Determination of the economic cutting of eucalyptus

The economic rotation of eucalyptus plantings was determined with respect to profit maximization according to the methods of economic analysis adopted in this study.

An estimate of the volume of wood in logs was realized by a system of production forecasting whose equations are available in Souza (2005). Table 3 presents the volumes forecast at various ages.

Age in years	Volume based on forecasts in m³/ha
5	100
6	120
7	140
8	160
9	180
10	195
11	203
12	208
13	212
14	215

Table 3. Volume of timber logs for poles of diverse ages from forestry logging.

The economic analysis used the methods of Net Present Value (NPV), Internal Rate of Return (IRR) and Equivalent Periodic Benefit (BPE), taking the alternative interest rate in the market of 10% per year. This is the rate of return required by the investors of this company for their investments, however, one expects a return above 10% per year, so that the system could be seen as a more profitable investment. According to Rezende & Oliveira (2008), the NPV, IRR and BPE can be calculated with the following formulae.

$$NPV = \sum_{j=0}^{n} R_j \left(1+i\right)^{-j} - \sum_{j=0}^{n} C_j \left(1+i\right)^{-j} \tag{1}$$

$$0 = \sum_{j=0}^{n} R_j \left(1+IRR\right)^{-j} - \sum_{j=0}^{n} C_j \left(1+IRR\right)^{-j} \tag{2}$$

$$BPE = \frac{NPV\left[\left(1+i\right)^{t} - 1\right]\left(1+i\right)^{nt}}{\left[\left(1+i\right)^{nt} - 1\right]} \tag{3}$$

where;

R_j = receipts at end of year j;
C_j = costs at end of year j;
n = age of eucalyptus cutting corresponding to the end of the cycle in the system;
i = alternative market rate of interest;
t = number of periods of calculation;

6. Market options for timber from the agroforestry system

The original proposal

The market conditions changed after the implantation of this agroforestry system in 1999. Originally, wood from the forestry plantation would be sold to sawmills and for firewood. The results of the original proposal are stated in Table 4, in which one can observe the analytic conditions for the most productive segment.

	Sale of logs		Sale of milled lumber (product 1 and charcoal)		Sale of milled lumber (product 2 and charcoal)	
	NPV (R$/ha)	BPE (R$/ha)	NPV (R$/ha)	BPE (R$/ha)	NPV (R$/ha)	BPE (R$/ha)
5	717.49	189.27	1.246,72	328.88	1,322.01	348.74
6	1,502.83	345.06	2.139,15	491.16	2,229.67	511.95
7	1,270.91	261.05	1.874,90	385.12	1,960.82	402.76
8	1,020.59	191.30	1.589,77	297.99	1,670.75	313.17
9	747.12	129.73	1.278,38	221.98	1,353.95	235.10
10	478.63	77.89	972,68	158.30	1,042.97	169.74
11	229.95	35.40	689,55	106.16	754.93	116.23
12	-1.13	-0.17	426,46	62.59	487.28	71.52

Source: Souza et al.,2007.

Table 4. NPV and BPE based on the sale of timber logs, milled lumber and charcoal, from diverse years of cutting.

The results indicate an optimal age for cutting at six years, with a maximum annual profit of R$511.95, derived from the sale of half the volume of timber for the production of charcoal and the other half from the production of milled lumber for a product of greater total value. However profitable, the system does not become more interesting than other eucalyptus plantings purely for the production of charcoal.

7. The actual market situation

The new market possibilities allowed the company to seek alternatives for the principal product of the system, timber. The timber of the agroforestry system do not attain sufficient diimensions for the use as poles for rural electrification until seven years of age. Thus, beginning with the eigth year, 30% of the timber can be used for poles and the remaining 70% sold as firewood.

At nine years, 60% of timber was sold for poles and 40% for firewood; in the 10th year 90% was sold for poles and 10% for firewood, in years 11 and 12 all the timber was sold for poles.

Table 5 shows the values of cash flow with the net value occuring between yerars 6 and 12.

The values presented in Table 5 represent the cash flow from the installation of the agroforestry system, beginning with a value of –R$ 2,181.08/ha until the final value when the income from the sale of timber is calculated. One notes that from the third year, the cash flow sign changes from negative (-R$ 776.47/ha) to positive (R$ 324.66/ha). The sale of soy and rice merely reduce the costs in the initial years; note that the cash flow becomes positive beginning with the sale of fattened cattle.

Year 6	Year 7	Year 8	Year 9	Year 10	Year 11	Year 12
-2,181.08	-2,181.08	-2,181.08	-2,181.08	-2,181.08	-2,181.08	-2,181.08
-408.59	-408.59	-408.59	-408.59	-408.59	-408.59	-408.59
-776.47	-776.47	-776.47	-776.47	-776.47	-776.47	-776.47
324.66	324.66	324.66	324.66	324.66	324.66	324.66
716.94	716.94	716.94	716.94	716.94	716.94	716.94
716.94	716.94	716.94	716.94	716.94	716.94	716.94
6,664.74	716.94	716.94	716.94	716.94	716.94	716.94
	6,934.74	716.94	716.94	716.94	716.94	716.94
		10,480.74	716.94	716.94	716.94	716.94
			14,145.54	716.94	716.94	716.94
				17,937.24	716.94	716.94
					19,528.74	716.94
						19,972.74

Table 5. Cash flow considering income and costs between years 6 and 12 after implantation of the Agroforestry System.

An analysis of investment viability used in this cash flow within the methods of VPL, BPE, and TIR, in order to determine the age of timber cutting which mazimized profits is shown in Table 6.

Year	NPV (R$/ha)	BPE (R$/ha.yr)	IRR (%)
6	1,746.59	401.03	20
7	1,947.83	400.09	20
8	3,646.46	683.50	24
9	5,090.67	883.94	25
10	6,311.21	1,027.12	26
11	6,516.74	1,003.33	25
12	6,287.25	922.73	24

Table 6. Values of NPV, BPE and IRR between the years 6 and 12 in the Agroforestry System.

Analyzing individually the columns of data in Figure 6, it seems that the major Net Present Value (NPV) occurs at 11 years with a value of R$6,516.74 / ha. The values of NPV increase according to the forecast increase in tree volume in every age. When this volume begins to increase at a diminishing rate, the NPV diminishes the rate of increase in value, reaching a maximum at 11 years and decreasing in year 12 to a value of R$6,287.25 / ha. Thus, based solely on the NPV, the age for the final cutting of the system to maximize profit is at 11 years.

When one looks at the Equivalente Periodic Benefit (BPE) column, the same dynamic is present; an increase in value until a point of maximum value followed by a soft decline. From the point of investment analysis with the use of BPE, the optimal moment for timber felling and the end of the cycle of the agroforestry system is at 10 years of growth. This represents a net annual profit of R$1,027.12 /ha.

8. Conclusion

In decision making between projects which have different time horizons, the Equivalent Periodic Benefits (BPE) method has priority over the Net Present Value (NPV). Thus, considering that each different age represents an alternative project with different planning horizons, the alternative with a greater BPE must be chosen. In this case, the project with cutting at 10 years is given more economic viability with respect to an annual return on capital invested of 26% (Internal Rate of Return, IRR), followed by the project with cutting at 11 years.

Comparing the results of the original proposl and the actual proposal, there is an increase in annual net profit for a hectare of the agroforestry system from R$511.95 up to R$1,027.12. In addition, actual market conditions allowed a useful life of 10 years to the project as opposed to the 6 years in the earlier circumstances where the wood was destined for the construction of pallets and charcoal.

Agroforestry systems have gained support among development agencies in agricultural activities and with institutions of learning and research. As a result, many rural producers in the savanna regions of Brazil dedicate part of their land to the activity which offers the better use of land as a factor of production.

9. References

Antunes, F.Z. Caracterização climática dos Estado de Minas Gerais: Climatologia agrícola. Informe Agropecuário, Belo Horizonte, v.12, n.138, p. 9-13, jun. 1986.

Bernhard-Reversat, F. Soil Nitogen Mineralization under a Eucalyptus Plantation and Natural Acacia Forest in Senegal. In: Forest Ecology and Management, 23(4):233-244. Madison, 1988.

Bhojvaid, P.P.; Timmer, V.R. & Singh, G. Reclaiming sodic soils for wheat production by Prosopis juliflora (Swartz) DC aflorestation in India. In: Agroforest Systems 39: 139-150. The Netherlands, 1996.

Cerri, C.C.; Volkoff, B. & Eduardo, B.P. Efeito do desmatamento sobre a biomassa microbiana em latossolo amarelo da Amazônia. In: Revista Brasileira de Ciência do Solo, 9(1): 1-4. 1985.

Daniel, O.; Couto, L.; Silva, E.; Passos, C.A.M.; Jucksch, I. & Garcia, R. Sustentabilidade em sistemas agroflorestais: indicadores socioeconômicos. Revista Ciência Florestal. v.10, n.1, p.161-177. Santa Maria, Jun. 2000.

Dube, F.; Couto, L.; Silva, M.L.; Leite, H.G.; Garcia, R. & Araújo, G.A.A. A simulation model for evaluating technical and economic aspects of an industrial eucalyptus-based agroforestry system in Minas Gerais, Brasil. Agroforestry Systems. 55: p.73-80. Dordrecht, 2002.

Ferreira Neto, P.S. Comportamento inicial do Eucalyptus grandis W. Hill ex. Maiden em plantio consorciado com leguminosas na Região do Médio Rio Doce, Minas Gerais. UFV. Viçosa, 1994. 92p. Dissertação de Mestrado.

Filho, J.M.O.; Carvalho, M.A. & Guedes, G.A.A. Matéria orgânica do solo. Informe Agropecuário, Belo Horizonte, 13(147): 22-24, 1987.

Franco, F.S. Sistemas agroflorestais: uma contribuição para a conservação dos recursos naturais na Zona da Mata de Minas Gerais. UFV. Viçosa, 2000. 147p. Tese de Doutorado.

Hoeflich, V.A. Pesquisa agroflorestal no contexto brasileiro. In: Seminário Eucalipto: Uma visão Global, 1995. Belo Horizonte. Anais... Belo Horizonte: 1995. p.159.

Kaur, B.; Guota, S.R. & Sungh, G. Bioamelioration of a sodic soil by silvopastoral systems in Northwestern India. In: Agroforestry Systems 54:13-20, The Netherlands, 2002a.

Kaur, B.; Gupta, S.R. & Singh, G. Carbon storage and nitrogen cycling in silvopastoral systems on a sodic soil in Northwestern India. In: Agroforestry Systems 54: 21-29, The Netherlands, 2002b.

Lima, D.G. Desenvolvimento e aplicação de um modelo de suporte à decisão sobre multiprodutos de povoamentos de eucalipto. Viçosa, UFV, Impr. Univ., 1996. 80p. (Tese M.S.).

Macdicken, N.K. & Vergara, N.T. Agroforestry: Classification and management. New York: John Wiley ans Sons, 1990. 382p.

Macedo, R.L.G. Sistemas agroflorestais com leguminosas arbóreas para recuperar áreas degradadas por atividades agropecuárias. In: Simpósio Nacional de Recuperação de Áreas Degradadas, 1, 1992. Anais... Curitiba:1992. P.288-297.

Martins, A.R.A. Dinâmica de nutrientes na solução do solo em um sistema agroflorestal em implantação. ESALQ. Piracicaba, 2001. 144p. Tese Doutorado.

Mielniczuk, J. Desenvolvimento de sistemas de culturas adaptadas a produtividade, conservação e recuperação de solos. IN: Congresso Brasileiro de Ciência do Solo, 21. Campinas, 1988. Anais..., Campinas, SBSC, 1988. p. 109-116.

Passos, C.A.M. Comportamento inicial do eucalipto (*Eucalyptus grandis* W. Hill ex. Maiden) em plantio consorciado com feijão (*Phaseolus vulgaris* L.) no Vale do Rio Doce, Minas Gerais. UFV. Viçosa, 1990. 64p. Dissertação de Mestrado.

Passos, C.A.M.; Couto, L. & Fernandes, E.N. Comportamento inicial de *Eucalyptus grandis* W. Hill ex. Maiden consorciado com milho (*Zea mays* L.) e feijão (*Phaseolus vulgaris* L.) no Vale do Rio Doce, Minas Gerais. In: Congresso Florestal Panamericano, 1, Congresso Florestal Brasileiro, 7. Curitiba. Anais... Curitiba, 1993. P.270-273.

Oliveira, A.D.; Scolforo, J.R.S. & Silveira, V. de P. Análise econômica de um sistema agro-silvo-pastoril com eucalipto implantado em região de cerrado. Revista Ciência Florestal. v.10, n.1, p. 1-19. Santa Maria, Jun.2000.

Rezende, J.L.P. & Oliveira, A.D. Análise Econômica e Social de Projetos Florestais. Editora UFV. Viçosa, 2008. 389p.

Santos, F.L.C. Comportamento do *Eucalyptus cloëziana* F. Muell em plantio consorciado com forrageiras na Região do Cerrado em Montes Claros, Minas Gerais. UFV. Viçosa, 1990. 83p. Dissertação de Mestrado.

Silva, J.M.S. Estudo silvicultural e econômico do consórcio de *Eucalyptus grandis* com gramíneas sob diferentes espaçamentos em áreas acidentadas. UFV. Viçosa, 1999. 115p. (Dissertação Mestrado).

Souza, A. N. Crescimento, produção e análise econômica de povoamentos clonais de *Eucalyptus* sp em sistemas agroflorestais. 2005. 203 p. Tese (Doutorado em Engenharia Florestal) Universidade Federal de Lavras, Lavras, 2005.

Souza, A.N.; Oliveira, A.D.; Scolforo, J.R.S.; Rezende, J.L.P. & Mello, J.M. Viabilidade Econômica de um Sistema Agroflorestal. Cerne, Lavras, v. 13, n. 1, p. 96-106, jan./mar. 2007

Effects of a Unilateral Tariff Liberalisation on Forestry Products and Trade in Australia: An Economic Analysis Using the GTAP Model

Luz Centeno Stenberg and Mahinda Siriwardana
*University of Notre Dame Australia; University of New England
Australia*

1. Introduction

Australia has 147.4 million hectares of native forest areas which includes 48.4 million hectares of closed forest and open forest (Bureau of Rural Sciences [BRS], 2010). Some 103 million hectares of native forest areas are either privately owned or leasehold while the balance is multiple use forest (9.4 million hectares), conservation reserves (22.4 million hectares) or other categories of public ownership (12.4 million hectares).

Negotiations on trade liberalisation and bilateral agreements between countries and regions suggest that further tariff reductions are inevitable. With the current trade negotiations under the Doha round, the global forestry sector as well as Australia's could be affected by the outcomes of the negotiations. Moreover, the increasing recognition of the importance of the forestry sector in terms of addressing climate change issues suggests that any policy affecting this sector can be significant in terms of its economic as well as environmental impact. Forest conservation and carbon emissions are two issues linked to a possible carbon trading scheme in Australia.

The chapter attempts to verify the findings of previous studies (Gan & Ganguli, 2003; Liu et al., 2005; Sedjo & Simpson, 1999). These studies suggest that further reductions in tariffs on forest products are likely to generate only very modest increases in worldwide trade and production. Moreover, the increased harvest pressures on forests due to tariff reduction should be small (Sedjo & Simpson, 1999). At present, the paper does not explicitly model land use or carbon sequestration. Sohngen et al. (2008) highlight the challenges to computable general equilibrium (CGE) modellers in capturing the full range of potential inter-relationships of the forestry sector to the rest of the economy such as land use changes, carbon sequestration and climate policy. Unlike previous studies, this paper highlights the Australian forestry industry as well as the emerging regions in forestry trade such as the Russian Federation and sub-Saharan Africa.

The chapter aims to examine the economic and potential environmental effects of tariff liberalisation on forest products and merchandise trade in Australia using the global trade analysis project (GTAP) general equilibrium modelling framework. The chapter is organised as follows: Section 2 provides recent developments in trade of Australian forestry products.

It also highlights Australia's trading partners and the leading countries in merchandise trade. Section 3 describes the theoretical framework employed in this study. Section 4 discusses the model simulations while Section 5 summarises the results.

2. Background

Australia is a net importer of traditional forest products. Australia does not have significant forest trade relationship with the US. However, the US economy is a major consumer of forest products internationally. In terms of forest products, Australia's main exports are woodchips and sawnwood and it mainly imports wood-based panel and paper and paperboard. Japan and China are the main destination of exports, where Japan is the biggest market for Australian woodchips (broadleaf and conifer) and China is the biggest market for recovered paper. It is anticipated, however, that China will be Australia's biggest market for forest products in the next few years.

Table 1 shows the top exporting and importing countries of world merchandise. Interestingly, the top four exporters are also the top importers of world merchandise. Moreover, nine countries dominate world merchandise trade. Interestingly, the Russian federation is amongst the top exporters of world merchandise. It is ranked third behind the United States and Canada and ahead of Brazil and China in terms of industrial roundwood production (Food and Agriculture Organization [FAO], 2009). The Russian Federation is also a major producer of wood-based panels behind China, the United States, Germany and Canada (FAO, 2009). The region is amongst the top five exporting and consuming countries for industrial roundwood, wood-based panels and sawnwood.

Exporters	Value (Bn $)	Share	Importers	Value (Bn $)	Share
Germany	1461.9	9.1	United States	2169.5	13.2
China	1428.3	8.9	Germany	1203.8	7.3
United States	1287.4	8.0	China	1132.5	6.9
Japan	782.0	4.9	Japan	762.6	4.6
Netherlands	633.0	3.9	France	705.6	4.3
France	605.4	3.8	United Kingdom	632.0	3.8
Italy	538.0	3.3	Netherlands	573.2	3.5
Belgium	475.6	3.0	Italy	554.9	3.4
Russian Fed	471.6	2.9	Belgium	469.5	2.9
United Kingdom	458.6	2.9	Korea, Republic of	435.3	2.7

Source: International Trade Statistics, WTO (2009)

Table 1. Leading Exporters and Importers in World Merchandise Trade, 2008

Table 2 shows the direction of trade for Australia. This suggests that Australia has a strong trade relationship with Asia than any other region in the world. Although Australia does not have significant forest trade relationship with the US, it is still one of its major markets in terms of the direction of trade.

Australia imported $3.8 billion worth of forest products in 2000-2001 (Australian Bureau of Agricultural and Resource Economics [ABARE], 2001). In 2002, Australia had a trade deficit in wood products of $1.7 billion (National Association of Forest Industries [NAFI] briefings 2010). The World Trade Organization calculates the average final bound duties and MFN applied duties for wood, paper, etc. in Australia for 2008-2009 at 7 per cent and 3.4 per cent, respectively.

Exports	Millions, US$	Imports	Millions, US$
Japan	41515	China, People's Republic of	32804
China, People's Republic of	28282	United States	25311
Korea, Republic of	14969	Japan	19120
United States	10249	Singapore	14977
New Zealand	8005	Germany	10701
India	10999	United Kingdom	8431
United Kingdom	7324	Malaysia	8788
Singapore	5536	Thailand	9367
Thailand	4516	New Zealand	7366
Indonesia	3839	Korea, Republic of	6276

Source: Asian Development Bank (www.abd.org/Statistics)

Table 2. Australia's Direction of Trade, 2008

According to the Department of Agriculture, Fisheries and Forestry (DAFF), Australia has traditionally carried a deficit in the trade of its forest and wood products. From 1997 to 2007, in terms of volume, Australian forest product exports increased by 74 per cent while its imports increased by 37 per cent (ABARE, 2009). In contrast, from 1999 to 2009, exports increased by 41 per cent while its imports increased by only 11 per cent (ABARE, 2011).

In terms of value, total exports of wood products in 2009–10 were $2.26 billion while imports were $4.2 billion. This constitutes a trade deficit of $1.9 billion. In 2008-2009, mainly due to the financial crisis, forest product exports decreased by 5.2 per cent and imports increased by 1.1 per cent (ABARE, 2009). Most of this deficit is in paper products as they account for around half ($2.2 billion) of Australia's imports in 2008–09 (DAFF, 2010).

Lower tariffs have arguably been accepted as beneficial to society's economic well-being. Tariff levels have come a long way since the General Agreement on Tariffs and Trade (GATT). However, protectionism especially on local employment from developed countries is resurfacing due to the financial crisis of 2008-2009. If tariff levels continue to decline then global merchandise trade liberalisation would boost Australia's agricultural exports by an estimated US$9 billion (in 2006 dollars) in 2020 (ABARE, 2007).

The Doha round of trade negotiations is considered to provide a major opportunity for developing countries. This trade negotiation started in November 2001 and emphasises on tariffs, non-tariff measures, agriculture, labour standards, environment, competition, investment, transparency and patents. As part of the series of negotiations since 2001 in Hong Kong after four years, trade ministers representing most of the world's governments reached a deal that sets a deadline for eliminating subsidies of agricultural exports by 2013. The effect of the Doha round on forest product's trade is of practical importance for this study. Unfortunately, the current negotiations on trade collapsed in

July 29, 2009. Informal negotiations are taking place in nine key sectors based on what has been dubbed 'the crucial mass' approach – where a certain number of countries representing a certain percentage of world production in a sector are required to participate in order to create a sectoral initiative (Smaller, 2005). These sectors include electronics, bicycles and sporting goods, chemicals, fish, footwear, forest products, gems and jewellery, pharmaceuticals and medical devices, and raw materials. The possible increase in forest products trade due to lower tariffs can have a significant effect on deforestation and as a consequence carbon trading.

Partial equilibrium models have been used in the past to analyse the effects of tariff reductions in the forest sector (Liu et al., 2005). These models cannot generally include the interactions of different sectors in the economy with the forestry sector. Since forest products can be processed to have a higher value-adding within an economy's production as well as consumption, changes in forestry production (and consumption) due to tariff reduction can have significant impacts on the whole economy. Using a global CGE model, such as GTAP, the changes in one sector of the world economy say, countries with higher endowment of forest products or countries that rely heavily on forest products, can be predicted and analysed. Industries and/or countries that are affected in a positive or negative way can be identified. The GTAP model has also been used to analyse the effects of tariff liberalisation on the forest sectors of Brazil, the European Community and the United States (Coelho et al., 2006; Francois et al., 2003; Tsigas, 2005).

3. Theoretical model and data specifications

The model used in the study is developed within the global trade analysis project (GTAP). The project is a global network of researchers and policy makers conducting quantitative analysis of international policy issues. The standard GTAP model is a multi-region (i.e. 113 regions), multi-sector (i.e. 57 sectors), computable general equilibrium model, with perfect competition and constant returns to scale. Each region has a single representative household. The share of aggregate government expenditure in each region's income is held fixed. There is a global banking sector which intermediates between global savings and consumption. International trade and transport margins are treated explicitly and bilateral trade is handled via the Armington assumption. Full documentation of the theoretical structure of GTAP is available in Hertel (1997).

The study uses GTAP database version 7. It contains complete bilateral trade information, transport and protection linkages among 113 regions for all 57 commodities for 2004. The database also includes energy data and OECD domestic support. In this study, the regions are aggregated to 25 regions selected to emphasise global trade on forestry products. There are 13 sectors (commodities) selected to place emphasis on the forest sector and the other sectors in the economy that depend on it (i.e. forestry, wood products and paper products) and they are summarised in Table 1A (see appendix). The regions are selected and grouped to identify the main players in forestry trade. Regions like Russian Federation and Sub-Saharan Africa are included in contrast to Liu et al. (2005) to highlight the relative importance and contribution of these countries. There are five factors of production: land, unskilled labour, skilled labour, capital and natural resources, where labour and capital are assumed mobile. There is no change in the parameters used within the standard GTAP data base. Both short-run and long-run closures are implemented, where capital is fixed in the

former and mobile in the latter. Tariffs are removed for the 16 regions which are considered
dominant in the global forestry trade as shown in Table 3.

Regions
Australia
New Zealand
China
Japan
Korea
Indonesia
Malaysia
Thailand
USA
Rest of N America
Latin America
Germany
United Kingdom
EU_25[1]
Sub-Saharan Africa[2]
Russia

Table 3. Tariffs are removed for 16 out of 25 regions

The GTAP model requires a large data base (like any macroeconomic model) to reflect the
underlying economic structure and dependency between regions. It provides a
global/macroeconomic framework on inter-linkages between and amongst regions. The
results from CGE analyses should be taken with caution and should not be relied on as the
sole source of information (Siriwardana & Yang 2007, p. 26). Hence, specific country effects
and microeconomic implications have to be tested using single-country CGE modelling.
Nevertheless, a global model such as GTAP can provide useful insights to potential
distribution effects of tariff liberalisation policies.

4. Results

On the macroeconomic level, the effects of a global trade liberalisation within the global
forestry sector are minimal. The short-run and long-run effects for most regions are similar
as shown in Table 4. Thailand, Sub-Saharan Africa and Russia would experience a reduction
in their terms of trade relative to the other regions in the model between -0.07 per cent and -
0.14 per cent. However, these countries together with Malaysia would experience an
increase in their real gross domestic products (GDP) relative to the rest of regions between
0.06 per cent and 0.55per cent. The gains in economic growth are higher in the long-run than
in the short-run for Thailand, Sub-Saharan Africa and Russia. In terms of welfare, some
regions benefit more in the short-run than in the long-run and vice versa. In the short-run,

---

[1] Excluding Germany and the United Kingdom (see appendix, Table 2A).
[2] For the list of countries included in this region refer to appendix, Table 2A.

amongst the countries, China and Japan would experience the most gains at around US$ 400 million. China, Thailand, Russia and collectively, Latin America, Sub-Saharan Africa and the EU_25 would gain more in the long-run at US$ 445 million, US$ 415 million, US$ 1.35 billion, US$ 722 million, US$ 594 million and US$ 1.03 billion, respectively. Australia is only slightly affected with modest increase in welfare.

Under Doha with the possible liberalisation of forestry tariffs, Australia can benefit since its major exports are agricultural goods which include forestry products. Table 5 shows changes in the Australian output and consumption of forest products compared to the other major traders of forest products as well as major traders of world merchandise in general. Forest products are classified into three groups such as forestry[3], lumber and wood products and pulp and paper products. Australian output of forest products is predicted to decline whereas the opposite is true for the consumption of forest products.

Amongst its major trading partners of agricultural products, Japan's output of forestry and lumber and wood products would decline by -0.42 per cent and -0.72 per cent, respectively but the output of pulp and paper as well as the consumption of forest products would increase by 0.19 per cent, 0.14 per cent, 0.97 per cent and 0.04 per cent. China would benefit both in the production and consumption of forest products with the exception of pulp and paper production which would decline by -1.45 per cent. In contrast, the United States of America would experience a slight reduction in the production of forestry and lumber and wood products and a slight increase in the consumption of forest products. Indonesia would increase its production of forestry output by 2.06 per cent, lumber and wood products by 3.97 per cent and pulp and paper output by 1.69 per cent.

The European Union is also a major market for Australian agricultural products. At the moment, there are 27 countries included in the EU market. In the model simulation, the 25 EU countries are grouped into one region (i.e. EU_25) with Germany and the United Kingdom treated as separate regions. Germany would experience an increase in outputs for lumber and wood products and pulp and paper products by 0.62 per cent and 0.40 per cent, respectively while the UK would experience a reduction in the output as well as the consumption of its forestry products and the opposite for pulp and paper products. Collectively, the output of forest products in EU_25 would increase between 0.19 per cent and 0.58 per cent.

Table 6 shows the changes in Australia's market prices of forest products. Australia's domestic price of forest products would decline more than the world prices of forestry, lumber and wood and pulp and paper products by -0.36 per cent, -0.47 per cent and -0.22 per cent. Amongst its top trading partners in agricultural products, only Indonesia would experience a relatively substantial increase in the market price of forestry products at 1.05 per cent and a reduction of -0.52 per cent in the market price of pulp and paper products although China's market price for pulp and paper products are reduced by -0.32 per cent.

Russia, Latin America and Sub-Saharan Africa would experience the most reduction in the market price of forestry products between -0.54 per cent and -1.13 per cent while

---

[3] The Food and Agriculture Organization classifies forest products into eight categories: wood fuel, industrial roundwood, sawnwood, wood-based panel, wood pulp, other fiber pulp, recovered paper and paper and paperboard. GTAP has three commodities that correspond to FAO's classification: forestry, wood products and paper products.

Malaysia and Taiwan[4] would experience the biggest increase in the market price of forestry products at 1.06 per cent and 3.24 per cent respectively. Korea would experience a reduction in the market price of lumber and wood products at -1.25 per cent while Thailand would experience the most reduction in the market price of pulp and paper products at almost -5 per cent. New Zealand and the rest of North of America (which includes Canada) would experience a similar reduction in the market prices of their forest products.

Region	Terms of Trade (%Δ)		Real GDP (% Δ)		Welfare ($US mn)	
	Short-run	Long-run	Short-run	Long-run	Short-run	Long-run
Australia	-0.04	-0.05	0.02	0.02	98.72	47.63
New Zealand	-0.05	-0.05	0.00	0.00	-10.39	-14.64
RO Oceania	-0.01	-0.02	-0.01	-0.03	-2.05	-6.21
China	-0.02	-0.02	0.04	0.04	458.24	445.09
Japan	0.01	0.02	0.01	0.00	407.69	241.88
Korea	0.01	0.02	0.04	0.03	261.65	208.30
Taiwan	0.03	0.05	0.05	0.01	214.90	104.42
RO East Asia	0.05	0.05	0.04	0.05	141.38	141.40
Indonesia	0.07	0.06	0.07	0.04	228.70	125.60
Malaysia	0.01	0.02	0.25	0.27	304.24	267.53
Singapore	0.02	0.02	0.04	0.05	64.51	70.87
Thailand	-0.09	-0.14	0.15	0.55	151.16	415.31
RO SEAsia	0.01	0.01	0.00	0.01	11.47	13.32
India	0.01	0.03	0.00	0.01	32.97	68.94
RO South Asia	0.00	0.01	0.00	0.00	5.31	9.43
USA	0.00	0.01	0.00	0.00	239.13	231.13
RO N America	-0.02	-0.03	0.00	0.01	-69.28	-81.86
Latin America	-0.07	-0.07	0.04	0.08	377.96	722.18
Germany	0.01	0.01	0.01	0.01	251.46	319.82
United Kingdom	0.01	0.01	0.01	0.00	288.48	87.77
EU_25	0.01	0.02	0.01	0.01	1141.94	1033.97
M East and N Africa	0.03	-0.01	0.01	-0.03	155.07	-274.70
Sub-Saharan Africa	-0.09	-0.12	0.06	0.19	168.21	593.76
Russia	-0.07	-0.11	0.06	0.28	362.92	1351.70
Rest of World	0.01	0.00	0.00	-0.02	-5.25	-198.26

Table 4. Effects of tariff liberalisation on terms of trade, real GDP and welfare

Table 7 shows the percentage changes in the value of forest products trade. In terms of forestry products exports, Korea and Taiwan would experience the highest increase at 14.64 per cent and 19.43 per cent, respectively while Indonesia and Russia would experience a similar increase in their imports of forestry products by 10.5 per cent and 9.65 per cent respectively. Latin America and Sub-Saharan Africa would experience the highest increase in forestry products imports by 15.08 per cent and 20.46 per cent respectively.

---

[4] It is interesting that Taiwan consistently shows up to have significant changes in its forestry sector. It might be worth while investigating why a country so small would experience relatively bigger changes in its production of forest products.

Region	Forestry		Lumber and Wood Products		Pulp and Paper	
	Output	Consumption	Output	Consumption	Output	Consumption
Australia	-0.59	0.22	-2.82	1.78	-0.71	0.43
New Zealand	-0.32	0.08	-1.29	1.34	-1.17	0.08
RO Oceania	-0.61	0.03	-1.34	-0.01	-0.15	-0.02
China	0.13	0.01	1.23	0.07	-1.45	0.23
Japan	-0.42	0.14	-0.72	0.97	0.19	0.04
Korea	-0.74	0.31	0.09	2.94	0.86	0.25
Taiwan	6.39	-0.36	5.18	0.08	2.44	0.10
RO East Asia	0.11	0.11	0.23	0.14	4.12	0.14
Indonesia	2.06	-0.16	3.97	0.11	1.69	0.23
Malaysia	2.88	-0.11	6.37	3.76	-0.23	1.59
Singapore	1.53	-0.30	2.29	0.14	1.08	0.18
Thailand	-0.50	0.45	-2.25	1.77	3.40	4.09
RO SEAsia	-0.07	0.01	-1.09	0.02	0.16	0.05
India	0.21	0.00	0.24	0.01	0.51	0.02
RO South Asia	0.03	0.00	0.07	0.01	-0.09	0.03
USA	-0.16	0.03	-0.30	0.14	0.04	0.01
RO N America	-0.53	0.12	-0.73	0.30	-0.24	0.11
Latin America	-0.75	0.25	-1.95	0.67	-2.47	0.73
Germany	0.09	-0.01	0.62	0.09	0.40	0.02
United Kingdom	-0.06	-0.01	-0.01	0.19	0.07	0.01
EU_25	0.19	-0.01	0.58	0.07	0.37	0.04
M East and N Africa	0.15	-0.06	0.48	-0.03	0.25	-0.02
Sub-Saharan Africa	-0.43	0.26	-8.99	0.96	-5.81	1.15
Russia	-1.17	0.84	-11.04	2.92	-3.99	1.20
Rest of World	-0.28	0.02	-0.57	-0.01	-0.12	0.01

Table 5. Percentage changes of Australian output and consumption of forest products

Japan and Korea would increase their exports of lumber and wood products by 30.16 per cent and 40.32 per cent while China, Taiwan, Indonesia, Malaysia, Latin America and the Middle East and North African countries would also increase their exports of lumber and wood products between 5 per cent and 8 per cent. In terms of imports of lumber and wood products, Latin America, Sub-Saharan Africa and Russia would experience an increase between 30 per cent and 37 per cent while China, Indonesia, Malaysia and Thailand would experience an increase of between 15.06 per cent and 17.94 per cent. Australia, New Zealand and Korea would experience a modest increase in their imports of lumber and wood between 7 per cent and 10 per cent. Thailand's export of pulp and paper products would increase by 33.55 per cent while East Asia (excluding Japan, Korea and Taiwan), Taiwan, Malaysia and India would increase their export of pulp and paper products between 10 per cent and 17 per cent. Japan and Korea would also experience an increase in their export of pulp and paper products by 7.51 per cent and 9.62 per cent, respectively while New Zealand would experience a decline of around -4.57 per cent. In terms of import of pulp and paper products, Thailand's imports would increase the most by 20.09 per cent. Russia, Latin America and Sub-Saharan Africa would increase their imports of pulp and paper products by 11.86 per cent, 14.75 per cent and 16.79 per cent, respectively. Australia, China, Korea, Indonesia and Malaysia would also increase their imports of pulp and paper between by 4 per cent and 10 per cent.

Region	Forestry	Lumber and Wood Products	Pulp and Paper
World Price	-0.04	-0.11	-0.09
Australia	-0.36	-0.47	-0.22
New Zealand	-0.22	-0.20	-0.13
RO Oceania	-0.23	-0.09	-0.06
China	0.07	-0.17	-0.32
Japan	-0.16	-0.14	-0.04
Korea	-0.26	-1.25	-0.20
Taiwan	3.24	0.03	0.00
RO East Asia	0.14	0.02	0.02
Indonesia	1.05	0.16	-0.52
Malaysia	1.06	-0.26	-1.28
Singapore	0.58	-0.03	-0.06
Thailand	-0.38	-0.49	-4.99
RO SEAsia	-0.03	-0.01	-0.14
India	0.11	0.04	0.00
RO South Asia	0.01	-0.03	-0.04
USA	-0.05	-0.05	-0.02
RO N America	-0.24	-0.24	-0.10
Latin America	-0.54	-0.57	-0.49
Germany	0.10	-0.01	0.01
United Kingdom	0.06	-0.08	0.00
EU_25	0.10	-0.02	0.00
M East and N Africa	0.10	-0.02	-0.03
Sub-Saharan Africa	-0.64	-0.97	-0.72
Russia	-0.75	-0.21	-1.13
Rest of World	-0.06	-0.02	-0.02

Table 6. Percentage changes in domestic and world market prices of forest products

Region	Forestry		Lumber and Wood Products		Pulp and Paper	
	Exports	Imports	Exports	Imports	Exports	Imports
Australia	2.78	-0.34	1.84	7.61	3.27	6.35
New Zealand	1.03	-0.29	0.13	10.35	-4.57	1.20
RO Oceania	-0.52	-0.33	-4.79	-0.17	-0.30	0.26
China	4.63	1.98	5.70	16.04	5.55	9.79
Japan	7.51	-0.32	30.16	2.76	7.51	0.78
Korea	14.64	1.17	40.32	9.83	9.62	4.85
Taiwan	19.43	3.74	7.33	2.55	17.00	1.99
RO East Asia	0.52	0.61	3.61	0.52	15.55	1.36
Indonesia	7.67	9.65	6.36	17.94	5.18	5.36
Malaysia	-4.71	3.01	8.41	15.06	11.90	5.86
Singapore	1.54	0.04	3.11	0.57	3.32	1.13
Thailand	6.13	-0.61	1.91	17.58	33.55	20.09
RO SEAsia	1.24	-1.62	-1.38	0.07	4.77	0.64
India	9.57	0.13	3.20	0.87	10.03	0.56
RO South Asia	1.17	-0.16	1.95	0.29	2.54	0.29
USA	0.38	0.34	-0.22	1.10	0.88	0.33
RO N America	0.93	0.31	0.13	3.83	-0.35	0.79
Latin America	4.38	15.08	5.42	29.89	1.85	14.75
Germany	-0.46	1.42	2.22	0.60	1.74	0.22
United Kingdom	0.72	0.46	3.51	0.49	0.85	0.13
EU_25	-0.18	1.04	1.87	0.84	1.66	0.43
M East and N Africa	2.78	0.45	6.53	0.18	2.75	0.20
Sub-Saharan Africa	4.66	20.46	0.51	37.10	-0.20	16.79
Russia	2.56	10.50	2.69	30.78	5.19	11.86
Rest of World	-0.30	0.07	-2.29	-0.08	-0.21	0.14

Table 7. Percentage changes in the value of forest products trade

5. Summary and conclusion

The interaction between economic activity and the environment are increasingly being recognised not only locally but internationally. Globalisation and the relevance of international trade suggest that increasing cooperation amongst countries is required. Trade liberalisation and climate change are issues that will continue to be in the political agenda for the next few years. With forestry included in the DOHA round of trade negotiations, the sector's effects on the domestic economy as well as its importance in managing climate change could reveal important policy implications.

The study analyses the effects of trade liberalisation on forestry products in Australia, amongst the leading exporters and importers of forest products, in particular, as well as global merchandise, in general. The study utilises the Global Trade Analysis Project (GTAP) model and its database, version 7 with 2004 data. There are 25 regions aggregated to emphasise global trade on forestry products and 13 sectors to emphasise the forest sector and the other sectors in the economy that depend on it. There are five factors of production namely, unskilled and skilled labour, capital, land and natural resources. There is no change in the parameters used within the standard GTAP data base. The study has not incorporated the role of the forestry sector in carbon sequestration.

Given that forest products only comprise a small proportion of world merchandise trade, it is expected that trade liberalisation would cause small changes in terms of trade, real GDP, production, consumption and prices of forest products in most countries. In the short-run, national welfare in China and Japan would increase substantially by more than $US400 million while the opposite is true for North America (excluding the United States). In the long-run, national welfare in China, Thailand, Latin America and Russia would increase between $US445 million and $US1.35 billion. Collectively, EU_25 and Sub-Saharan Africa would experience the highest increase in welfare in the long-run by $US1.03 billion and $US594 million, respectively. It seems that Asian countries, Latin America, Russia, the EU as well as Sub-Saharan Africa would gain the most with a tariff reduction on forest products namely forestry, wood and paper products.

As a caveat, the study does not explicitly model land use or carbon sequestration. It also cannot capture the full benefits of sustainability issues in forestry such as rotation periods and forest cover. However, there are recent attempts in the literature to address this short-coming (Sohngen et al., 2008). It is apparent that forestry and hence the forestry sector generally can have environmental benefits such as biodiversity, low salinity, low soil erosion, etc. The incorporation of all non-monetary benefits requires a substantially rich environmental data set and modelling methodologies. The specific costs and benefits of localised industries and/or regions should be explored further via case studies to ensure success of any policy attempting to balance economic and sustainable issues.

6. Appendices

No.	Sector	Commodities
1	Agriculture	Paddy rice, Wheat, Cereal grains nec, Vegetables, fruit and nuts, Oil seeds, Sugar cane and sugar beet, Plant-based fibers, Crop nec, Cattle, sheep, goats and horses, Animal products nec, Raw milk, Wool and silk-worm cocoons, Meat, Meat products, Processed rice
2	Forestry	Forestry
3	Fishing	Fishing
4	Mining and Extraction	Coal, Oil, Gas, Minerals nec
5	Manufacturing	Vegetable oil and fat, Dairy products, Sugar, Food products nec, Beverages and tobacco products, Textiles, Wearing apparel, Leather products, Petroleum and coal products, Chemical, rubber and plastic products, Mineral products, Ferrous metal, Metals nec and Metals products, Motor vehicles and parts, Transport equipment nec, Electronic equipment, Machinery and equipment nec, Manufactures nec
6	Wood Products	Wood Products
7	Paper products	Paper products and publishing
8	Construction	Construction
9	Public Service	Electricity, Gas manufacture and distribution, Water
10	Trade	Trade
11	Sea Transport	Sea Transport
12	Air Transport	Air Transport
13	Other Services	Transport nec, Communication, Financial services, Insurance, Business services, Recreation and other services, Public Admin, Defence, Health, Education and Dwellings

Table 1. A. Sectoral aggregation

EU 25	Sub-Saharan Africa
Austria	Nigeria
Belgium	Senegal
Cyprus	Rest of Western Africa
Czech Republic	Central Africa
Denmark	South Central Africa
Estonia	Ethiopia
Finland	Madagascar
France	Malawi
Greece	Mauritius
Hungary	Mozambique
Ireland	Tanzania
Italy	Uganda
Latvia	Zambia
Lithuania	Zimbabwe
Luxembourg	Rest of Eastern Africa
Malta	Botswana
Netherlands	South Africa
Poland	Rest of South African Customs
Portugal	
Slovakia	
Slovenia	
Spain	
Sweden	
Bulgaria	
Romania	

Table 2. A. Region Composition

7. References

ABARE (2011). Australian Forest and Wood Products Statistics, September and December Quarters 2010, May, Commonwealth of Australia, Canberra

ABARE (2009). Australian Forest and Wood Products Statistics, March and June Quarters 2009, November, Commonwealth of Australia, Canberra

ABARE (2007). Australian Commodities, March quarter 07.1, Commonwealth of Australia, Canberra

ABARE (2001). Australian Forest and Wood Products Statistics, March and June Quarters, October, Commonwealth of Australia, Canberra

ABS (2010). International Trade in Goods and Services, 5368.0, January, Australian Bureau of Statistics, Commonwealth of Australia, Canberra

BRS (2010). Bureau of Rural Sciences, Department of Agriculture, Fisheries and Forestry, 25.05.2011, Available from:

<http://adl.brs.gov.au/data/warehouse/pe_brs90000004181/forestsAtGlance2010
_ap14.pdf>

Coelho, A. M.; Lima, M. L., Cury, S. & Goldbaum, S. (2006). *Impact of the proposals for tariff reductions on non-agricultural goods (NAMA)*, 14.01.2010, Available from: <http://www.anpec.org.br/encontro2006/artigos/A06A121.pdf>

DAFF (2010). Australia's Forest Industries, Department of Agriculture, Fisheries and Forestry, 25.05.2011, Available from: <http://www.daff.gov.au/forestry/national/industries>

FAO (2009), Forest products (2003-2007), Food and Agriculture Organization of the United Nations, Rome

Francois, J.; van Meijl, H. & van Tongeren, F. (2003). *Economic implications of trade liberalization under the Doha round*, 14.01.2010, Available from: <https://www.gtap.agecon.purdue.edu/resources/download/1397.pdf>

Gan, J. & Ganguli, S. (2003). Effects of global trade liberalization on U.S. forest products industries and trade: A computable general equilibrium analysis. Forest Products Journal, Vol. 53, No. 4, pp. 29-35

Hertel, T. W. (ed.) (1997). Global trade analysis: modeling and applications, Cambridge University Press, Cambridge

Liu, C.; Kuo, N. & Hsue, J. (2005). Effects of tariff liberalization on the global forest sector: application of the GTAP model. International Forestry Review, Vol. 1, No. 3, pp. 218-226

Low, K. & Mahendrarajah, S. (2010). Future directions for the Australian forestry industry, Australian Bureau of Agricultural and Resource Economics, Issues Insights 10.1, Canberra

Narayanan, B. & Walmsley, T.L. (eds.) (2008). *Global trade, assistance, and production: the GTAP 7 data base*, Center for Global Trade Analysis, Purdue University

National Association of Forest Industries (2010). International forestry issues, NAFI Briefings, 25.05.2010, Available from: <http://www.nafi.com.au/briefings/index.php3?brief=27>

Sedjo, R. A. & Simpson, R. D. (1999). *Trade liberalization, wood trade flows, and global forests*, Discussion Paper 00-05, Resources for the Future, December, Washington, D.C

Siriwardana, M. & Yang, J. (2007). Effects of Proposed Free Trade Agreement between India and Bangladesh. South Asia Economic Journal, Vol. 8, No. 1, pp. 21-38

Smaller, C. (2005). *In the name of development: whose ambitions for the Doha round?*, Institute for Agriculture and Trade Policy, Geneva Office, Geneva Update 19th August, 2005,22.01.2009, Available from: <http://www.s2bnetwork.org/.../Geneva%20update%20august%202005.pdf?id=49>

Sohngen, B.; Golub, A. & Hertel, T. W. (2008). *The role of forestry in carbon sequestration in general equilibrium models*, GTAP Working Paper no. 49, Center for Global Trade Analysis, Purdue University, West Lafayette, Indiana

Tsigas, M. E. (2005). An analysis of US regional implications of tariff reductions under the Doha Round of negotiations, Paper presented to the 8th Annual Conference on Global Economic Analysis, Lübeck, Germany, 9-11 June

Varian, H. R. (1992). *Microeconomics analysis* (3rd edition), WW Norton and Company, New York

World Trade Organization (2009). International Trade Statistics, 14.12.2011, Available from: <http://www.wto.org/english/res_e/statis_e/statis_e.htm>

8

Strategic Cost Management Practices Adopted by Segments of Brazilian Agribusiness

Marcos Antonio Souza[1] and Kátia Arpino Rasia[2]
[1]*Universidade do Vale do Rio dos Sinos – Unisinos*
[2]*Universidade Federal de Rio Grande – Furg*
Brazil

1. Introduction

The perception that the globalization event creates an increasingly fierce competition between companies is already known by managers and customers (Shank & Govindarajan, 1997). Like in the other segments, in agribusiness the tougher competition has been one of the inducing vectors, among others, in the search for more effectiveness and efficiency in the use of resources, as a way of ensuring business continuity.

In agribusiness, factors such as scarcity of resources, emergence of new technologies in equipment and supplies, coupled with changes in the eating habit, may also change consumer's expectations if the products become less attractive and/or more expensive. This dynamics involves and guides the competitiveness of agribusiness companies, needing to anticipate trends relating to these issues and taking actions to increase the demand of their products and achieve higher contribution margins (Cyrillo, 2010). In general, globalization impacts the business in terms of both threats and opportunities, whether domestic or foreign. The growth of agribusiness in Brazil led to the expansion of the performance frontiers, making a strong presence in foreign markets, with Brazilian products gaining international prominence and competing favorably with products from other countries (Gasques et al, 2004).

The management process of the organizations demands information from all kinds, financial and non-financial quantitative and qualitative. Among the quantitative information, the ones from cost assume strategic aspect in the analysis of competitiveness and have to face the competition by organizations. From this perspective it is important to understand and analyze the valuable activities of the organization with the external environment, called by Porter (1989) as value chain analysis, unique for each company. This proposed analysis allows to identify the strengths, weaknesses and opportunities for improving business performance based on their processes and costs. For Shank & Govindarajan (1997), cost management can lead to an effective control of spending and can ensure an advantage against the competition. Costs affect all organizations' areas and activities and are determined by the set of management decisions.

From the 80s, new cost management practices emerged in response to criticism about the usefulness of the available techniques and used ones by the traditional managerial accounting. Those were shaped by a more complex operational environment in which there was need for the information to reflect the changes . (Johnson and Kaplan, 1987; JOHNSON, 1992; Bacic, 1994). The commonality in the criticisms, besides the proposition and discussion on specific practices, is the finding that the cost management should also consider environmental factors external to firms, creating a new knowledge field and professional practice called Strategic Cost Management – SCM.

The adoption of strategic cost management practices by the companies has been the subject of several studies at national (Reckziegel, Souza and Diehl, 2007, Marques et al, 2003, Ferreira, Silva and Batalha, 2010; Muniz, 2010, Souza, Collaziol and Damacena, 2010) and international level (Guilding, Cravens and Tayler, 2000; Bowhill and Lee, 2002; Dekker and Smidt, 2003, Waweru, Hoque and Uliana, 2005; Cinquini and Tenucci, 2006; Cadez and Guilding, 2007; Zoysia and Herath, 2007; Noordin, Zainuddin and Tayler, 2009; Angelakis, and Thério Floropoulos, 2010). Based on the findings of these studies, it is observed that the adoption of the strategic cost management practices, when segregated in countries, present higher frequency of use in developed countries like Japan, Italy and United States of North America.

Still, the analysis of the researches' findings shows the lack of studies in the agribusiness companies, despite their social and economic relevance. Thus, the research problem that guides this study is: What best strategic cost management practices are used by segments of agribusiness companies in Brazil?. The overall objective is to identify, in Brazilian agribusiness segment, which practices in strategic cost management are used. The specific objectives are: (a) to identify the practice of cost management highlighted by the literature, (b) to identify the potential benefits, limitations and disadvantages of adopting these practices, (c) to verify the existence of a relationship between the use of SCM practices and the level of competition, (d) to examine the perception of the practices' benefits in those companies with international operations, (e) to identify the degree of difficulties noticed in adopting the surveyed practices.

The study contributes to a complex business segment and still lacking in empirical investigations on the cost management practices. It also meets the recommendations of Callado and Callado (2006), that by highlighting the importance of agribusiness emphasize the paucity of academic studies focused on cost management in firms of this segment. Specifically, the authors show the need of the analysis of cost practices, an important tool for generating information for decision making. Besides this introduction, form the text other four parts: (i) in section two the theoretical framework, (ii) the methodology in section three, (iii) in the fourth section results and the discussion of research findings and (iv) final. Finally, in references are listed the sources used in the research's development and theory.

2. Theoretical framework

2.1 Strategy applied to business management

Miles and Snow (1978) identified four types of strategy that are distinguished by the corporate behavior of firms: (a) Prospector: corresponds to pioneering companies that want innovation, (b) Defender: used by companies with tight organizational controls of efficiency and quality, (c) Analyzer: combines prospectors' and defenders' features, and (d) Reactor: only reacts to the environment, as if it didn't have its own strategy, and seeks new products or markets only when it feels threatened by competitors. The authors emphasize that the chosen strategy must be adapted to the chosen market, with a particular configuration of technology, structure and process, consistent with its market strategy.

Porter (1989) sustains that there are only three successful and internally consistent strategies to achieve better performance than other firms. These generic strategies are: (a) Cost leadership: the company becomes a low-cost competitor in its industry; (b) Differentiation: the company seeks to be unique in its industry, observing some dimensions valued by buyers; (c) Focus: the company is based on the choice of a narrow competitive environment of an industry, selects a segment or group of segments in the industry and tailors its strategy to serve them. For the author these are the viable approaches to deal with competitive forces. However, companies must adopt only one of them, otherwise they would be stuck in the middle ground and without a defense strategy.

To Anthony and Govindarajan (2008), although the definitions on strategy differ, there is agreement that the strategy sets the direction and the plans for achieving the goals. A company develops its strategies combining its core competencies with market opportunities, while observing the risks and weaknesses. In the process of defining the strategies, the environmental analysis seek to identify opportunities and threats. Porter (1986) identifies the existence of five basic competitive forces: (i) the bargaining power of suppliers, (ii) of buyers, (iii) the threat of new entrants, (iv) the threat of substitute products and (v) rivalry among competing firms.

Authors from other areas such as marketing: like Kotler (1998) who defines the typologies in which businesses can be fitted due to its chosen strategy as: (i) leaders, (ii) challengers, (iii) followers and of market niches. A company is a leader when it has relevant participation in the product market and maintains its leadership by changing prices and launching new products. The challengers are those of lower ratings in an industry and they can attack the leader in an aggressively bid. The followers track, copy or improve a product to launch it and get high profits, since it did not incur expenses related to innovation. The ones of market niche create, expand and protect their portions of space.

Because of its operational characteristics, the agriculture and stock-breeding is under pressure by society groups concerned about the impact of its activities on the environment. In this sector, the concern with the environmental impacts that come from its activities has grown. For Marques et al (2003) the negative impacts result in the reduction of biodiversity, erosion and contamination of soil, silting and contamination of water sources, and they possibly cause changes in regional climate. To mitigate these effects, companies must implement environmental strategies. The social benefits offered by companies under the implemented environmental strategy, impact positively over its image. Recognition by society over environmental and socially correct practices of companies can contribute to the brand development and strengthening and its reputation, impacting directly on its sustainability (Thorpe and Prakash-Mani, 2003).

2.2 The management model as competitive edge

It is typical of a competitive environment that companies constantly seek to adapt. Also, despite competing in the same segment, organizations adopt and implement different models for managing their business. Although sometimes strategies are similar, only a few organizations can achieve their goals. One explanation for this may be because the used management model in the organization does not comply with the requirement imposed by the situation and the environment.

About the management model, Nascimento, Reginato and Lerner (2007) argue that the construction of a well-defined, organized and coordinated structure, to harmonize the performance of activities guided by pre-established conduct rules, may represent difference in terms of competition. The management model, when formalizing the management process - which generally comprises the steps of planning, execution and control - structures and approximates the foundations of strategic cost management. It will be in the management process, in the planning stage - strategic and operational - that strategic goals and strategies of organizational activities will be defined. In order to create interaction and synergy between parts, the management model should establish how the flow of information will happen, the channels of communication, internal controls (among other purposes, to evaluate the performance) and also the formalization of decision making and its models. Therefore, the success of strategic cost management is related to the adequacy of the model management, process management and decision-making process to the needs of the organization.

Frost (1999) points out that managers need support to achieve the highest performance and also have a good understanding of how the performance can be quantified and communicated. In the everyday of organizations, managers are pressured by consumer demand, increased competition and the ever-shrinking time. We live the time that everything has to be done better, faster and cheaper. In this context, it is vital for managers that the company translates its strategy into measurable goals through an appropriate measurement model, readily available by information systems.

Fischmann and Zilber (2008), addressing the issue of evaluation, emphasize that performance indicators emerge as one of the tools that can assist in defining the strategic planning and consequently the determination of business strategies. Thus, according to the authors, it is possible to check the property with which decisions were taken. It is in this environment that the strategic cost management plays an important role, providing information to managers to support the management process and decision making.

2.3 Agribusiness

The term *agribusiness* first came at Harvard University, USA, coined by Professors Davis and Goldberg (1957), based on study from the input-output approach and conceptualized as the total sum of all operations involved in the manufacture and distribution of farm supplies. In this approach, agribusiness is set around the business of agriculture and is the basis of food production. As agriculture could not be considered in isolation, the authors considered it as part of an extensive network of economic agents, starting with the production of raw materials, industrial processing, storage and distribution of agricultural products and derivatives.

Over the time, changes in the way of life bring out changes in consumer eating habits. Zuin and Queiroz (2006) found that the people's short available time for meals triggered the need for rapid preparation of food. As a result, there is need and demand for ready-to-eat or semi-ready-to-eat products, causing changes in products, production processes and forcing companies to use new production technologies. On the other hand, the failure to adapt to new corporate standards and expectations demanded by the consumer, can result in reductions in volumes and lower profit margins.

In the U.S., Jackson and Mitchell (2009) observed the increasing vertical integration of agribusiness food chain. However, they noticed negative effects of the near price monopoly featured by companies that have the power inside the chain. In this scenario, the authors believe that the power imbalances in food production and speculation in agricultural commodity markets should be analyzed for representing an obstacle to the operation of the food chain.

The dynamics of agribusiness is driven by the pursuit for competitive advantage, mainly occurring through vertical integration, (Silva and Batalha, 2010, Jackson and Mitchell, 2009; Callado and Callado, 2009; Azevedo, 2010). Vertical integration is defined by Porter (1989) as the division of activities between a company and its suppliers, channels and buyers, so a company can buy components rather than manufacture them. Then, a way the company can differ from others is assuming a greater number of buyer activities. For Loturco (2008) the growth of Brazilian agribusiness is in the process of modernization, with products competing in national and international markets. Brazil is asserting itself as a major supplier of food and raw materials of natural origin [commodities] in the international market (Lima et al (2009). For the author, the coverage and social and economic impacts of Brazilian agribusiness is impressive. Around 5000 cities depend directly on the agribusiness, contributing with 26.5% in the formation of the country's gross national product. Sobral (2008) adds that agribusiness in Brazil is growing due to its favorable climatic conditions and prices of major agricultural commodities in great demand international market.

Brazil, according to IICA (2009), takes the first place in the production of biofuels, representing 36% of world production. The Brazilian coffee production is also outstanding, reaching 34% of the world's, reaching the first place, followed by Vietnam (14%), Colombia (9%), Indonesia (5%), Ethiopia (5%), India (4 %) and Mexico (3%). Still according to data from IICA (2009), Brazil is the largest producer of oranges in the world context and accounts for 33% of production. The U.S. ranks second in world production of oranges, followed by China (12%), EU (11%), Mexico (8%) and Egypt (7%). Brazil is number two in world production of soybeans, with 26%, followed by Argentina (20%) and China (7%). According to CNA (2010), Brazilian agriculture and livestock breeding ended 2009 with production equivalent to R$ 718 billion, showing a decrease of R$46.6 billion compared to the previous year. Such reduction is equivalent to a loss of 6% in the share of agriculture and livestock breeding in the shaping the country's Gross National Product. In the National Food and Nutritional Security Council - CONSEA (2008), discussions are that the expansion of biofuel production can affect food production in Brazil, especially in the state of São Paulo. According to this group, there is need of public policies to harmonize food energy production, so that the crops won't lose space for the production of biofuels.

2.4 Strategic cost management

Simmonds (1981, p.26), in a seminal article on the subject states:

Strategic Management Accounting can be defined as the provision and analysis of management accounting data about a business and its competitors for use in developing and monitoring the business strategy, particularly relative levels and trends in real costs and prices, volume, market share, cash flow and the proportion demanded of a firm's total resources.

The author also states that SCM is not the redefinition of business, planning or marketing functions inside the company. Also, it is emphasized that the management accounting skills are essential. The use of traditional management accounting techniques combined with contemporary ones is adopted by Shank and Govindarajan (1997).

An important component of management information systems of any organization is the accounting system. Assuming the importance of management information required by managers according to Simmonds (1981), it is pertinent the alert from Johnson and Kaplan (1992): an inefficient accounting system can undermine both the development of superior products or the improvement of processes and marketing efforts. Kaplan and Norton (1997) and Shank and Govindarajan (1997), in general, are the most often mentioned authors when the topic is seeking competitive advantage with the use of cost information, among others. Also, it has been credited that as consequence of the criticisms over accounting (in particular by Johnson and Kaplan, 1992), there were new practices, approaches and procedures for costs considered more appropriate to the new demands of management. The main criticism - widely accepted and disseminated by a majority of authors - is based on the argument that the practices of cost accounting and management, still in use today, were developed to a reality of business and technology which no longer exists.

The SCM approaches by other authors, in general, are in accordance to the proposal of Simmonds (1981). Bacic (2009) sustains that management accounting should be considered within a framework that recognizes the impact of competition and strategy, noting the criteria and business's needs. In the same way, to Blocher et al (2007) since strategic factors are growing in importance for management, cost management has transformed its traditional role of cost of product and operational control into a broader and strategic focus. Thus, for the authors, strategic cost management is the development of cost management information in order to facilitate the primary function of management: strategic management. As expected from an evolutionary process, over the years specific cost management practices were developed and incorporated into SCM.

2.5 Analyzed strategic cost management practices

The practices of strategic cost management lead to information that can contribute so that the company can succeed in assuring competitive advantage. For Hansen and Mowen (2001), strategic cost management uses cost information to develop and identify superior strategies, capable of producing a sustainable competitive advantage. The analyzed SCM practices were classified according to three factors discussed in the study of Cinquini and Tenuci (2006), shown in Figures 1, 2 and 3.

In Figure 2 are listed the practices of Factor 2 (Processes and Activities) and Factor 3 (Clients).

In Figure 3 are listed the practices related to Factor 4 (Competitors).

The external analysis of costs comes from the recognition that only internal efficiency is no longer sufficient for the effectiveness of the company. It includes two approaches of data collection and analysis: (1) competitors, about best practices in processes, product attributes, accounting data, or costs, (2) it involves costs that focus on the processes and activities of customers and suppliers, constituted by the sequence of actions throughout the production workflow, from acquisition of raw materials to product delivery to consumers.

SCM Practices	Main Characteristics
Cost of Product Life Cycle - CPLC	The CPLC considers all the necessary steps of the product, from design to delivery and installation of the finished product. It evaluates the cost of products while designing, producing, distributing, consuming and disposal phase.
Total Cost of Ownership - TCO	It analyzes the cost of purchasing properties or services from a particular provider. It represents the total of all costs of properties/services, from its acquisition to its final consumption and disposal.
Environmental Costs	Costs incurred because there is poor environmental quality or because it can exist. They can be classified into four categories: (i) prevention costs, (ii) costs of detection, (iii) internal failure costs and (iv) external failure costs.
Kaizen costing	It means continuous improvement. It is a management technique from which managers and staff are committed to a program of continuous improvement in quality and other critical factors of success.
Intangible costs	Practice of hidden cost analysis, the result of structural items and inefficiency of management; deriving from the existence of intangible factors and resulting from the formation of intangible assets.

Source: Based on Cinquini and Tenucci (2006).

Fig. 1. SCM Practices, according to Factor 1: Quality

3. Methodology

The approach of this research is exploratory because it provides familiarity with the subject - strategic cost management in agribusiness segment firms - and the related problem (Menezes and Silva, 2001). It is also descriptive by highlighting the characteristics of a given population or phenomenon or establishment of a correlation between variables (Gil (2002). The classification is quantitative for the usage of statistical techniques for analysis and interpretation of data. The qualitative aspect is a result of the performed analysis based on quantitative data (Roesch, 2005). This research is also a survey and data were collected via structured questionnaire with closed questions and Likert scale to capture the intensity of the responses. The assertions in the questionnaire concerning to SCM practices were based on the study of Cinquini and Tenucci (2006). As usual, pre-test was conducted with teachers and controllership professionals, whose suggestions and contributions have improved the instrument.

	SCM Practices	Main Characteristics
Factor 2: Processes and Activities	Value Chain Analysis	Practice of monitoring opportunities in bonds with customers and suppliers. It divides the chain into relevant strategic activities so that one can understand the behavior of costs and sources of differentiation.
	Activity Based Management (ABM)	Practice developed for the funding and management of the activities that consume resources; it allows the identification, reduction or even elimination of activities that do not generate value to the customer.
	Target costing	It determines the cost for a product based on a certain competitive sale price, and that the product achieves the wanted profit. It uses Value Engineering to reduce costs based on the manufacture alternatives.
	Logistics cost	Analysis of costs of supply, purchase, distribution and storage of inputs and outputs.
	Analysis of the Determinants of Cost	Considered as central points of cost management, they represent the cause of costs and precede the effective execution of operations. In general, they are related to the facilities, technology and complexity of the activities used in the processes of activities related to production of properties and services.
Factor 3: Performance Analysis	Indicators and Non-Financial Metrics	Indicators allow the understanding, the comprehensive analysis of economic and financial situation of the company and can be applied to all strategic aspects of analyzed costs in practices of strategic cost management.

Source: Based on Cinquini and Tenucci (2006)

Fig. 2. SCM Practices, according to Factor 2: Processes and Factor 3: Performance.

	SCM Practices	Main Characteristics
Factor 4 External Analysis of Costs	Cost Analysis of Competitors	It collects data to appreciate the value chain of competitors, transforming them into useful information to decision-making.
	Interorganizational Cost Management, GIC	It is the exchange of information between chain companies to establish improvements to processes, through partnerships. It uses the Open-Book Accounting (OBA) for opening the company's cost information in order to reduce costs and optimize results.

Source: Based on Cinquini and Tenucci (2006)

Fig. 3. SCM Practices, according to Factor 4: Competitors.

The sample consists of investigated companies, those belonging to the segments (1) sugar and alcohol and (2) wood, pulp and paper. The sampling used was non-probabilistic and the sampling frame were the companies listed (Babbie, 2005) in the Yearbook of Agriculture Exam 2008. After identifying the companies, potential respondents were contacted and 120 emails were sent with the link to the questionnaire. After a few weeks, the companies were contacted again and encouraged to answer. With the return of 34 questionnaires, data collection phase was completed. The return rate was 28.3% - the average of other surveys of the area. After the collection, the data were tabulated and summarized by using MS-Excel spreadsheets and statistical analysis done by SPSS software. In processing the data, the techniques used were the descriptive statistics - frequency distribution, mean, median and standard deviation - and measures of correlation (Pearson coefficient and ANOVA).

4. Results and discussion

4.1 Description of companies

The companies in Brazilian agribusiness that took part in this research are part of the segments (i) sugar and alcohol and (ii) wood, pulp and paper. In order to make a description of the respondent companies, they were asked for specific information. The distribution of companies by revenue indicates that 38.2% of the respondents have annual revenue of up to R$ 500 million, 29.4% in the range of R$ 501 million to R$ 1 billion and 32.4% over R$ 1 billion. As regarding the origin of firms, there is a predominance of Brazilian with 82.3%, most of them (53.6%) of sugar and alcohol sector. All six international companies operate in the sector of wood, pulp and paper. In relation to the target market, most companies (94.1%) operate in the foreign market via exportation. For a significant number of companies in the sugar and alcohol segment, exportation revenues are around 10% of its total income. In this revenue edge for the foreign market, companies of the wood, pulp and paper sector account for 32.4% of the sample.

Most companies have a long period of existence, some over 100 years. From 1961 to 1975 was the creation of 30% of all enterprises. Regarding the location of firms, the prevailing states are São Paulo (35.2%), Paraná (23.5%), Alagoas (11.8%) and Minas Gerais (8.9%). In São Paulo there is equal division among segments. Among Paraná's companies prevail the wood, pulp and paper sector (85%). The surveyed Alagoas's companies operate only in sugar and alcohol sector. The perception of surveyed managers about the type of competition faced by the companies showed that 44.1% of them operate in highly competitive markets and 55.9% in medium competitive markets. In the segment of sugar and alcohol, for the surveyed managers, the highly competition is due to the increase of the demand for biofuels. Analysis undertaken by CONSEA (2008) points out that the increased demand for biofuels has led to expanding the frontiers of sugar cane crops, reducing the areas for food or forcing the displacement to other regions.

4.2 Analysis of SCM practices

Table 1 shows that the SCM practices frequently used by companies are (i) the determinants of costs, (ii) value chain analysis, (iii) indicators and non-financial metrics, (iv) target costing, (v) standard cost and (vi) logistics costs.

The practice of analyzing the determinants of cost is often used by all firms surveyed. These findings are justified by the particularities of agribusiness, leading to greater complexity in management (Vilckas and Nantes, 2006). With the technical knowledge on production processes, managers need to use management tools to plan the productive activities and add value to products. The value chain analysis is frequently used by 89% of companies surveyed. This finding is in accordance to the statement of Oliveira and Pereira (2008), for whom the management of agribusiness companies must observe the value chain to differentiate themselves from competitors. Thus, the use of this practice, besides providing a number of reviews, defends the adoption of strategies of differentiation (Porter, 1989).

Strategic Cost Management Practices	Degree of Usage of SCM Practices	Degree of Benefits for SCM Practices	Degree of Difficulty in Adopting SCM Practices
Intangible costs	2,26	2,56	4,00
Analysis of the Determinants of Cost	4,59	4,56	3,06
Value Chain Analysis	4,06	4,41	3,09
Cost of Product Life Cycle	2,47	3,71	3,35
Activity Based Management (ABM)	2,26	3,50	3,59
Environmental Costs	2,59	4,06	3,65
Indicators and Non-Financial Metrics	4,15	4,47	2,88
Target Costing	4,50	4,62	2,82
Standard Cost	4,09	4,24	2,85
Logistics Cost	4,35	4,59	2,85
Total Cost of Ownership - TCO	2,06	2,91	3,79
Kaizen Costing	2,06	3,12	3,74
External Analysis of Costs	2,79	3,56	3,65

Source: Research Collected Data

Table 1. Usage, Benefits and Difficulties in Adopting SCM Practices.

The indicators and non-financial metrics are significantly used by 88% of the companies investigated. For Nakagawa (1991) a management information system which does not incorporate non-financial indicators is weak, which does not occur with significant proportion of respondent companies. Queiroz (2006) states that the information systems of agribusiness companies must demonstrate the large performance of operations in financial and nonfinancial format.

The practice of target costing is used by 94% of companies, indicating that they consider the price charged by the market as a benchmark for setting their costs (Rocha (1999). The practice of standard costing in 89% of companies shows that its use and target costing's are complementary, not antagonistic. These findings suggest that there are benefits with the use of standard costing in the control stage, after the planning phase of the product with support the target costing, as proposed in the study of Carastan (1999). Although the literature stresses the importance of ABC/ABM to the analysis and value creation in activities (Kaplan and Cooper, 1998, Blocher et al, 2007), only 15% of companies surveyed use these practices to analyze and measure activities of the production process.

Diagrammed in Figure 4 is a comparison between the average level of use, benefits and difficulties noticed by managers in relation to the adoption of the practices of SCM, based on a scale from 1 to 5 1 to 5 (1 = no or no one, 5 = always or a lot, as appropriate).

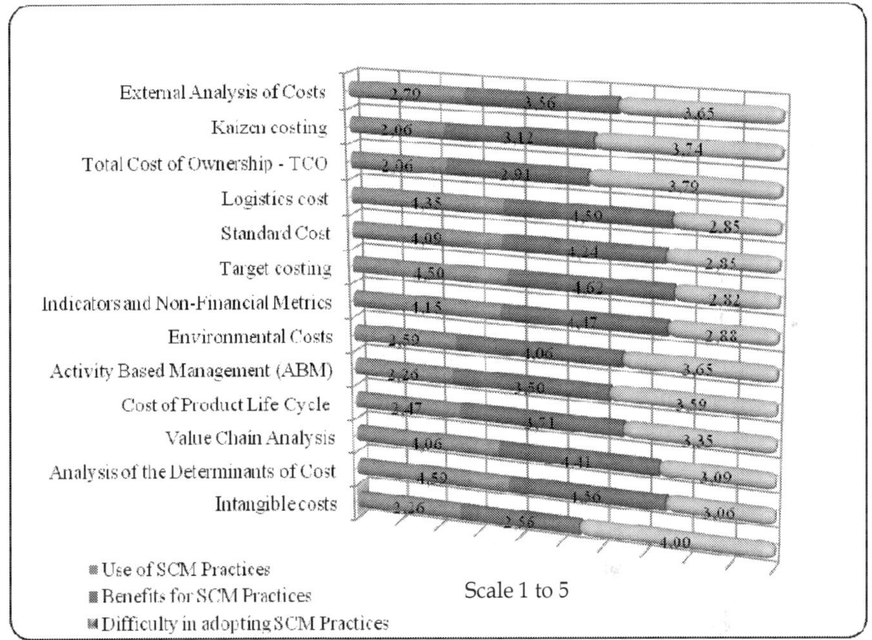

Source: Research Collected Data

Fig. 4. Average Degree of the Usage, Benefits and Difficulties in Adopting SCM Practices.

As shown in Figure 4, seven SCM practices have low adoption and great difficulties in implementation and usage (external analysis, kaizen costing, TCO, environmental costs, activity based management, life cycle costs and intangible costs). Despite the low utilization, the potential and the benefits are recognized, however, the exception is the practice of intangible costs. The low use of the environmental cost practice, ponders Gonçalves (2008), reveals the weakness of management given the international demands on environmental issues. The low adoption of the practice of activity based management, despite the benefit emphasized by literature, suggests that it is not noticed by managers. The same occurs with other practices such as life cycle costs, TCO, kaizen and analysis of external costs.

From the perception of 97% of respondents there are benefits in the usage of non-financial indicators. The benefits of using the standard cost reaches 88% in the total "high" and "very high". Blocher et al (2007) believe that these two practices may be associated by the reason that non-financial measures should be associated with costs for the control of strategic activities. The shown benefits by kaizen costing are noticed by 41% of respondents, although the adoption for this practice is only frequently used by 15% of companies. However, for the target costing both the noticed benefits and the frequent usage by businesses have high percentages. Eldenburg and Wolcott (2007) consider that the kaizen costing is similar to the target costing by setting goals to reduce costs. Still, the cost reduction goals are established to manage the downward trend in sale prices over the life cycle of products, given the value chain perspective. It is appropriate, in this context, the use of the kaizen costing (production stage) with the target costing (planning stage).

The benefits of the logistics costs practice are noticed by 97% of respondents as high and with the same percentage of frequent usage. The practice of environmental costs is often used by only 18% of respondent companies. These findings suggest that there are difficulties in its use. It is known that sanitary restrictions may affect competitiveness of firms and the intangible benefits for the adoption of environmental management. Environmental certifications value products since their preparation does not occur at the expense of environmental resources. The adoption of the practice of environmental cost management, in this background of requirements and benefits, might enable the companies highlight the investments made in benefits of society in this aspect. The low adoption of important SCM practices, though readily recognized its importance, can be credited to a number of difficulties encountered by companies. Among the difficulties, there is lack of qualified personnel and restructuring of procedures and system costs for companies. The main identified benefits are cost reduction, improvement in decision making and competitiveness.

The Pearson correlation coefficients, Table 2, indicate that SCM practices are positively related to the level of competition. However, as the coefficients are between 0.4 and 0.69, the existing association is moderate (Gageiro and Pestana, 2005).

SCM Practices	Level of Competition	Competitive Strategy	Capital Structure
Cost of Product Life Cycle	0,549**	0,040	-0,091
Activity Based Management (ABM)	0,596**	0,057	-0,090
Environmental Costs	0,628**	0,043	-0,147
Total Cost of Ownership - TCO	0,475**	0,296	-0,321
Kaizen costing	0,593**	0,133	-0,329

Source: Research Collected Data
(**) Significance (p value <0.005)

Table 2. Correlation - SCM Practices x Competition, Strategy and Capital Structure

The correlation between the SCM practices and the level of competition occurs in the low used practices by companies. For these practices, as evidenced in Table 1, there are greater difficulties for their implementation, despite the perception of potential benefits arising from their use. Such findings suggest the company that additional efforts justify the use of these practices. The Pearson correlation test identified no significant correlation between the use of the practices of SCM and capital structure (open / closed). The coefficients, negative, below 0.329 for all analyzed practices and statistically insignificant (p value <0.005). Regarding SCM practices and Competitive Strategy, the coefficients of correlation - positive - less than 0.296 and statistically not significant (p value <0.005) indicate no correlation. The lack of connection between these factors can be explained by similarity of firms on the size and experience, but can also be a feature of the segment.

The ANOVA test, Table 3, was conducted to determine whether there are differences between the intensity on the noticed benefits arising from the practices of SCM and the different magnitude of exportation, having as proxy the "percentage of exports on turnover" of companies (size). The results for the SS show the main effects, with p value <0.001, indicating that the probability of occurrence by chance is less than 0.01%. The F value, which tests the equality of variances, is represented by the ratio between AS (x) and AS(error). The importance of calculating the F value is due to the analysis of the significance of the variances between the means.

SCM Practices x Percentage of Exportation without Turnover		Sumo f squares (SS)	Degrees of freedom (DF)	Average of Squares (AS)	F	Sig.
Activity Based Management (ABM)	Between Groups (x)	13,567	5	2,713	5,088	0,002
	Inside Groups (error)	14,933	28	0,533		
	Total	28,500	33			
Target costing	Between Groups (x)	6,963	5	1,393	5,518	0,001
	Inside Groups (error)	7,067	28	0,252		
	Total	14,029	33			
Logistics Cost	Between Groups (x)	4,802	5	0,960	4,949	0,002
	Inside Groups (error)	5,433	28	0,194		
	Total	10,235	33			

Source: Research Collected Data

Table 3. ANOVA Test: Benefits x Percentage of Exportation without Turnover.

Based on scale of F Distribution, the critical value for 5 DF (numerator) and 28 DF (denominator) corresponds to 2.56 (F). Since the F values exceed this limit: 5.088 for the ABM; 5.518 for target costing, and 4.949 for the logistics costs, then the differences for these groups in terms of exportation variable is significant for a p value <0.005.

The differences in the ANOVA for companies that export, indicate relatively higher average about the perceived benefits. This suggests that management emphasizes logistics costs control as consumer markets are more distant. The benefit of target costing is outstanding in the exporting companies, given the limit of allowable costs to compete in these markets. The use of ABM is justified as it enables the implementation of other SCM practices: the use of the life cycle cost practice makes it necessary to check the activities of design, production, distribution, consumption and disposal. If it is the practice of environmental costs there will be are control activities (prevention, assessment), and the related lack of control ones (internal and external failure) analyzed.

Practices ABC/ABM, according to literature, are also needed in coordinating activities in the value chain for demonstrating the links among the needs, identifying opportunities for resource optimization and quality improvement (Porter (1998). In the analysis of perceived difficulties was applied Pearson's correlation test, as shown in Table 4. It was identified some relation in the degree of difficulty perceived by managers for implementation of the SCM practices.

The correlation coefficient of the difficulty level of implementation of kaizen costing indicates a strong association (between 0.70 and 0.89) with the level of difficulty of the TCO practice; other practices have a moderate association (between 0.40 and 0.69). For Pestana and Gageiro (2005), R (linear association) less than 0.20 is very low; between 0.20 and 0.39 low; between 0.40 and 0.69 moderate; between 0.70 and 0.89 high and finally, between 0.90 and 1.00 very high. For all variables, the significance (p value <0.005) indicates that the probability of the associations being by chance is smaller than 5%. As example of the analysis regarding the level of competition, the results indicate practices of little use: the environmental cost, ABC/ABM and kaizen costing.

Difficulties in Implementing SCM Practices	Target costing	Analysis of the Determinants of Cost	Indicators and Non-Financial Metrics	Cost of Product Life Cycle	Total Cost of Ownership - TCO
Standard Cost	0,644**	-	-	-	-
Indicators and Non-Financial Metrics	0,493**	-	-	-	-
Environmental Costs	0,479**	-	-	-	-
Logistics cost	0,661**	0,494**	0,478**	-	-
Value Chain Analysis	-	0,555**	-	-	-
ABC/ABM	-	-	-	0,663**	-
Kaizen Costing	-	-	-	-	0,846**

Source: Research Collected Data

Table 4. Pearson's correlation test – Noticed Difficulties in SCM Practices.

4.3 General Interpretation of data analysis

An overview of the analysis' findings shows that there is a balance between use and nonuse of the SCM practices surveyed. From the total of 13 practices surveyed, six of them (46.2%) are well used and the remaining have limited use. At first glance this ratio suggests an overall unfavorable scenario, however, the adoption identified in this study is greater than in other national surveys and lower than the international ones.

Although there is present emphasis on cost management literature, it draws attention the low usage of ABC and ABM. In this context, it is plausible the recent simplification of the ABC under the guise of TDABC. Literature also tends to emphasize that it is antagonistic the use of standard cost practices and target costing (Shank and Govindarajan, 1997). However, the survey results show the opposite, suggesting that the surveyed companies have managed to separate the utility of practices in the stages of planning and cost control, making them complementary. Finally, special attention should be given to the results of the correlation tests [Pearson and ANOVA], which did not identify common characteristics of companies with the degree of correlation that indicates a situation with force to leverage the use of SCM practices listed in this research.

5. Closing remarks

The focus of the study was to investigate the use, the importance of perceived benefits and difficulties of implementation by agribusiness companies of SCM practices, treated in the literature as the most appropriate ones to help managers in the management process of organizations. In summary, the findings on practices of SCM show heavier usage of six practices among the 13 listed. The benefits, limitations and disadvantages from the use of practices deserve to be highlighted for the paradoxical situation of high importance, low utilization and high difficulty degree of implementation found by some. The level of competition has no importance in the adoption of practices, nor the capital structure (open/closed) has relation to the type of strategy used.

Although the findings of this research cannot be generalized, it is displayed two major impacts. The impact to the academy is the challenge to rethink the contents of the subjects related to cost, managerial and strategic accounting. It is understood that the dissemination of knowledge must occur as an undergraduate, as addition for the student when in full professional activity. On the other hand, the impact for practitioners is to seek to overcome the conceptual shortcomings over SCM practices, to act as disseminators of knowledge in their organizations and show the benefits to the organization for its use.

Finally, during the study it was evidenced the need for more specific and accurate studies in terms of non-use of SCM practices given strong indications in literature. We believe that the findings of this exploratory study provide the foundation for conducting researches in the form of multiple case study and cross-case analysis. The results of investigations with these approaches may indicate - with more direct and objective responses to deal with issues, sometimes paradoxical - relation to the benefits and difficulties in implementation and use of SCM practices addressed in this research.

6. References

Angelakis, George; Theriou, Nikolaos; Floropoulos, Jordanis. (2010). Adoption and benefits of management accounting practices: evidence from Greece and Finland. Advances in Accounting, v. 26, n. 1, p. 87-96.

Anthony, Robert, N.; Govindarajan, Vijav. (2008). Sistemas de controle gerencial. 12ª ed. São Paulo: Atlas.

Azevedo, Paulo F. (2010). Comercialização de produtos agroindustriais. In: BATALHA, Mario Otávio. Gestão agroindustrial. 3ª ed. São Paulo: Atlas, p. 63-112.

Babbie, Earl. (2005). Métodos de pesquisas de survey. Belo Horizonte: Ed. UFMG.

Bacic, Miguel Juan. (1994). Escopo da gestão estratégica de custos em face das noções de competitividade de estratégia empresaria. In:CONGRESSO BRASILEIRO DE GESTÃO ESTRATÉGICA DE CUSTOS, 1. Anais Eletrônicos... São Leopoldo: ABCustos.

Bacic, Miguel Juan. (2009). Gestão de custos. Curitiba: Juruá.

Banco Nacional De Desenvolvimento Econômico E Social - Bndes. (2010). Porte de empresas. Disponível em:
http://www.bndcs.gov.br/SiteBNDES/bndes/bndes_pt/Navegacao_Suplementa r/Perfil/porte.html. Acesso em: 25 jan 2011.

Batalha, Mario Otávio; Silva, Andre Lago. (2010). Gerenciamento de sistemas agroindustriais: definições, especificidades e corrente metodológicas. In: BATALHA, Mario Otávio. Gestão Agroindustrial. 3ª ed. São Paulo: Atlas, p. 1-60.

Blocher, Edward J.; Chen, Kung H.; Cokins, Gary; Lin, Thomas W. (2007). Gestão estratégica de custos. São Paulo: McGraw-Hill.

Bowhill, Bruce; Lee, Bill. (2002). The incompatibility of standard costing systems and modern manufacturing. Journal of Applied Accounting Research, v.6, n.3, p.1-24.

Cadez, Simon; Guilding, Chris. (2007). Benchmarking: the incidence of strategic management accounting in Slovenia. Journal of Accouting & Organizational Change, v.32, n.2, p.126-146.

Callado, Antônio A. C.; Callado, Aldo L. C. (2006). Mensuração e controle de custos: um estudo empírico em empresas agroindustriais. Sistemas & Gestão Revista Eletrônica, v. 1, n. 2, p.132-141.

Callado, Antonio A. C.; Moraes Rodolfo A. (2009). Gestão de custos no agronegócio. *In:* CALLADO, Antonio A. C. Agronegócio. 2ª ed. São Paulo: Atlas, p. 20-28.

Carastan, Jacira T. (1999). Custo-meta e custo-padrão como instrumentos do planejamento empresarial para obter vantagem competitiva. *In:* CONGRESSO BRASILEIRO DE CUSTOS. 6. São Paulo. Anais eletrônicos... São Paulo: ABCustos.

Cinquini, Lino.; Tenucci, Andrea. (2006). Strategic management accounting: exploring links with strategy. Disponível em: http://www.unisanet.unisa.edu.au/Resources/.../ Cinquini%20&%20Tenucci.pdf. Acesso em: 20 dez 2010.

Confederação Nacional Da Agricultura E Pecuaria Do Brasil - Cna. 2010. Site institucional. Disponível em:< http://www.canaldoprodutor.com.br/> Acesso em: 15 dez 2010.

Conselho Nacional De Segurança Alimentar E Nutricional - Consea. 2008. Exposição de motivos. Disponível em: <https://www.planalto.gov.br/Consea/exec/index.cfm>. Acesso em: 06 Jan. 2011.

Cyrillo, Fabio. (2010). Novas demandas do varejo. In: CONGRESSO NACIONAL DE GESTÃO DO AGRONEGÓCIO, 1. Chapecó. Anais eletrônicos... Chapecó: Agrogestão.

Davis, John.; Goldberg, Ray. (1957). The genesis and evolution of agribusiness. *In:* Davis, J.; Goldberg, R. A concept of agribusiness. Harvard University, p.4-6.

DE ZOYSA, Anura; HERATH, S. Kanthi. (2007). Standard costing in Japanese firms. Industrial Management & Data Systems, V.107, N.2, P. 271-283.

Dekker, Henri; Smidt, Peter. (2003). A Survey of the adoption and use of target costing In Dutch firms. International Journal Of Production Economic, v. 84, n.3, p.293-305.

Eldenburg, Leslie G.; Wolcott, Susan K. (2007). Gestão de custos: como medir, monitorar e motivar o desempenho. Rio de Janeiro: LTC.

Fischmann, Adalberto A.; Zilber, Moisés A. (2001). Utilização de indicadores de desempenho como instrumento de suporte à gestão estratégica. RAM, v.1, n.1, p. 9-25.

Frost, Bob. (1999). Performance metrics. Strategy and Leadership, v. 27, p.34-36.

Gasques, José G.; Rezende, Gervásio C.; Verde, Carlos M. V.; Conceição, Júnia Cristina P. R.; CARVALHO , João C. S.; SALERNO, Mario S. (2004). Desempenho e crescimento do agronegócio no Brasil. Disponível em:http://www.ipea.gov.br/portal/ images/stories/PDFs/TDs/td_1009.pdf Acesso em: 20 jan 2011.

Gil, Antonio Carlos. (2002). Como elaborar projetos de pesquisa. São Paulo: Atlas.

Gonçalves, José A. (2008). Os bons exemplos que vem do campo. *In:* Anuário Exame Agronegócio: Ranking das 400 maiores empresas do agronegócio. São Paulo: jun, p.54-57.

Guilding, Chris; Cravens, Karen S.; Taykes, Mike. (2000). An international comparison of strategic management accounting practices. Management Accounting Research, v.11, n.1, p. 113-135.

Hansen Don R; Mowen, Maryanne M. (2001). Gestão de custos. São Paulo: Pioneira.

Instituto Interamericano De Cooperação Da Agricultura - Iica. Caderno de estatísticas do agronegócio. 2009. Disponível em: http:HTTP://www.iica.org.br/Docs/Publicacoes/Agronegocio/CadernoEstatistic as_03-009.pdf   Acesso em: 10 dez 2010.

Jackson, Anthony; MITCHELL Eve. (2009). Food sovereignty: time to choose sides. Encyclopaedia Britannica. Disponível em: http://media.web.britannica.com/ ebsco/pdf/37/37221176.pdf   Acesso em: 04 mai 2010.

Johnson, H. Thomas; KAPLAN, Robert S. (1987). Relevance lost: the rise and fall of management accounting. Boston: HBSP.

Johnson, H. Thomas. (1992). Relevance regained: from top-down control to bottom-up empowerment. New York: Free Press.

Kaplan, R. S.; Norton, D. P. (1997). A Estratégia em ação. Rio de Janeiro: Campus.

Kaplan, Robert & Cooper, Robin. (1998). Custo & desempenho. São Paulo: Futura.

Kotler, P. (1998). Administração de marketing São Paulo: Atlas.

Lima, Fátima M. S.; Abrantes, Luiz A. A.; Correia, Laura F. M.; Brunozi Jr., Antonio C. (2009). Políticas públicas de inovação tecnológica na cadeia agroindustrial do leite: o efeito da instrução normativa n° 51 nos produtores rurais da Microrregião de Viçosa, MG. In: ENANPAD, 33. Anais... São Paulo, ANPAD.

Loturco, Roseli. (2008). Os obstáculos para as multinacionais do agro. *In:* Anuário Exame Agronegócio: Ranking das 400 maiores empresas do agronegócio. São Paulo: p.21-23.

Marques, J.; Skorupa, L. A.; Ferraz, José M. G.; Bacellar, A. A. (2003). Indicadores de sustentabilidade em agro-ecosystems. Jaguariúna: Embrapa Meio Ambiente.

Miles, Raymond E.; Snow, Charles C. (1978). Organizational strategy, structure and process. New York: Mc Graw-Hill.

Muniz, Luciani Silva. (2010). Práticas de gestão estratégica de custos adotadas por empresas brasileiras. Dissertação (Mestrado em Ciências Contábeis). Unisinos. São Leopoldo.

Nakagawa, Masayuki. (1991). Gestão estratégica de custos. São Paulo: Atlas.

nascimento, auster m.; reginato, luciane; lerner, Daiane F. (2007). Avaliação de desempenho organizacional. *In:* NASCIMENTO, Auster Moreira; Reginato, Luciane (Org). Controladoria: um enfoque eficácia organizacional. São Paulo: Atlas, 2009.

Noordin, R. Zainuddin, Yuserrie; Tayles, Michael. (2009). Strategic management accounting information elements. Management Accounting Journal, v.4, n.1, p.17-34.

Oliveira, Deyvison L.; Pereira, Sidinei A. (2008). Análise do processo decisório no agronegócio. Revista Eletrônica Gestão e Sociedade, v.2, n.4, p.1-24.

Pestana, Maria H.; Gageiro, João N. (2005) Análise de dados para ciências sociais: a complementaridade do SPSS. Lisboa: Edições Silabo, Lda.

Porter, Michael E. (1986). Estratégia competitiva. Rio de Janeiro: Campus.

Porter, M. E. (1989) Vantagem competitiva. Rio de Janeiro: Campus.

Queiroz, Timóteo Ramos. (2006). Ferramentas de controle da inovação na propriedade rural. *In:* Agronegócios gestão e inovação. (Org) Zuin, Luís Fernando S.; QUEIROZ, Timóteo R. Cap.10. São Paulo: Saraiva.

Reckziegel, Valmor; Souza, Marcos A.; Diehl, Carlos A. (2007). Práticas de gestão adotadas por empresas estabelecidas nas Regiões Noroeste e Oeste do Paraná. Revista Brasileira Gestão de Negócios, v. 9, n. 23, p.14-27.

Rocha, Welington. (1999). Contribuição ao estudo de um modelo conceitual de sistema de gestão estratégica.Tese (Doutorado em Controladoria e Contab.) – FEA/USP, São Paulo.

Roesch, Sylvia Maria Azevedo. (2005). Projetos de estágio e de pesquisa em administração – guia para estágios, trabalhos de conclusão, dissertações e estudos de caso. São Paulo: Atlas.

Shank, J.K.; Govindarajan, V. (1997). A revolução dos custos. R. Janeiro: Elsevier.

Silva, Andre L.; Batalha, Mario O. (2010) Marketing estratégico aplicado ao agronegócio. *In*: Batalha, Mario Otávio. Gestão agroindustrial. São Paulo: Atlas, p. 1-60.

Silva, E. L.; Menezes, E. M. Metodologia da pesquisa e elaboração de dissertação. 3°. ed. Florianópolis: Laboratório de Ensino a Distância da UFSC, 2001.

Simmonds, K. (1981) Strategic management accounting. Management Accounting, april, p. 26-9

Sobral, Eliane. Mais dinheiro no campo. (2008). *In*: Anuário Exame Agronegócio: Ranking das 400 maiores empresas do agronegócio, p. 90-92. São Paulo.

Souza, Marcos A.; Collaziol, Elisandra; Damacena, Cláudio. (2010). Mensuração e registro dos custos da qualidade: uma investigação das práticas e da percepção empresarial. RAM, v.11, n. 4, p.66-97.

Thorpe J.; Prakash-Mani, K. (2003). Developing value the business case for sustainability in emerging markets. Greener Management International, v.44, p. 17-32.

Vilckas, Mariângela; Nantes, José F. D. (2006). Planejamento e agregação de valor nos empreendimentos rurais. *In*: ZUIN, Luiz F. S.; QUEIROZ, Timóteo R. Agronegócios gestão e inovação. São Paulo: Saraiva, p.167-188.

Waweru, N. M.; Hoque, Z.; Uliana, E. (2005). A Survey of management accounting practices in South Africa. National Journal of Accounting, Auditing and Performance Evaluation, v.2, n.3, p.226-263.

Zuin, L. F. S.; Queiroz, T.R. (2006). Agronegócios gestão e inovação. São Paulo: Saraiva.

9

Forestry and Life Cycle Assessment

Andreja Bosner, Tomislav Poršinsky and Igor Stankić
Faculty of Forestry, University of Zagreb
Croatia

1. Introduction

Ever since recognizing the effect of global warming and ozone degradation, scientists are trying to develop new methods of environmental impact assessment to ensure environmental protection. Life Cycle Assessment (LCA) is such a method and with its fully comprehensive approach better nature conservation is possible. Sustainability can be achieved through minimal consumption of renewable and non-renewable materials, energy saving, reuse and recycling, emission control, etc. However, the cost aspect is still the most important factor in today's world and therefore achieving sustainability is made even more difficult. LCA according to FAO is a useful tool for comparing the environmental aspects of specific products as it enables the ecological comparison of two or more products made of different raw materials but used for the same purpose. LCA can help in measuring environmental aspects and potential impacts of a product through its entire life-cycle from raw material acquisition (as the beginning of the life-cycle) to manufacturing, use, recycling, reuse and final disposal (as the end of the product's life-cycle). The ISO/EN 14040 defines LCA as a technique for assessing the environmental aspects and potential impacts associated with a product by: 1) compiling an inventory of relevant inputs and outputs of a system; 2) evaluating potential environmental impacts associated with those inputs and outputs; 3) interpreting the results of the inventory analyses and 4) impact assessment in relation to the objectives of the study (Figure 1).

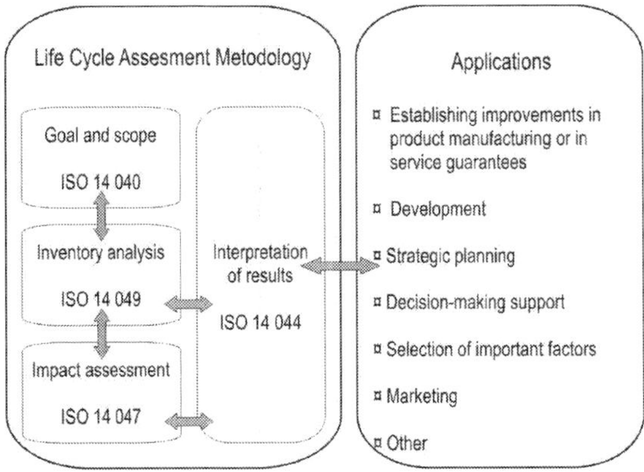

Fig. 1. Phases in the LCA according to ISO standards (source: www.iso.org)

Goal and scope of the LCA explain why the study is being done, set functional units (for example 1 m^3 or 1 m^2 or 1 tonne are usually taken for wood products) and give system boundaries to determine which unit-processes of LCA will be considered in the study (data quality requirement, precision and representatives of data). Inventory analysis indicates relevant input data (used energy, raw and supporting materials etc.) and output data (releases to air, water, land and manufacture of side products, by-products etc.). To avoid high complexity of the inventory analysis, processing machines, manpower, buildings and land-use for biological production or transportation are not included. Impact assessment reveals the potential environmental repercussions and their significance. Impact categories can be defined form local to global. Interpretation of results combines all the findings together and final conclusions and recommendations can be made.

Other ISO/EN standards deal with LCA as well:

- ISO 14020: 2000 (Environmental labels and declarations – General principles)
- ISO 14021: 1999 (Environmental labels and declarations – Self-declared environmental claims, Type II environmental labelling)
- ISO 14024: 1999 (Environmental labels and declarations – Type I environmental labelling – Principles and procedures)
- ISO 14025: 2006 (Environmental labels and declarations – Type III environmental declarations – Principles and procedures)
- ISO 14040: 2006 (Environmental management – Life cycle assessment – Principles and framework)
- ISO 14044: 2006 (Environmental management – Life cycle assessment – Requirements and guidelines)
- ISO/TR 14047: 2003 (Environmental management – Life cycle impact assessment – Examples of application of ISO 14042)
- ISO/TS 14048: 2002 (Environmental management – Life cycle assessment – Data documentation format)
- ISO/TR 14049: 2000 (Environmental management – Life cycle assessment – Examples of application of ISO 14041 to goal and scope definition and inventory analysis).

According to COST ACTION E9 Memorandum of Understanding (Life cycle assessment of forestry and forest products) Europe currently produces and consumes the roundwood equivalent of 400 million m^3 of wood products. Forests, moreover cover more than 25% of the world's land area with approximately 600 billion m^3 of standing stock. Annually 3.6 billion m^3 of wood is used as firewood (55%) or as different products (45%). Because wood is a renewable energy source, CO_2 neutral and recyclable, it represents the most important renewable raw material and fuel and a very important carbon sink. Even more importantly, wood can be permanently available if sustainable forest management is made obligatory. The main reasons of performing LCAs by COST Action E9 of forestry and forest products are:

- To enable comparison between different materials, provided that products are used for the same purpose;
- To obtain quantified and reliable information of benefits of wood products and their environmental impact, so that industry and policy makers can use such information;
- To highlight unknown or uncertain areas of environmental impact of wood products;

- To improve production and recycling techniques by minimizing steps of high environmental impacts or choosing different processing routes to reduce such high environmental impact.

However, certain problems when dealing with LCA in forestry arise:

- Forestry uses considerable areas of land;
- Life cycle of forest products can vary from relatively short (for example paper) to very long (for example structural timber);
- Long production chains which start in the forest and end with disposal or burning for energy;
- Forest products have complex relationships between products, by-products and waste.

Complexity of the LCA study is shown in figure 2.

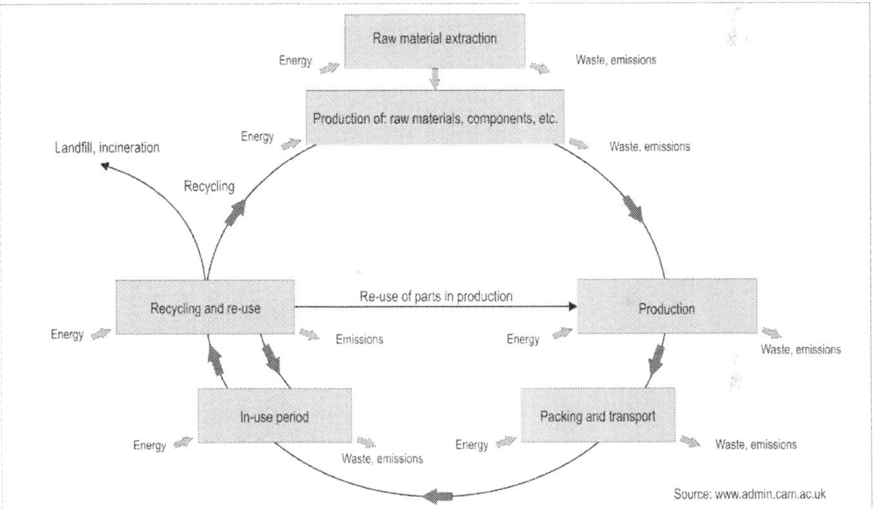

Fig. 2. Product lifetime

2. Challenges for LCA in forestry

The carbon cycle starts with the biosynthesis in the forest and ends by releasing CO_2 into the atmosphere during combustion or biodegradation. The most important material and energy cycles of wood as a renewable raw material should be therefore considered in LCA studies. Production of timber requires large area of land, but different kinds of land use should be also considered and compared, for example timber production in forests in comparison to agriculture raw material production, renewable forest material and fuel production in comparison to the exploitation of non-renewable materials and fuels and of land use in industry. Different forest management systems should also be compared and studied. Complexity of LCA studies for forestry and relating industries is shown in Figure 3 by Jungmeier (2003) in his System analysis of forestry, forest products and recovered wood (COST Action E9 and E31).

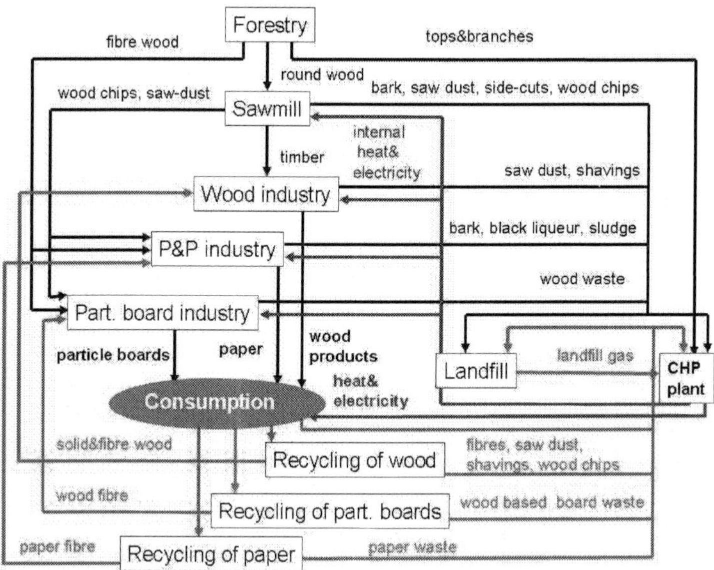

Fig. 3. Complexity of forestry products by Jungmeier (2003)

2.1 Space-time aspect

Many scientists believe that time effect should be excluded from LCA studies. However, in forestry social and time aspects are important, not just because forests affect on soil stability, water quality, biodiversity, wildlife habitat, but also because of the most important carbon cycle. Frühwald (1995) emphasises that forests are a part of the environment and therefore should be considered as an impact category of its own. Furthermore, forests are a substantial part of the global ecosystem and are negatively influenced not only by poor or inadequate management, but also by polluted air, acid rain etc. LCA is a method mainly focused on industrial production and to include/expand it to wood production and make it possible for forestry it should consider land use (biological production) and site conditions of raw material production. Only a serious impact assessment can characterize a substance as an environmental burden. If the emissions of a substance (such as CO_2) is interpreted as a burden, consequently a reduction of this environmental burden by fixing this substance in the biological growth process should be stated in the inventory and should be interpreted as a benefit (Thoroe and Schweinle, 1995).

Even though in industrial production it is sufficient to concentrate on flows between different phases of a product lifetime, in forestry that is insufficient and stocks should also be considered. To make this possible, raw material acquisition i.e. forest production should be standardized. Standard argument for exclusion of beneficial effects forests have on the environment is a lack of methods for measuring such effects (Thoroe and Schweinle, 1995).

To understand and quantify the role of the forest ecosystem in the carbon cycle it is necessary to quantify both the net annual carbon fluxes (Mg C ha^{-1} yr^{-1}) and the total carbon content (Mg C ha^{-1}) of representative forest ecosystems, thereby including the carbon fluxes and stocks in the soil (Nabuurus and Mohren, 1993). Virtanen and Nilsson (1993) made a life cycle analysis of waste paper recycling, where CO_2 fixation was 0.7 to 0.9 kg for each kg of biomass (logs including bark and water content). Calculations of net CO_2 uptake for 1 kg of biomass were 0.46 kg CO_2-equivalents, which gave 0.92 kg CO_2-equivalents per kilo of usable wood (assuming 50% of water content). Grasser (1994) makes CO_2 uptake calculations in a stack-wood chain. The chain stars with wood formation, and the CO_2 uptake is based on a carbon content of 0.5 kg per kilo of oven dry matter, which means 1.833 kg of CO_2 uptake per kilo of usable wood (including bark). To provide a kilo of usable wood (including bark), 1.2 kg of wood needs to be cut, of which 0.2 kg will stay in the forest as wood waste and decay and the bounded CO_2 within will be set free again during its biological decomposition. This represents 17% of the total growth. Lox (1994) calculates CO_2 uptake of produced cardboards and states that biological decomposition of wood waste if different. 34% of the total growth will stay in forest, which means 0.5 kg of wood waste per kilo of usable wood (including bark) because only cellulose and hemi cellulose fraction will decompose, which means 2.2 kg of CO_2 uptake per kilo of usable wood. Different ways to calculate the CO_2 uptakes have been mentioned and De Feyter (1995) states that the CO_2 should not be seen as a credit, but as an implementation of the carbon neutrality of wood when looking at its lifecycle. The emitted carbon can be present in the form of CO_2, CO, CH_4 etc. and those emissions have a different behaviour towards environmental effects which can result in different CO_2-equivalents.

When comparing recycling and non-recycling of paper in LCA, most analysis show a higher contribution to global warming of recycling systems. This is because recycling systems use more fossil fuels, however recycling of paper holds carbon longer in fixed rotation, but this aspect cannot be taken into account in LCA because LCA still has a static approach and no time aspects.

Wollenman (2006) investigates aims at balancing carbon sequestration and emission fluxes of forest land use to evaluate methodology for a close-to-nature and plantation regime. The regimes have a positive sequestration capacity of about 85 kg C m^3 (eucalyptus plantation) and about 180 kg C m^3 (close-to-nature beech forest) respectively. Assuming a given area of forest cover a division of so called ecological labour between plantation forest and forest reserve provides a net carbon sink capacity of 0.18 kg C m^{-2}a^{-1} compared to 0.14 kg C m^{-2}a^{-1} for close-to-nature regime. These results clearly indicate that management regime determines the net carbon sink capacity and that so called ecological labour division between intensively and unmanaged forests offer opportunities to maximize carbon sequestration.

Forest roads provide accessibility to forests, and their construction is one of the largest investments in forestry itself. Multi-purpose use of forest roads creates problems involving the process of building and maintaining forest roads in the life cycle assessment of forest products. Lifetime of roads is difficult to define, since it depends on many factors and generalizes their depreciation over time. Mroueh et al (2000) in a study of life cycle assessment of road construction analyze numerous factors (environmental loadings, Table 1) that need attention:

- Consumption of energy and fuels (including energy consumed by the machine, or the vehicle during the processing of raw materials and energy contained in organic materials);E

- Emissions and dust;
- Environmental noise pollution (Table 2);
- Substances released into the soil depending on the type of building material (for example sulphate compounds, mercury, arsenic, chromium, copper, vanadium, etc.).

Impact category	Environmental loading	Results, unit
Resource use	Use of natural resources	t/construction selected[*]
	Industrial by products	t/construction selected[*]
	Energy	kWh/construction
	Fuels[**]	m^3/construction
	Land use	Verbal estimation of the significance of land use
Effluents to soil and waters	Leaching of metals (e.g. As, Cd, Cr, Cu, Mo, Ni, Se, Pb, Zn)	mg/m^2 of construction selected
	Leaching or migration of compounds from materials[***]	mg/m^2 of construction selected
	CI, SO_4 [***]	mg/m^2 of construction selected
Emissions to air	CO_2	kg/construction selected
	NO_X	
	SO_2	
	VOC	
	CO	
	Particles	
Wastes	Inert waste	t/km
Other loadings	Noise	Noise x time/km

[*]Only constructions that meet the same performance requirements and are designed for the same site are compared in the assessment.
[**]Fuel means diesel oil used to power machines and vehicles or oil used as a raw material in industry. Can be presented separately or included in the energy consumption.
[***]The substance included in the analysis is selected according to the material used.

Table 1. Environmental loadings in the life cycle assessment of road construction

The same authors according to the study of Häkinnen and Mäkelä (1996) estimate the environmental burdens that arise during maintenance and repair of roads in Finland during the period of 50 years. The frequency of repairs is determined in advance adopted strategy (Table 3).

Water-soluble substances present in the materials can be carried away by run-off water into the environment and into the groundwater. The amount of leaching depends on the composition of the material, the amount of water passing through material and the manner in which it is laid (Ranta et al. 1987). Pollutants leaching out of alternative constructions over a hundred years are presented in table 4 (Mroueh et al., 2000).

Some of the most important factors, such as COD emissions to water and land, have been absent from this study and applicable calculation methods requires more extensive further research. The creation of procedures for calculating the environmental loadings would be important for assessing the impacts of material selections and the total loadings of road usage.

Machine type	Noise level range	Average noise level
	dBA	
Drilling rig	98 – 101	100
Blasting	125 – 136	130
Hydraulic hammer	87 – 92	90
Conveyor belt	84	84
Crushing plant	100	100
Hydraulic excavator	82 – 100	89
Earth moving machines	91	91
Lorry	84	84
Bulldozer	80 – 89	84
Road roller	84 – 101	92
Asphalt layer	74 – 89	81
Road grader	85 – 89	87

*At a distance of 7 m from the source

Table 2. Noise levels of working machines* (Mroueh et al. 2000)

Environmental loadings	Construction	Maintenance
CO_2, kg/km	263 000 – 562 000	33 900
SO_2, kg/km	280 – 610	4,1
NO_X, kg/km	2600 – 3800	140
CO, kg/km	600 – 1100	20
Violatile organic compounds (VOC), kg/km	550 – 980	210
Fuel consumption, l/km	63 000 – 100 000	18 200
Energy consumption, kWh/km	790 000 – 1 .470 000	183 300

Table 3. Environmental loadings caused by road construction and maintenance (Mroueh et al. 2000)

The classic LCA includes a static approach that ignores the dimension of time and flows are determined through emissions that occur in a certain production period, while the duration of the period is not taken into account. Forestry on the other hand just by using dynamic models predicts the production and thus the time dimension is completely necessary for modelling. Furthermore, the accumulation of substances in the classical analysis of life cycles are ignored and only the flows of goods between production systems and the environment are taken into account, while in forestry supplies represent the basic information for modelling. It should be noted that the static analysis of the life cycle for forestry has not yet been performed because of the mentioned dynamic character of the forestry production systems in general. Karjalainen et al (2001) reported that in the short term analysis it is easier to implement dynamic models in the study of forestry especially in view of the circulation of substances (CO_2, NH_3, NO_3, etc.).

Substance	Ash FA1	Ash FA2/FA3	Concrete CC1	Concrete CC2	B-F slag BFS
			mg/m^2		
Sulphate	692000	446000	546000	791000	-
Chloride	84600	44600	13800	20800	-
Arsenic	84	69	-	-	<0.1
Barium	7.7	7.7	-	-	-
Cadmium	0.06	0.05	-	-	0.4
Chrome	250	140	92	131	<0.1
Copper	4.4	2.7	51	65	<1
Mercury	0.3	0.2	-	-	-
Molybdenum	1260	1260	-	-	<0.1
Nickel	19	12	-	-	<0.1
Lead	0.15	0.15	-	-	4
Selenium	45	28	-	-	-
Vanadium	615	615	-	-	5
Zinc	3	2	-	-	<10

Table 4. Pollutants leaching out over hundred years

Domino effect caused by land use begins with the influence of vegetation on the soil and ground, the influence of the nutrients and water in the soil, soil compaction (the probability of erosion, and thus the loss of fertile soil), until the end of the product life cycle, through which one can see the total impact of products on the environment (Guinée et al 2006).

Cowell and Clift (2000) proposed how to solve the problem of soil erosion in the life cycle assessment. They believe that soil washout is a category of impact, the consequence is the erosion of fertile soil (kg/ha/yr), and indication of the erosion is presented with global supplies of fertile soil. Global supplies of fertile soil are determined as the ratio of total upper layer of soil in the world (tonnes) and of an annual loss of the upper layers of soil due to erosion (t/yr).

2.2 Forest products

The definition of waste in the context of life cycle analysis is given in the standard ISO 14040 ("output from the production process that no longer can be used"), but in forestry waste can get a new meaning. A typical example of waste transferring to product in forestry is in the extraction of forest residue for energy purposes (branches and tree tops for example which are a by-product of wood in the timber assortment production, and usually represent waste in wood cuttings). In case the forest residue is left to decompose in the forest, i.e. additional profit will not be made; it represents a loss or a waste.

As well as the term "waste" is more complex when dealing with forestry, complications arise in use of other terms for products. A group of standards, ISO 14000, gives definitions for a main product (a product or a co-product which depends on economics and on the conversion of production process itself), a final product (product which requires no additional transformations prior to its use), a by-product (the product is not the intended output of a process and there is a market for the product; economic value is relatively low), a co-product (any of two or more products from the same unit process; production of this product is the intended output of the process with a defined market and economic value is relatively high) and a waste (any output from the production system which is disposed of and will not go further processing). It is clearly defined that a product is a physical things with measurable characteristics such as size, colour, taste, etc., resulting in the production process or services offered to consumers and requested by the consumer. Definitions of other sub-products depend entirely on the current market conditions. In wood processing many "waste streams" are used for internal heat or energy, which can also cause problems in the allocation procedure and definition of system boundaries. Wood itself, as an output, can be referred to as "kg absolute dry" or in "kg with X% moisture content dry base" (Karjalainen et al., 2001).

Apart from wood products there are many other products which can be gained from the forest and also have market value. Up till 2001 and COST E9 2001 workshop (Energy, carbon and other material flows in the life cycle assessment of forestry and forest products) the non-wood products have not been mentioned in the LCAs on wood products and no environmental impacts have as yet been allocated. Compared to the mass of wood produced and harvested in the forests, the amount of non-wood products is generally bellow 1% (Schwaiger and Zimmer, 2001). Schwaiger and Zimmer (2001) made a survey of reported non-wood products in tonnes per year according to the national data-bases and UN-ECE/FAO study (Table 5).

State	Resin	Mushrooms	Game and other hunted animals	Berries	Reindeer (*Rangifer tarandus*)	Cork	Chestnuts	Others
			t/yr					
Austria	-	-	-	-	-	-	-	-
Denmark	-	-	-	-	-	-	-	-
Finland	-	1408	5857	8441	2200	-	-	311*
France	-	8200	-	1000	-	4000	-	600
Germany	-	-	-	-	-	-	-	-
Greece	6140	-	-	-	-	-	-	-
Ireland	-	-	-	-	-	-	-	-
Italy	-	2614	-	496	-	9204	69852	22658
Norway	-	1200	7787	25000	-	-	-	1500
Slovenia	-	800	840	600	-	-	-	3000
Sweden	-	8500	17000	20700	-	-	-	26163
Switzerland	-	735	1597	-	-	-	-	525

*Finland uses 30.1 million m³ of peat for energy and 1.6 million m³ for horticultural and bedding peat out of forests.

Table 5. Reported non-wood products in tonne per year.

It is obvious that many data is not available.

2.3 Data availability

Life cycle analysis has a static character, which represents a problem of applicability in forestry, because forestry is determined by works in space and time and thus has a dynamic character. There is a big difference between the duration and physical activity between the chain cycles of industrial products and production chains in forestry so there is a the need to develop a modelling approach that will enable the creation and collection of representative data for the accurate analysis.

The modelling can use these types of data (Karjalainen et al., 2001).:

- Empirical models - modelling is based on an analysis of existing data. Inventory based modelling is based on knowledge of the past and present situation.
- Process based models - development modelling with prediction of future events, taking into account any changes in the production process. Nonlinear effects can be taken into consideration by using corresponding productivity models or other process models.
- Hybrid models

For modelling it is essential that complex dynamic models are simplified in an appropriate way and as such are used in the analysis of the life cycle. Depending on forest management authors propose the rotation (management) period to be determined in two ways:

- For even aged forests a modelled period (determined by the length of rotations) is proposed;
- For selection forests it is proposed to use a period of 120 years.

Berg (1998) presented how different ways of data collection can give considerable differences in results, which may or may not be dependent on physical factors. Data were produced from two kinds of sources:

1. Statistics regarding actual work volume (area in hectares, cubic meters of harvested timber or tonne per kilometre of made wood transport).
2. Operational data or work study data about production efficiency and fuel consumption per work or time unit.

Another approach was also tried and data were collected from vast forest area during one-year work operations. Records were kept of all recourses allotted to that area in order to deliver the timber that was transported from the area. During collection, data was sorted according to four origins:

1. Measurements of actual consumption of fossil fuel and actual delivery of timber.
2. Measurements and constants (for example actual wood volume produced with actual machine hours, but with a measured constant of fuel/machine hour).
3. Local averages of data concerning equipment of the specific area.
4. General averages

Author claims that discrepancies occurred due to several reasons. Gross data did not cover all the measurements performed, problems occurred between boundary operations and secondary transport, logging and stand treatment appeared and there was different data quality depending on data origin. Data from contractors were seldom available or not certified according to ISO standards. Available emission factors in literature were not appropriate with respect to fuels and engines used.

Berg and Karjalainen (2000) revealed the importance to depend on the origin and quality of data. In their study which was based on data of environmental loads from forest operations in Sweden and Finland in the late 1980s and early 1990s, the authors have compared records of CO_2 emissions from logging operations, transport of timber to industry and some silvicultural activities. Even though harvested volumes in both countries were similar, together with operational conditions (terrain difficulty and level of mechanization in thinnings were the only differences) and machine types used in logging, mixed origin of data caused mistakes in calculating energy consumption. Two types of data were available, either from time-studies or data from follow-up routines.

Athanassiadis (2000) reported of difficulty to get specific data from the industry due to confidentiality problems. Author in his study summarizes the results from four separate studies on energy and resource consumption and emissions during the life cycle phases of harvesters and forwarders in the cut-to-length harvesting systems. Energy input during operations was 82 MJ/m^3, 11% was due to energy consumption during the production phase of the fuel. Exhaust emissions varied considerably depending on the kind of fuel that was examined (rapeseed methyl ester, environmental class 1 and environmental class 3 diesel fuels) and on whether emissions produced during the production phase of the fuels were taken onto consideration. It was also estimated that 35 l/1000 m^3 of chainsaw oil was used for felling and crosscutting while hydraulic oil spillage from both harvesters and forwarders was 20 l/1000 m^3. 52% of the forwarder's mass was replaced during its operational lifetime, 56% of mass for single-grip harvester and 50% of mass for two-grip harvester. About 6% of the machinery's life cycle energy consumption was due to activities connected with the production of the vehicles; raw material acquisition and intermediate processing, fabrication of individual components, assembly of the vehicles and associated transports.

McManus et al. (2004) analyzed the life cycle of mineral oil and BIO (rapeseed) oil used in mobile hydraulic systems (Table 6). The authors note that the use of BIO oil is a relatively new process and that it is difficult to obtain accurate input data. In most emissions occurring in the production of both types of oil, emissions of pollutants at production of BIO oil exceed those in production of mineral oil, except in greenhouse gas emissions. Authors further conclude that although it is not always better to use the BIO (negative impact on parts of the hydraulic system - seals and hoses), production of mineral oil is by itself volatile because mineral oil is a derivative of non-renewable resources.

Schwaiger and Zimmer (2001) collected data from 12 European countries (data from Croatia was included from the study of Beuk et al. 2007) participating in the COST Action E9 "LCA of Forestry and Forest Products" (Figures 4, 5, 6). The study is restricted to the forest operations "harvesting" and "extraction" due to the lack of available data for all other forest processes such as "scarification", "stand establishment", "tending and seeding", "clearing" or "use of pesticides", "forest road construction and maintenance", "delimbing", "debarking" etc. Quantity and quality of information available varied considerably. Data, such as felling, increment, forest area, etc., are available in all countries and performance data on forest operations were not often available (e.g. transport distances were assumed ...). In countries such as Austria, Finland, Switzerland and Sweden larger amounts of data were available at the national level. However, it should be noted that in none of the analyzed countries secondary forest products (in the lifecycle analysis) were not mentioned and no differences was made between the timber harvested in thinnings or in final fellings (clear cuts).

Category	Total people emission equivalents	
	Mineral oil	BIO rapeseed oil
Greenhouse gases	$2.73*10^{-4}$	$2.30*10^{-5}$
Ozone-depleting gases	$9.61*10^{-12}$	$4.59*10^{-10}$
Acidification	$3.41*10^{-5}$	$2.91*10^{-5}$
Eutrophication	$9.89*10^{-6}$	$2.68*10^{-5}$
Heavy metals	$9.23*10^{-6}$	$6.90*10^{-6}$
Carcinogens	$1.49*10^{-10}$	$5.99*10^{-9}$
Winter smog	$1.91*10^{-5}$	$1.03*10^{-5}$
Summer smog	$8.96*10^{-10}$	$2.67*10^{-5}$
Pesticides	0	$1.48*10^{-5}$
Energy	$3.73*10^{-5}$	$3.89*10^{-5}$
Solid waste	0	0

Table 6. Normalized data for the production of 1 kg of mineral and rapeseed BIO oil

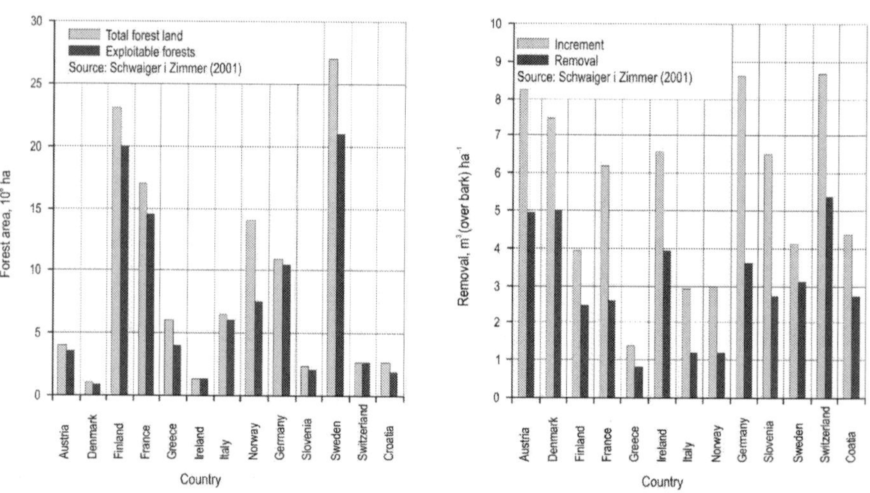

Fig. 4. Data of forest land, increment and removal

Authors claim that data availability depended on forest ownership (Table 7) because data from private and company forests were often incomplete.

State	State forests	Federal forests	Corporation forests	Private forests	Company forests
			%		
Austria	17.0	3.5	9.8	69.7	–
Denmark	40.0	–	–	–	60.0
Finland	33.2	–	4.7	54.3	7.8
France	10.2	–	16.2	73.6	–
Germany	4.3	29.6	26.1	40.0	–
Greece	–	–	–	–	–
Ireland	–	–	–	35.0	65.0
Italy	7.6	–	27.4	59.9	5.1
Norway	7.0	2.0	–	91.0	–
Slovenia	32.6	–	–	64.9	2.4
Sweden	3.4	–	8.0	49.2	39.4
Switzerland	5.0	1.0	62.0	29.0	3.0
Croatia	76.0	19.7	–	4.3	–

Table 7. Distribution of forest ownership

In order to create standardized and representative data suitable for comparison, harvesting was divided to felling and timber transport (Figure 5). Timber transport was further divided to timber extraction and remote transport (Figure 6)

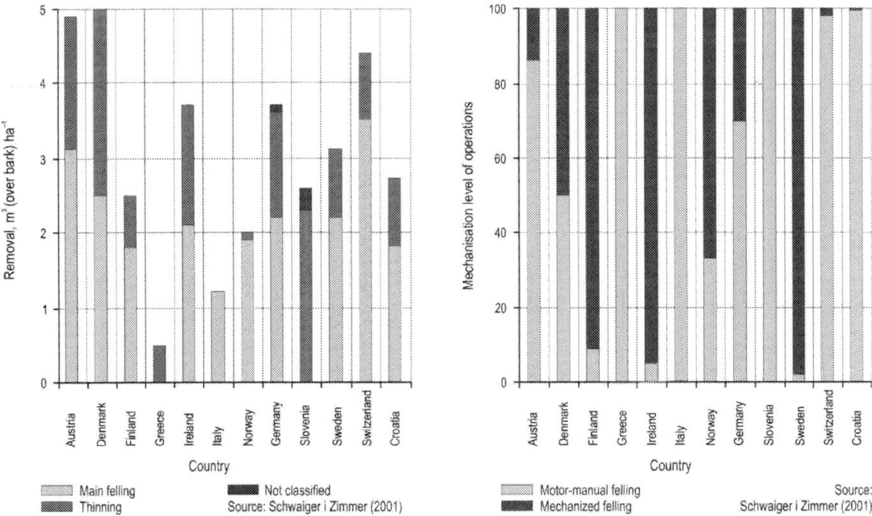

Fig. 5. Removal structure and mechanization level of felling and processing

Mechanization level of timber harvesting and processing in these countries should be compared with respect to the proportion of tree species in the growing stock, type of forest management, terrain (the influence of terrain slope, surface barriers and the load-bearing capacity of forest vehicles) and to the relationship between incomes gained in timber felling (Figure 4).

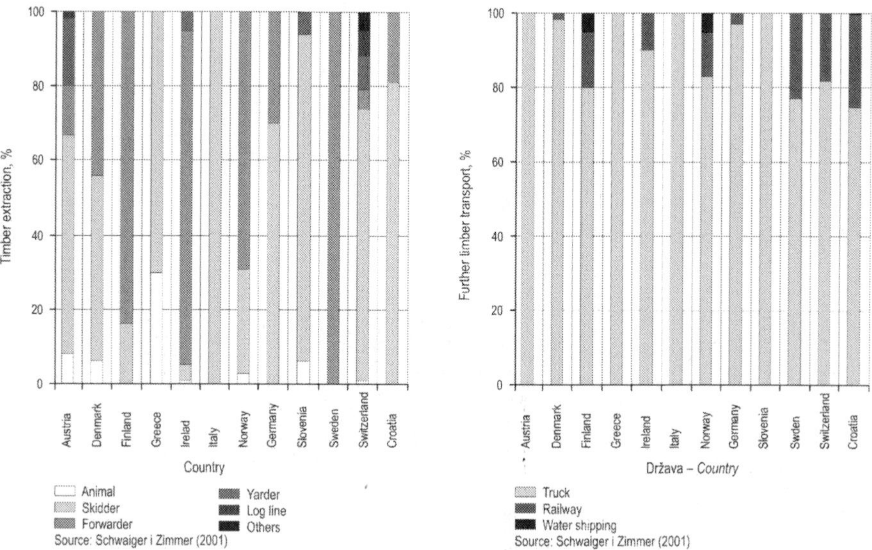

Fig. 6. Structure of timber extraction and further transport

On the choice of extraction vehicles used, the above-mentioned terrain factors and certainly the level of road openness was crucial which has led to significant performance differences in the primary timber transport between European countries (Figure 5).

Remote (secondary) timber transport for all European countries mainly relies on the transport of timber by trucks (> 75%).

Shares of transport by rail and water in some European countries are related to the existence of a network of railroads and their affiliation with the international transport corridors, or the existence of navigable waterways.

In analyzing life cycle, except general information related to individual forestry (Figures 3, 4 and 5), the environmental assessment is required and the data on productivity, fuel consumption and the amount of discharged pollutants are an integral part of each harvesting systems (Table 8).

Transport of timber with trucks significantly affects the life cycle analysis, because fuel consumption and exhaust (GHG) emissions depend on the distance of transport and vehicle type. It should be noted that legal restrictions vary considerably within European countries (Table 9), and that the total number of axles affect considerably on the permissible weight of the vehicle and trailer. Fuel consumption per kilometre for trucks depends on the type and age of the vehicle, its total weight and the distance of timber transportation. However, due to lack of data on the structure of passed kilometres, depending on the type of road (public, forest roads), the greatest attention was on relative fuel consumption due to the volume of transported wood (Table 9).

Means of work (producer)	Reference	Productivity	Fuel consumption		Exhaust gasses		
					CO_2	N_2O	CH_4
		m^3/h	l/h	kg/m^3	g/kg fuel		
Felling and processing							
Chainsaw (Stihl 026)	Fedrau (2000)	4.00	1.50	0.28	3150	0.02	6.91
Chainsaw (Stihl 036)	Fedrau (2000)	8.00	2.40	0.23	3150	0.02	6.91
Chainsaw (Stihl 026/036)	Fedrau (2000)	6.00	2.00	0.24	3150	0.02	6.91
Harvester (Timberjack 1270)	Knechtle (1997)	13.00	11.30	0.77	3455	2.20	5.23
Timber transport							
Animal (horse)	FPP(1991)	1.50	–	–	–	–	–
Skidder (Mahler Unifant)	Fedrau (2000)	7.00	6.00	0.64	3455	2.20	5.23
Forwarder (Timberjack 810B)	Knechtle (1997)	17.00	9.80	0.43	3455	2.20	5.23
Yarder (Sincrofalke)	Winkler (1997)	6.00	7.20	0.90	3455	2.20	5.23
Log line (Leykam)	Trzesniowski (1989)	1.50	–	–	–	–	–
Truck (MAN)	Frischnecht (1995)	–	4.00		3180	0.10	0.20

Source: Schwaiger i Zimmer (2001)

Table 8. Productivity, fuel consumption and amount of exhaust gasses in harvesting operations

Country	Fuel consumption	Allowed weight
	$kg/m^3 * 100$ km	t
Austria	1.47	38
Denmark	1.28	48
Finland	1.16	56
Germany	1.42	40
Greece	1.47	24
Ireland	1.43	40
Italy	1.42	44
Norway	1.28	50
Slovenia	1.42	40
Sweden	1.14	60
Switzerland	1.42	40

Table 9. Fuel consumption and total allowed weight of trucks

The total amount of pollutants disposed into the environment during forest operations is directly linked with the consumption of fossil fuels. Table 10 shows the estimated fuel consumption in European countries in different harvesting operations.

Country	Felling and processing	Timber extraction	Further transport	Total
	kg fuel/m³			
Austria	0.31	0.59	3.68	4.60
Denmark	0.51	0.46	1.28	2.30
Finland	0.73	0.36	2.08	3.20
Germany	0.40	0.58	1.42	2.40
Greece	0.28	0.45	5.89	6.60
Ireland	0.75	0.48	1.98	3.20
Italy	0.24	0.64	0.86	1.70
Norway	0.60	0.47	1.41	2.50
Slovenia	0.24	0.62	0.99	1.90
Sweden	0.77	0.43	2.08	3.30
Switzerland	0.25	0.58	1.22	2.00

Table 10. Estimated fossil fuel input for harvesting operations

Authors conclude that the forest data availability in European countries differs significantly. While data of forest land use (ha), distribution of tree species (%), growing stock, increment and harvested volume (m³) are available in all countries, specific data and information on different forest operation are rather small.

Heinimann et al. (2006) state that since the introduction of ISO 14 030 environmental impact awareness became more present, but there is still nevertheless a small number of prepared and/or published studies with forestry topics. The same author explores the environmental benefits of certain harvesting operations with respect to energy consumption, carbon dioxide emission, as well as carbon sequestration. Determines the energy consumption in the range of 0.12 to 0.62 MJ/kg, which is only 0.5 – 3.7% of energy gained form forest biomass production. The CO_2 emissions amounted to 0.005 to 0.032 kg/kg of forest biomass, for which from 0.005 to 0.018 kg of forest biomass must be produced which would lead to carbon sequestration in trees. Although it was previously thought that felling and processing of wood with a chainsaw (motor-manual cutting) is more environmentally friendly, it was proved by the author that completely mechanized felling and processing (harvester-forwarder system) of short wood has the same impact on the environment (Figure 7).

Data on the life cycle of machines used in forestry can also be obtained from some manufacturers. Timberjack (John Deere Group member) in its "environmental label" provides information about the environmental impact of harvester Timberjack 770 and Timberjack 1410 forwarders throughout their life cycle (Figure 8), which is divided into these five periods:

- Extraction of raw materials - the impact on the environment during the manufacture of steel, cast iron, rubber ...
- Production of the vehicle - the impact on the environment in the development and preparation of the vehicle (determined by ISO 14001);
- Use of vehicles - dumping pollutants into the environment due to the consumption of fuels and lubricants;
- Vehicle maintenance and repairs - the environmental impact of disposal and recycling of old oil, tires and other vehicle parts;

- Use of materials at the end of the life cycle - materials recovery and recycling with reuse of parts and storage of non-recyclable materials.

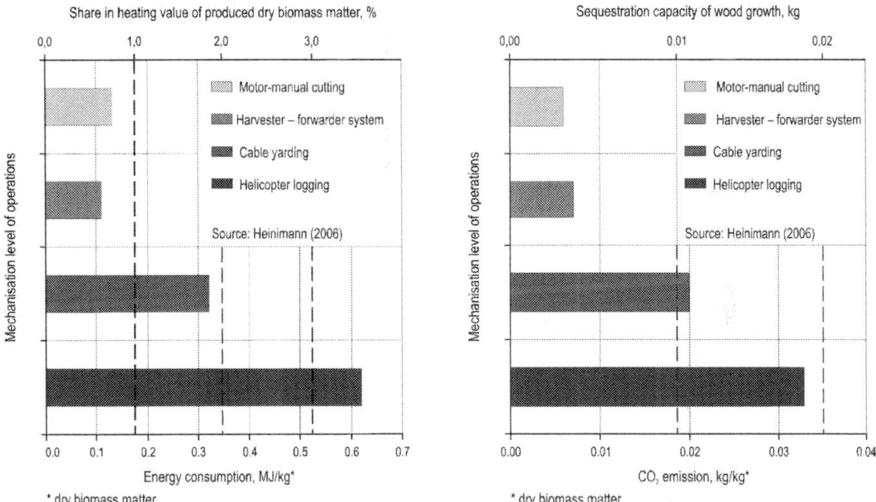

Fig. 7. Environmental suitability of some harvesting technologies

The greatest impact on the environment throughout the life cycle of both vehicles had a period of vehicle use (Figure 8). Period of vehicle use should be perceived by the two parameters known in the cost calculations of machine work: 1) normal use of the vehicle corresponds to the number of operating hours during which the operating cost per hour does not increase due to increased maintenance costs (for the harvester and forwarder it is usually 10 000 operating hours), 2) during the obsolescence of the vehicle which corresponds to the greatest time period in years when the use of the vehicle is still economical, and after which comes technological obsolescence (corrosion, fatigue) and reduction of operational safety (for the harvester and forwarder it is usually 10 years).

In the period of vehicle use, the biggest impact on the environment is made due to consumption of fossil fuels (Table 11). The most important pollutants are vehicle exhaust gasses, carbon dioxide (as the most important greenhouse gas that affects global warming), nitrogen oxides (as one of the causes of acid rain), and as well as sulphur oxides (cause respiratory diseases).

The environmental benefit of each product is presented with the possibility for recycling or reuse of materials at the end of its life cycle. Recycling must be primarily technologically feasible, and economically viable. The rate of vehicle recycling at the end of its period of use is determined by the ratio of the mass of materials that can be recycled and the total mass of the vehicle. According to EU Directive 2000/53/EC, the rate of recovery at the end of the period of use of vehicles must be at least 80% (i.e. 85% when reusable parts of the machine are taken into account). Till the year 2015 the prescribed recycling rate will be 85% (i.e. 95% when reusable parts of the machine are taken into account).

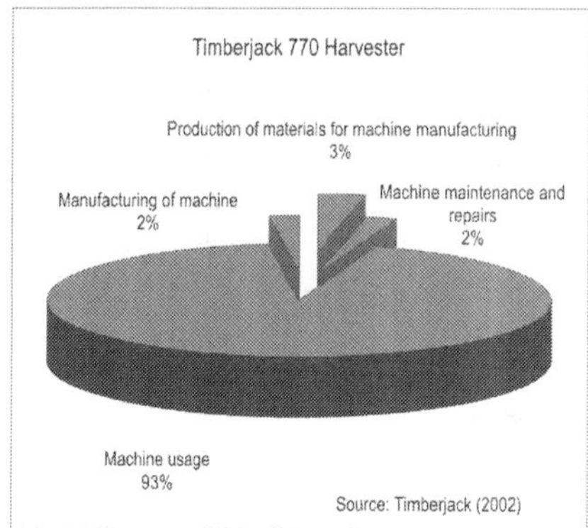

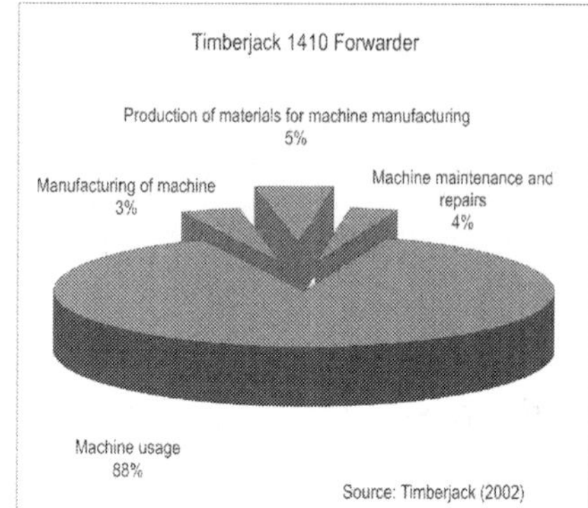

Fig. 8. Environmental impact throughout different life cycle phases of harvester and forwarder

Vehicle type	Gaseous and particulate emission							
	CO_2		NO_X, NO_2		SO_X, SO_2		Particulates	
	Total	During usage	Total	During usage	Total	During usage	Total	During usage
	kg							
Harvester	499 000	451 400	5428	5 293	624	523	284	256
Forwarder	509 100	451 400	4356	4 206	713	523	232	209

Source: Timberjack (2002)

Table 11. Pollutions during harvester and forwarder life cycle

At the end of the period of use, the harvester waste (un-recyclable materials) is less than 8% of the total mass, and forwarder waste is even smaller, only 4% of the total vehicle's mass (Figure 9).

The VTT Technical Research Centre of Finland applied the LCA method in a study to determine the most significant environmental impacts of the Timberjack 1490D slash bundler. In the LCA study of the slash bundler, environmental impact is divided into five phases:

1.   Production of materials,
2.   Manufacturing of the machine,
3.   Usage,
4.   Maintenance and repairs,
5.   Post-use disposal.

The total emissions of carbon dioxide is 727 001 kg where 88% of the total carbon dioxide emissions occur during entire life cycle of the slash bundler. Nitrogen oxide emissions amount to 4881 kg during the slash bundler's life cycle (96% during usage phase). 1402 kg of sulphur oxides are generated (84% during usage phase) and 246 kg of particulate emissions are generated during the usage phase. Slash bundler is built on forwarder chassis, to which is attached the bundler component and the automation that drives its usage, so the recyclability rate is the same as for forwarders (Fig.10).

Økstad (1995) states that common experiences for paper and pulp industry in Norway with LCA projects in gathering data, from the production process and the raw materials, is much

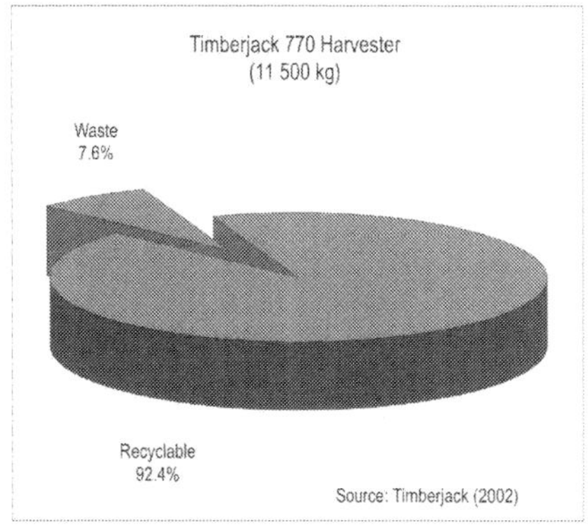

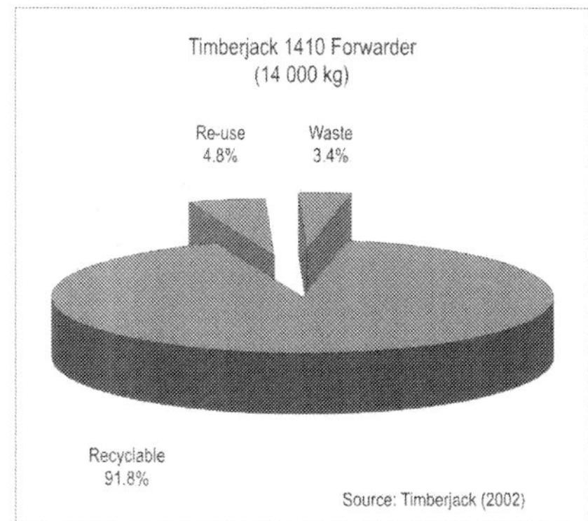

Fig. 9. Harvester and forwarder recyclability rate

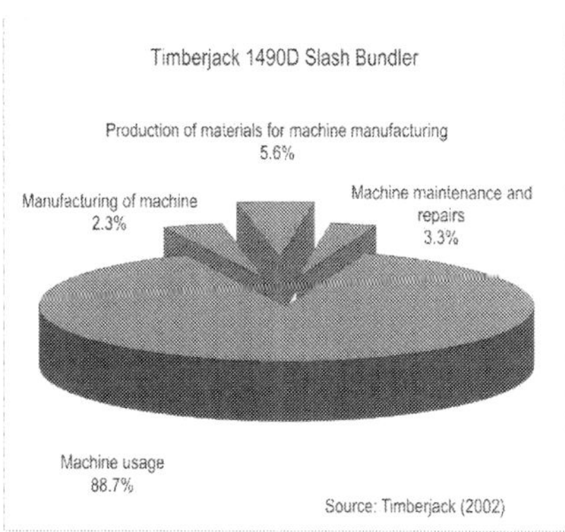

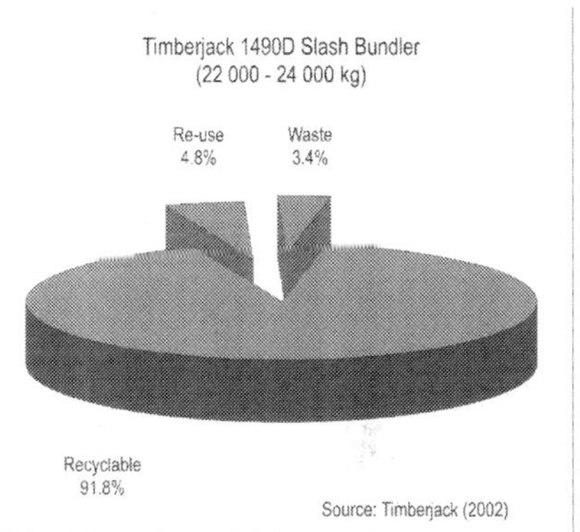

Fig. 10. Environmental impact throughout different life cycle phases of a slash bundler and
its recyclability rate

larger task than assumed. Even if there is a good knowledge of the total emissions and the
energy demand of the factory, the processes are often integrated and there are problems
allocating different emissions to the products. In addition to the internal problems of finding
the correct data from the manufacturing processes there are problems to procure data from
the processing of raw materials (forest products and chemicals). Author further claims that
some pulp and paper companies have established their own expertise in the field of LCA on
the corporate level and have their own LCA strategy:

- LCA is seen as a useful tool in the development of the new products;
- LCA may be used in the process to select the most environmentally friendly product. It
  is emphasised that it may be difficult to get a just result from such comparisons because
  different system boundaries in the assessment will lead to different answers;
- LCA may be used as a tool to identify the points in the production process which gives
  the highest environmental impacts and where the potential improvement is the highest;

- Having carried out LCA of their own products gives the industry much better "cards on hand" in the communication with local and national environmental authorities.

Some companies have taken the challenge and initiated their own work to collect data and carry out LCA of their activities and products, while others have a more reluctant attitude towards the method and have the opinion that the method is not fully developed to be used as a tool for industry. Author concludes that LCA is a useful tool in the development of new products and it may be used in the process of selecting the most environmentally friendly product or for highlighting points in the production process which give the highest environmental impact. In some cases LCA will be required in the licensing process of discharge permits.

3. Conclusion

Life cycle assessment is a very complex tool in the process of assessing the environmental impact of a product. To be successfully developed, high quality data sources are required that will contain all the inputs and outputs of different materials used, energy consumption and the amount of pollutants and emissions produced during:

1. Extraction of raw materials,
2. Product production,
3. Time of product usage,
4. Possible recycling and re-use,
5. Final disposal of waste.

For forestry, the life cycle assessment is particularly difficult due to the long production process of raw materials, primarily wood (length of rotation period), and because of the high spatial and temporal impact forests have on the environment. As the life cycle assessment is a static and not a dynamic tool of environmental impact, many experts believe that the spatial and temporal aspects should be excluded from the assessment itself. However, the life cycle assessment related to forestry and its products makes such exclusion impossible. Forests not only have an important social impact, but they store large amounts of carbon (carbon sequestration), protect soil and water resources, ensure biodiversity, produce oxygen, etc., thus affecting the global climate. These are beneficial functions of forests that can not and should not be ignored when dealing with life cycle assessment in forestry. On the other hand, the question of how to include all forest management activities (from silviculture to forest protection and harvesting) could be involved in the life cycle assessment of forest products.

Life cycle assessment is a good tool for determining the effects of each product (and manufacturing processes in which the product is being produced) on the environment, especially today when there is more and more awareness of environmentally friendly technologies, renewable energy and eco-efficiency. The greatest use of life cycle assessment tool in forestry, from the aspect of timber harvesting, is in comparing different harvesting systems for selecting environmentally friendly versions.

Due to the lack of quality data sources, and conversion between a static to a dynamic character of the life cycle assessment, requires further research and improvement of this otherwise useful, but very complex tool in assessing the environmental acceptability of products and production processes.

4. References

Anon., (1997). Memorandum of Understanding for the implementation of a European Concerted Research, *COST Action E9, Life cycle assessment of forestry and forest products.* Technical annex, April 11, 1997, Bruxelles, Belgium.

Athanassiadis, D., (2000). Resource consumption and emissions induced by logging emissions induced by logging machinery in life cycle perspective. *Dissertation,* ISBN 9157658773, Umeå, Swedish University of Agricultural Sciences, 88 pages.

Berg, S., (1998). An inventory of forest operations for LCA of forest products. *International Conference of life cycle assessment in agriculture,* Agro-Industry and Forestry, 3-4 December 1998, Brussels, Belgium.

Berg, S. and Karjalainene, T. (2000). Comparision of emissions from forest operations in Finland and Sweden. *Presentation at COST Action E9,* Mid-term Seminar in Espooo/Finland, Dipoli 27-29 March 2000.

Beuk, D., Ž. Tomašić, D. Horvat, (2007). Status and development of forest harvesting mechanisation in Croatian state forestry. *Croatian journal of forest engineering,* 28(1): 63–82.

Cowell, S. J., & Clift, R. (2000). A Methodology for Assessing Soil Quantity and Quality in LCA. *Journal of Cleaner Production,* Vol. 8, 321-3, ISSN: 0959-6526.

De Feyter, S., (1995). Handling of the Carbon Balance of Forest in LCA. *Proceedings of the International Workshop: Life-Cycle Analysis – A Challenge for Forestry and Forest Industry,* May 3–5, 1995, Hamburg, Germany, pg. 37–43.

Frühwald, A., (1995). LCA – A Challenge for Forestry and Forest Products Industry. *Proceedings of the International Workshop: Life-Cycle Analysis – A Challenge for Forestry and Forest Industry,* May 3–5, 1995, ISBN:952-9844-16-6, Hamburg, Germany, pg. 9–14.

Guineé, J., L. van Oers, A. de Koning, W. Tamis, (2006). Life Cycle Approaches for Conservation Agriculture, Part I: A definition study for data analysis. *CML report 171,* June 2006, Universiteit Leiden, ISBN-10: 90-5191-148-3, Leiden, Netherlands, pg. 1–67.

Grasser, C. (1994). Ökoinventare für Energiesysteme. *Grunlagen für den ökologische Vergleich van Energiesystemen und der Einbezug van Energiesystemen in ökobilanzen für die Schweiz.* Bundesamt für Energiewirtschaft.

Häkkinen, T. and Mäkelä, K. (1996). Environemtal adaption of concrete. Environmental impact of concrete and asphalt pavements. *VTT Research Notes 1752.* ISBN 951-38-4507-4, Espoo, Finland, 61 pages.

Heinimann, H. R., A. Wollenmann, R. Knechtle, (2006). Life Cycle Approaches for Conservation Agriculture, Part I: A definition study for data analysis. *CML report 171,* June 2006, Universiteit Leiden, ISBN-10: 90-5191-148-3, Leiden, Netherlands, poster presentation.

Jungmeier, G., (2003). System Analysis of Forestry, Forest products and Recovered Wood. *International Conference »Efficient Use of Biomass for Greenhouse Gas Mitigation«.* COST Action E9 and E31, PPT, September 30 – October 1, Österund, Sweden.

Karjalainene, T., T. Apnesth, P. Esser, L. Finér, G. Jungmeier, B. Košir, K. E. Kvist, J. M. Roda, H. Schwaiger, S. Berg, B. Zimmer, J. Welling, (2001). Identification of Problems using Case Studies. *Energy, Carbon and Other Material Flows in the Life Cycle Assessment of Forestry and Forest Products.* Achievements of the working group 1 of the COST action E9, European Forest Institute, Discussion paper 10, ISBN: 952-9844-92-1, Joensuu, Finland, pg. 9–21.

Lox, F. (1994). Milieubalans van vouwkarton. Vrije Universiteit Brussels, in opracht van Pro Carton België, draft, april 1994.

Mcmanus, M. C., G. Hammond, C. R. Burrows, (2004). Life-cycle Assessment of Mineral and Rapeseed Oil in Mobile Hydraulics Systems. *Journal of Industrial Ecology*, 7(3–4): 163–177, ISSN: 1530-9290.

Mroueh, U. M., P. Eskola, J. Laine-Ylijoki, K. Wellman, E. M. M. Juvankoski, A. Ruotoistenmäki, (2000). Life Cycle Assessment of Road Construction. *Finnra report*, 17/2000, ISBN 951-726-633-2, Helsinki, Finland, 59 pages.

Nabuurus, G., J. and Mohren G., M., J. (1993). Carbon fixation through forestation activities. *A study of the carbon sequestration potential of selected forest types*. Institute for Forestry and Nature Research (IBN-DLO), Arnhem/Wageningen.

Økstad, E., (1995). Experience with LCA in the pulp, paper and packaging industry in Norway. *Proceedings of the International Workshop: Life-Cycle Analysis – A Challenge for Forestry and Forest Industry*, May 3–5, 1995, Hamburg, Germany, pg. 141–149.

Ranta, J., Wahlström, M., Rouhomäki, J., Häkkinen, T. and Lindroos, P. (1987). Desulphurisation products of coal combustion plants. *VTT Research Notes 741*. Espoo, 140 pages.

Schwaiger, H., B. Zimmer, (2001). A comparison of Fuel Consumption and Greenhouse Gas Emissions from Forest Operations in Europe. *Energy, Carbon and Other Material Flows in the Life Cycle Assessment of Forestry and Forest Products*. Achievements of the working group 1 of the COST action E9, European Forest Institute, Discussion paper 10, : 952-9844-92-1, Joensuu, Finland, pg. 33–53.

Timberjack, (2002*): Green forest machines for sustainable development – Environmental Declaration*. Timberjack – A John Deere Company, Tampere, Finland, 12 pages.

Thoroe, C., J. Schweinle, (1995). Life Cycle Analysis in Forestry. *Proceedings of the International Workshop: Life-Cycle Analysis – A Challenge for Forestry and Forest Industry*, May 3–5, 1995, ISBN: 952-9844-16-6, Hamburg, Germany, pg. 15–24.

Virtanen, Y. and Nilsson, S. (1993). Environmental impacts of waste paper recycling. International Institute for Applied Systems Analysis, Earthscan Publications Ltd, ISBN 1853831603, London, UK, 166 pages.

Wollenmann, R., (2006). Balancing Carbon Emission and Sequestration Fluxes of Forest Land Based on LCI – approach.). Life Cycle Approaches for Conservation Agriculture, Part I: A definition study for data analysis. *CML report 171*, June 2006, Universiteit Leiden, ISBN-10: 90-5191-148-3, Leiden, Netherlands, poster presentation.

www.iso.org

www.admin.cam.ac.uk

10

Wildlife Forestry

Daniel J. Twedt

U.S. Geological Survey
Patuxent Wildlife Research CenterVicksburg, Mississippi
USA

1. Introduction

Wildlife forestry is management of forest resources, within sites and across landscapes, to provide sustainable, desirable habitat conditions for all forest-dependent (silvicolous) fauna while concurrently yielding economically viable, quality timber products. In practice, however, management decisions associated with wildlife forestry often reflect a desire to provide suitable habitat for rare species, species with declining populations, and exploitable (i.e., game) species. Collectively, these species are deemed priority species and they are assumed to benefit from habitat conditions that result from prescribed silvicultural management actions.

Early wildlife conservation efforts largely focused on controlling indiscriminate slaughter of wildlife by restricting the season or sex of harvested species (Graham, 1947). Subsequent conservation efforts targeted increased protection of populations through creation of sanctuaries or reserves (Knight, 1999), such as national parks and wildlife refuges, within which harvest of wildlife was prohibited or greatly restricted. Typically parks and wildlife reserves were located in areas with abundant wildlife populations, and therefore, little emphasis was placed on active management to improve or maintain suitable habitat.

In his description of a "land ethic" in which people are members of the natural community, Leopold (1949) indicated the appropriateness of habitat alteration and management to aid the continued existence of wildlife species. His philosophical principles were succinctly stated as "*A thing is right when it tends to preserve the integrity, stability, and beauty of the biotic community. It is wrong when it tends otherwise.*"

Anthropogenic alteration of forest habitat for the benefit wildlife species has long been practiced, particularly with regard to intentional use of fire (Bonnicksen, 2000, Ford et al., 2002). Even so, active manipulation of forest structure to purposefully enhance wildlife habitat has been slow to evolve. Most early efforts regarding habitat alteration focused on providing suitable escape cover, foraging habitat, or distinct habitat features (e.g., nest sites) for game species. Within forested landscapes, improving wildlife habitat often meant increasing local heterogeneity and providing more forest edge habitat. However, increased knowledge of the possible negative effects associated with forest edges, such as greater predation and nest parasitism rates, has called into question the practice of providing increased heterogeneity within forested landscapes. At the same time, a desire to provide habitat for specific species that may require specialized habitat conditions, as is the case for some endangered and threatened species as well as other charismatic megafauna with widespread public support, has prompted development of habitat management prescriptions designed to improve forest habitat conditions for specific species (Wilson et al., 2007).

Partly in response to what has been perceived as management for single-species, but also to advocate a desire to prevent future population declines of common species, recent emphases in wildlife conservation have been placed on the development of comprehensive, eco-regional, conservation plans whose scope encompasses multiple species or the entirety of a species-group. This approach gained momentum with the North American Waterfowl Management Plan (1986) which identified conservation actions intended to return waterfowl populations to their former abundance and distribution. Subsequently, this basic concept has been expanded to other species groups such as songbirds (Rich et al., 2004), reptiles and amphibians (Bailey et al., 2006), and shorebirds (Brown et al., 2001), as well as other species groups. Although these conservation plans differ markedly with regard to their recommendations for habitat management, many of these conservation plans advocate use of prophylactic management prescriptions to enhance or maintain suitable habitat conditions for priority wildlife species. Indeed, most conservation plans recognize the need for alteration of habitat and landscapes via management actions so as to attain the desired distribution, abundance, and viability of priority wildlife species.

Before managers can provide appropriate forest habitat for priority wildlife, the landscape and site characteristics that contribute to viable populations of these species must be identified. Preferably, those forest habitat characteristics identified can be quantified via standard measurement protocols. If possible, the desired or acceptable range of values should be determined, and either quantitatively or qualitatively stated, for each of these characteristics. In addition, a threshold value for each characteristic should be determine, that if exceeded would justify prescription of management actions to return site characteristics within the desired range (Fig. 1).

Because wildlife forestry silviculture does not target optimal production of forest products, many forest management practitioners (i.e., foresters) have been reluctant to adopt these alternative silvicultural practices. Indeed, these concerns may be justified, as desired stand conditions may include less then fully stocked stand densities, retention of some economically mature trees, maintaining species diversity which includes less merchantable tree species, and preservation of less vigorous (decadent) trees. However, at present we lack credible, long-term economic data to suggest that development and maintenance of quality forest habitat for priority wildlife species will result in a sacrifice in timber production or quality. Regardless of these overarching financial concerns, silvicultural management actions prescribed to enhance desirable forest habitat conditions for priority wildlife species are expected to be commercially viable. Maintaining the long-term commercial viability of prescribed silviculture is import, because without sufficient financial returns, needed management actions will not be undertaken.

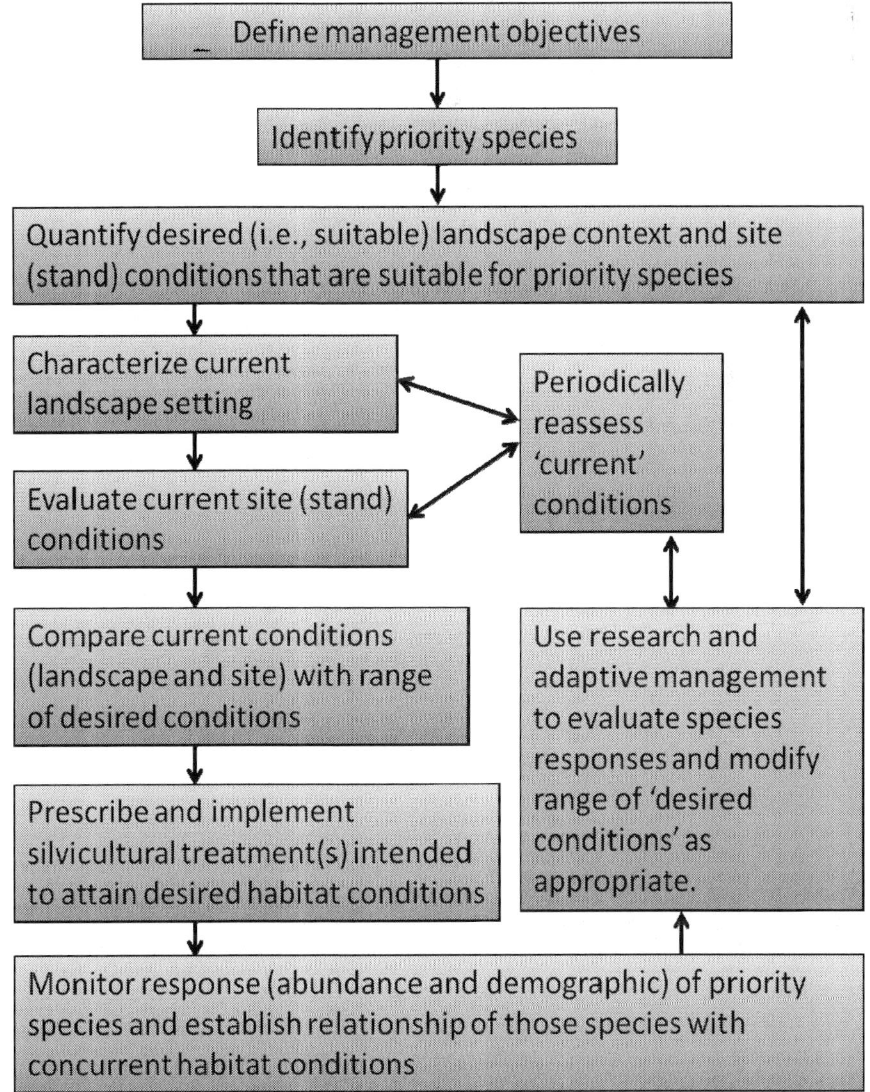

Fig. 1. Characterization of wildlife forestry silvicultural methodology.

2. Forest management

Silviculture is the manipulation of the establishment, growth, composition, and quality of forest vegetation to achieve management objectives. Historically a primary objective on managed forest lands has been to maximize financial returns from production of forest products such as timber and wood pulp. Production-oriented silviculture has focused on reducing the length of production cycles between stand regeneration and harvest while concurrently maximizing wood product yields. Reduction of production cycles (i.e., rotation length) tends to lessen the time regenerating forests are in early seral habitats, and consequently the size and age of shrubs and there their likelihood of persistence within subsequent seral stages is reduced (Hagar, 2004). Often production-forestry objectives favored a limited number of tree species due to their high economic value or growth potential. Concomitantly, to increase growth and yield of desired trees, other species with lesser economic value, as well as congeneric competitors, were culled. These actions result in increased spatial homogeneity within managed stands. The apogee of this trend towards increased spatial regularity is expressed within planted plantation forests. Climate, soils, species, and other factors affect the optimal length of production cycles within production-oriented forests (Lindenmayer et al., 1999). Thus, rotation periods within plantations vary markedly from rotations of up to 150 years to as few as 10 years. Notably, recent developments and increased advocacy of short-rotation woody coppice systems for use in biomass production have further compressed the length of production cycles to <4 years (Tubby and Armstrong, 2002).

Even in non-plantation forest management systems, historical forest management practices, especially those focused on timber or pulpwood production, have often resulted in relatively homogeneous mature forest conditions. This homogeneity has been exacerbated by environmental changes that limit natural disturbances within forest stands (e.g., fire suppression). When these homogenous forests are managed using even-aged silviculture, during which all or most trees are harvested and the entire stand regenerated, successional 'boom and bust' conditions are dispersed throughout landscapes. The forest conditions resulting from even-aged management (e.g., clear cuts) are temporarily favorable for some species, specifically benefiting species that exploit early-successional forest habitat (Askins, 2001, Hunter et al., 2001). Even so, these early-successional habitats are transitory and are followed by an extended period of stand development (stem exclusion and understory reinitiation; Oliver and Larson, 1996) during which the homogeneous, closed-canopy forest structure with sparse understory vegetation offers attraction (forage or cover) primarily for common (non-priority) forest wildlife species. Indeed, the abundance of homogeneous, closed canopy forests with sparse understory cover that have resulted from historical silvicultural practices may have contributed to why species for which these habitat conditions are suitable are considered common and are not priority species in need of management actions to increase their abundance. In addition, the altered (often depauperate) species compositions that have resulted from selective harvest and culling during past management practices may amplify the potential for catastrophic damage from outbreaks of insects or diseases or limit the seasonality of mast crops that are exploited by wildlife.

In contrast to the relatively short periodicity of production-oriented silviculture and the periodic boom-bust cycles associated with even-aged silviculture, natural successional processes within forests may require hundreds, or potentially a thousand, years to achieve complete replacement via small-scale disturbance or as the duration between catastrophic stand-replacement events. Owing to differences in their successional development, structural and ecological characteristics within production-oriented managed forests and in forests managed using even-aged silviculture are manifestly different from those same characteristics within natural forests (Seymour and Hunter, 1999). Ecological differences between managed and natural forest stands are exacerbated over time, as natural stands approach climax or 'old-growth' conditions while intensively managed forests are repeatedly regenerated.

To mitigate the detrimental effects of forest management on priority wildlife species, recommendations have been made to increase the length of production cycles, thereby allowing increased time for development of heterogeneous forest structure (Kerr, 1999). Unfortunately, small increases in rotation lengths provide little opportunity for increased complexity of forest structure within these stands – and periodic, complete harvest and stand regeneration thwart retention of forest structure over time. Similarly, forest managers have argued that temporal and spatial dispersion of clear-cuts throughout landscapes will provide continual availability of wildlife habitat – albeit at different locations. Although early-successional habitat may be continuously afforded via such dispersion, few of the other benefits conferred by increased forest structural heterogeneity, such as emergent or senescent legacy trees, are realized.

Increased recognition of the structural and ecological deficiencies within production-oriented forests has spurred interest in and development of alternative silviculture (Franklin, 1989). Alternative silviculture has been especially valuable for stands where management objectives are not solely financial but where stewardship objectives include preservation of regional biodiversity, enhancement of wildlife habitat, maintaining landscape aesthetics, or providing increased recreational opportunities.

Alternative silvicultural methods developed for use in many forest types have been based on identification and quantification of natural disturbance regimes, with subsequent implementation of silvicultural methods intended to emulate these disturbances (Mitchell et al., 2002, 2006; Palik et al., 2002). For many forest types, emulation of natural disturbance regimes includes small-scale disturbances implemented via single-tree and patch-cut harvests (Franklin et al., 2007, North and Keeton, 2008). If successfully implemented, emulation of natural disturbance regimes should promote natural stand development and succession. Unfortunately, natural stand development and concurrent successional changes may conflict with landowner objectives if succession results in changed species composition or reduced financial returns. Moreover, strict adherence to silvicultural regimes that mimic natural disturbance may not be necessary to maintain biodiversity or to provide desired wildlife habitat (Palik et al., 2002).

Another common alternative silviculture has been inspired by the desire to achieve forest structure reminiscent of old-growth forest (Bauhus et al., 2009). Although the structure of old-growth forest varies among forest types (Hayward, 1991), common attributes generally include increased vertical heterogeneity via a multi-layered canopy, increased horizontal heterogeneity associated with canopy gaps and different regeneration cohorts, the presence of large trees that are often in older-age classes, abundant snags (standing dead trees) or senescent trees, and large diameter downed woody debris (Bauhus et al., 2009, Keeton, 2006). Myriad ecological benefits, including increased biodiversity, have been ascribed to old-growth forests (Lindenmayer and Franklin, 2002). As such, forest reserve lands have been designated (e.g., wilderness areas and parks) which afford unfettered, natural development of these characteristics. However, on lands with timber production as an objective, even if not the sole objective, implementation of production cycles that span hundreds of years is not likely to occur. In addition, merely increasing the length of production cycles may provide little enhancement of biodiversity if the forest structure favored by production-oriented silviculture is maintained (Carey, 2006). Furthermore, where forest reserves are small or located within inhospitable landscapes, full ecological benefits of old-growth forests may not be attainable or sustainable (Kneeshaw and Gauthier, 2003).

Fortunately, it is likely that many of the benefits conferred by old-growth forests, or at least specific species associated with old-growth forests, are positively related to the heterogeneous structure (vertical and horizontal) and other structural attributes found within these forests, not necessarily with the prolonged existence of these forests (Beggs, 2004, Carey, 2003b). As such, alternative silvicultural methods have been developed which retain or enhance structural heterogeneity within managed forests, thereby promoting structural attributes of old-growth forests and encouraging greater biodiversity (Bauhus et al., 2009, Garman et al., 2003; McClellan, 2004). Most often, increased heterogeneity and other desired attributes have been achieved via retention of living or dead trees (or other forest elements). These retained structures have been referred to as legacy elements (Franklin et al., 1997).

Based on the concept of retained legacy elements, various alternative silvicultural methods have been advocated to enhance structural complexity within managed forests (Whitman and Hagan, 2003). Some of these methods have being designated as green-tree retention harvests (Zenner, 2000), variable-retention harvests (Aubry et al., 2004. Maguire et al., 2007), dispersed retention harvests (Heithecker and Halpern, 2006), variable density thinning (Aukema and Carey, 2008), and active intentional management (Carey et al., 1999, Carey, 2006). The legacy elements retained under these different silvicultural regimes may differ markedly. Retained elements may differ in quantity (e.g., % canopy cover or % basal area retained; Fig. 2a,b) but they may also differ in their composition, as different canopy species are preferentially harvested or specific regenerating species favored for retention. In addition, the quality of retained elements may differ via selection of retained trees with differing perceived fates – from robust, canopy-emergent trees through decadent, dead or dying trees. Moreover, retained elements may differ in their dispersion (e.g., evenly dispersed or aggregated; Fig. 2c,d) and the spatial juxtaposition of retention (e.g., clumped or linear).

Wildlife forestry silviculture exploits the versatility of legacy retention by combining specific retention and removal of canopy to achieved habitat conditions that are deemed suitable for priority wildlife species. The resultant forest stand may exhibit a mixture of tree species, size classes, and decadence within a heterogeneous distribution of aggregated retention and canopy gaps (Fig. 3).

2.1 Landscape considerations

Although specific silvicultural methods have most often been developed for and are implemented within forest management units (a.k.a. forest stands), a similar ecologically based approach has evolved with regard to maintaining or restoring landscapes surrounding managed forest stands (Lindenmayer et al., 2002, Lindenmayer & Hobbs, 2007). Intact forested landscapes are conducive to maintaining floristic and wildlife diversity, but also provide tangible social and economic benefits (Poudyal et al., 2010).

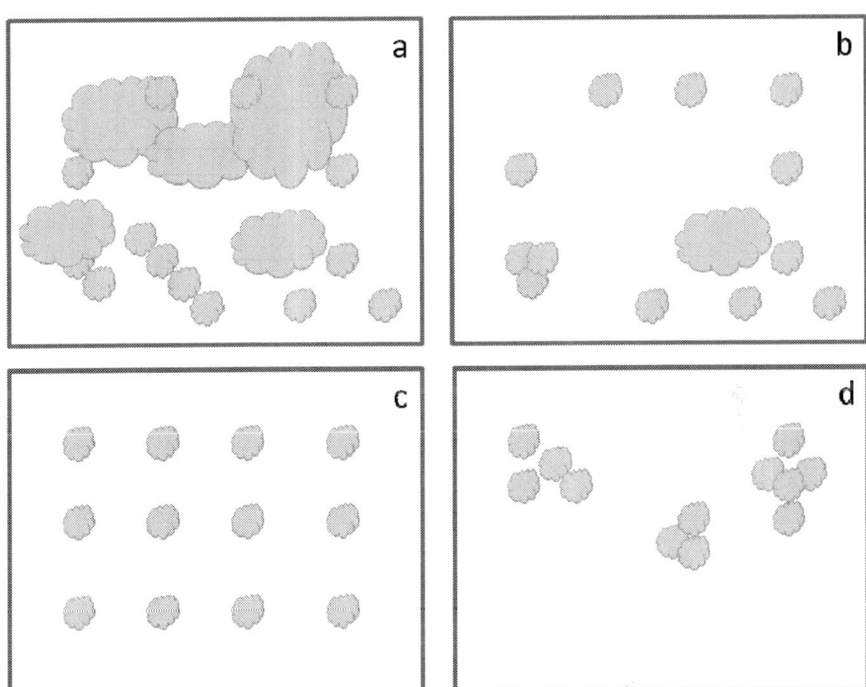

Fig. 2. Potential distribution of legacy elements after application of alternative silviculture, differing in quantity of retained canopy cover (a, b) or dispersion of retained elements (c, d).

Loss of diversity within landscapes is often a reflection of the increased homogeneity within forest stands. Although forest structure within a landscape ranges from regenerating stands to mature merchantable stands, these stands may be monotypic in species composition and physical structure. In addition, ensuring appropriate size of habitat patches and maintaining connectivity among these patches is critical to sustaining viable wildlife populations within landscapes (Franklin and Forman, 1987). Although the appropriate patch size is dependent upon the species under consideration, increased fragmentation tends to disrupt dispersion of fauna and flora and when habitat area is reduced without appropriate connectivity, extirpation of some species can result (MacArthur and Wilson, 1967). Finally, maintaining landscapes capable of sustaining viable wildlife populations requires that evaluation, planning, and management decisions reflect the temporal processes inherent in the ecosystem. Four landscape conditions have been recognized as contributing to ecosystems that support sustainable wildlife populations: 1) diversity within the landscape, 2) connectivity among landscape elements, 3) appropriate habitat patch sizes, and 4) sufficient time to achieve ecological functions (Silva, 1992). The impact of the degradation of forest landscapes was succinctly summarized by Franklin (1989) as: *"In general, we have tended to forget that what is good for wood production is not necessarily good for other organisms or processes in a forest ecosystem. Fully stocked young forests,... are the most simplified stage of forest development in terms of structure and function, and the most impoverished in terms of biological diversity.. . . . Simplification--genetic, structural, landscape and temporal — reduces ecosystem resilience...[Thus] the key to retaining resilience must be in maintaining ecological complexity or diversity".*

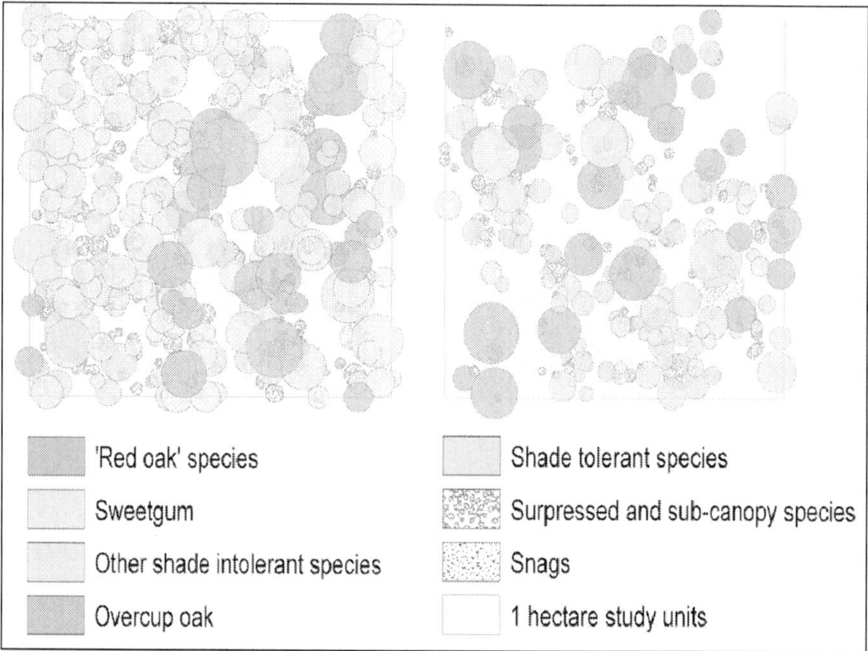

Fig. 3. Canopy cover accounted for by individual tree crowns within bottomland hardwood forests on an untreated control stand (left) and on a stand subjected to wildlife forestry variable-retention silvicultural treatment (right), 6 years after treatment (Twedt & Wilson, 2007).

As when considering disturbance within the context of forest stand development, landscape conditions have often been influenced by large-scale disturbance events, particularly fires (Kaufmann et al., 2003, Van Willgenburg and Hobson, 2008). Forest landscapes that result from management based on historical fire regimes, when compared with landscapes with extensive reserves and traditional silvicultural management, had increased area of late-successional habitat, more overstory structure in young forests, and larger forest patches (Cissel et al., 1999). Similarly, when compared with human altered landscapes, natural landscapes were composed of larger habitat patches which formed the landscape matrix, had greater connectivity among patches, the range of variation in patch size and complexity was greater, as was the likelihood of encountering 'rare' habitat types (Mladenoff et al., 1993).

Despite some understanding of desirable characteristics of forested landscapes, the appropriate landscape area remains nebulous. In part this is because an appropriate landscape area is dependent upon the species being considered, landscape characteristics, desired response, and the management context (Karau and Keane, 2007; Mayer and Cameron, 2003). As an example, the landscape appropriate for a grizzly bear is vastly different from the appropriate landscape for a shrew.

Even with knowledge of the appropriate landscape area and characteristics, manipulation of landscape conditions is usually beyond the purview of individual landowners. This is true even for large industrial or governmental landowners. Nevertheless, landscape conditions influence the ecological benefits conferred by forest stands, including the distribution, abundance, and viability of wildlife species within forest stands. In boreal forests, stand structure accounted for 54% of variability in bird and small mammal communities but landscape characteristics accounted for 29% (St Laurent et al., 2008). Similarly, bird species abundance within coniferous dominated landscapes was positively related to the proportion of broadleaf forest, with landscape thresholds for individual species ranging from 1% to 25% canopy cover. A strong negative relationship was noted between landscape broadleaf forest and avian population trends, such that the species most associated with broadleaf forests had greatest population declines. Therefore, maintaining or restoring broadleaf vegetation is important within coniferous dominated landscapes (Betts et al., 2010).

Because of landscape influences, providing desired forest conditions within forest stands may be insufficient for some priority species if these stands are located within inhospitable landscapes. Moreover, for long-term sustainable populations, most sites, even those with suitable habitat conditions, must have spatial and temporal linkages that facilitate dispersal, colonization, and interchange within metapopulations (Hanski, 1998). Towards that end, desired landscape conditions for priority wildlife species are usually composed of large forest patches as the preponderance of the landscape matrix, have a wide range of variation in patch sizes and complexities, and provide extensive connectivity among forest patches.

2.2 Why wildlife forestry?

The application of wildlife forestry silviculture is in many respects the logical extension of the various alternative silviculture methods whose objectives are to maintain or increase forest biodiversity (Table 1). Although forest management targeting increased biodiversity is a laudable objective in and of itself, managers of public conservation lands (e.g., National Wildlife Refuges and State Wildlife Management Areas) or private lands under conservation easements may have legal, or self-appointed, mandates to manage and conserve priority wildlife species rather than the more general goal of improving biodiversity. On other private lands, the economic and recreational values associated with silvicolous wildlife, including fee hunting, hunt-club memberships, wildlife viewing and other eco-tourism activities, have income potential that may equal or exceed that of extractive timber harvest. Hence, ensuring sustainable populations of wildlife may be essential for maintaining financial revenues from these forest lands.

In addition, land ownership patterns of private forest lands are changing. Changing ownership demographics and associated alteration of management objectives of these forest land owners have resulted in smaller landownership holdings. For owners of small landholdings, maintaining continuous forest cover and the aesthetic considerations associated with forests may be of paramount importance (Kaetzel et al., 2011).

Legacy characteristic	Wildfire	Wind	Insect / disease	Flood	Clearcut	Wildlife forestry
Live trees	Sparse	Variable	Variable	Sparse	Sparse	Variable
Snags	Abundant	Variable	Abundant	Abundant	Sparse	Variable
Downed woody debris	Variable	Abundant	Variable	Variable	Sparse	Variable
Understory development	Abundant	Abundant	Variable	Variable	Abundant	Variable
Spatial heterogeneity	Variable	Variable	High	Variable	Low	High
Duration of alterted state	Variable	Variable	Variable	Variable	Long	Variable

Table 1. State of habitat features following different distubances and prescribed wildife forsetry silviculture (expanded from Swanson et al., 2011).

Even on large land ownerships, such as industrial forest lands or lands managed by timber consortiums, there are increased public expectations to provide sustainable, 'green' forest products (Barneycastle, 2001). Concurrently, corporate efforts to maintain positive environmental images have prompted greater reliance on alternative silviculture, such as wildlife forestry. Finally, even on lands where implementation of traditional silviculture may be the preferred management option, exurban encroachment may constrain traditional forest management practices (Egan and Luloff, 2000).

2.3 Priority species

What are 'priority' species? For some groups of wildlife species, conservation planners have established priority rankings based on objective criteria such as current population abundance, distributions, availability of suitable habitats, perceived threats to habitat, etc. (Mehlman et al., 2004). For example, for North American landbirds, Partners in Flight has established ecoregional priority rankings (Panjabi et al., 2005). For other wildlife groups, priority status may be conferred based on species being listed as threatened or endangered by national or regional governmental agencies (Greenwald et al., 2006) or listed as a species of concern by private conservation organizations (Butcher, 2007). Regardless of the underlying rationale for their selection, priority wildlife species within managed forests differ among regions and forest types. Moreover, the priority species targeted for habitat improvement via prescribed management actions ultimately are determined by the landowner's objectives.

Why favor these species over others? All forest management, including no active management (a.k.a., passive management or benign neglect), influences the abundance and composition of wildlife species. For many priority species, insufficient habitat, or an overabundance of habitat with unsuitable conditions, likely contributed to their designation as a priority species. Therefore, where modification of forest habitat will promote habitat conditions conducive to supporting sustainable populations of these species, it behooves managers to prescribe silvicultural actions that will result in increased habitat for these species.

3. Desired habitat conditions for priority species

Because priority species differ among regions and among forest types, quantification of specific habitat characteristics suitable for all forests and all species is not practical. Even so, empirical evaluations of wildlife species responses to different silvicultural treatments provide insight into the general habitat conditions that benefit different species groups.

Amphibians – Within landscapes, amphibians benefit from protection of existing forest, avoiding further forest fragmentation, and disturbances that mimic natural processes (Kingsburg & Gibson, 2002). Within stands, seasonally available water, especially vernal pools, should be protected from disturbance (e.g., siltation) by retaining forest canopy (≥75%) immediately surrounding pools and maintaining >50% canopy for up to 120 m from seasonal waters (Calhoun & deMaynadier, 2004). Amphibians tend to benefit from retention of sufficient canopy throughout the forest stand to provide a partially shaded forest floor. Their abundance also tends to be positively associated with a deep litter layer and abundant woody debris on the forest floor (Maguire et al., 2005). Within southern pine forests (e.g., longleaf pine [*Pinus palustris*]), thinning and burning are positively related to increased abundance of priority (open-pine adapted) amphibian species (Steen et al., 2010).

Reptiles - The effects of silvicultural treatments on reptiles have been mixed, often exhibiting no discernable effect. Despite inconclusive studies, development of heterogeneous canopies, with numerous gaps enabling light to reach the forest floor, enhances thermoregulation opportunities for reptiles. Herpetofauna (reptiles and amphibians) may exhibit different regional responses to silviculture. Similarly, species-specific response to silviculture may result in species responding positively, negatively, or not at all (Russell et al., 2004). Thus, more research is needed on the effects of thinning and retained structural elements on reptiles and amphibians (Jones et al., 2009; Verschuyl et al., 2011).

Birds: The intensity of harvest, otherwise stated as the magnitude of retention, influences species composition of bird communities. More canopy retention favors retention of mature-forest species, whereas more intense harvest tends to favor early-successional species and forest generalist species (Schieck et al., 2000). In northern hardwood forests, retention of, 20 m² ha⁻¹ basal area maintained habitat for mature-forest birds yet afforded habitat for birds using early-successional habitat (Homes and Pitt, 2007). Similarly, in South American hardwood forests, combined dispersed and aggregated retention permitted colonization by early successional birds yet retained species characteristic of old-growth forests within retained aggregates (Lencinas et al., 2009). In Tasmanian eucalypt forest, long-term, aggregated retention harvests better sustained mature forest bird communities, post-harvest, compared with clearcut or dispersed retention harvests (Lefort and Grove, 2009). In young, conifer-dominated stands, thinning promotes diversity of breeding birds — a variety of thinning intensities and patterns, ranging from no thinning to widely spaced residual trees, will maximize avian diversity at the landscape scale and provide structural diversity within and among stands (Hagar et al., 2004). In southern pine forests, bird abundance and richness were positively related to volume of coarse woody debris and density of snags (Jones et al., 2009). In Midwestern oak-hickory forests, thinning harvest and prescribed fire increased avian nest survival (Streby and Miles, 2010). However, responses to variable retention harvests are species specific and temporally dynamic: Response of some species is immediate but short-lived, whereas other species exhibit a deferred (maybe several years post-treatment) but long-lasting response (Hagar and Friesen, 2009, Twedt and Somershoe, 2009).

Bats – Many bat species avoid spatial clutter: Thus bats tend to be more abundant in forests with open structure (harvested or old-growth) compared to closed-structured mid-successional forest (Loeb and O'keefe, 2006; Patriquin and Barclay, 2003; Menzel et al., 2005).

Forest structure appears more important than forest composition for foraging bats (Loeb and O'Keefe, 2006). Increased species and structural diversity of woody species, including providing broadleaf trees and shrubs in corridors and surrounding water features, tend to enhance foraging opportunities for bats. Conversely, forest type (or tree species) may be an important determinant of roosting location or behavior (Barclay and Kurta, 2007; Perry et al., 2007). Specifically, roost trees of bats are generally tall with large diameters and are located in stands with open canopies and high snag densities (Kalcounis-Rüppell et al., 2005). Compared to trees used by foliage-roosting bats, cavity-roosting bats use trees within stands that have more open canopies and trees that are closer to water than are random trees (Kalcounis-Rüppell et al., 2005).

Deer – Within landscapes, decreased road density and lessening of associated traffic reduces the negative impacts of disturbance. Silviculture should promote structurally complex, uneven-aged stands, that include forest openings (1 – 20 ha), and provide structural heterogeneity within and among stands (Nelson et al., 2008). Resulting habitat should provide understory forage and cover while retaining species diversity of trees.

Bear – Habitat selection is largely influenced by food abundance (Costello and Sage, 1994). A diversity of tree and shrub species producing hard mast (acorns and nuts) as well as soft mast (berries and fruits) is beneficial. Thinning, group selection, or patch cuts should be used to promote regeneration and recruitment of mast producing species, as well to increase understory food production, escape cover, and bedding areas. Care should be exercised to retain some large trees within each stand and to protect potential den trees during harvest.

Small Mammals – Silviculture that retains legacy structures, variable-density thinning, and management for increased forest decadence combined to support 3 species of squirrel in Douglas-fir forests (Carey, 2000). In southwestern conifer forests, thinning and burning treatments had positive effects on most small mammals. Even so, effects of silviculture on small mammals are species-specific, such that a positive response in abundance of one species may be offset by negative response in abundance of another species.

Effective strategies to achieve habitat for different species across spatial and temporal scales should include developing: 1) measurable objectives, 2) integrated conservation goals and silvicultural prescriptions, 3) clear and practical guidelines, effective training, and communication programs, and 4) a monitoring and an adaptive management process to evaluate and improve results (Munks et al., 2009).

3.1 Desired landscape conditions

Landscape conditions deemed suitable for priority species are highly dependent upon which species have been designated as priority species within the landscape under consideration. Furthermore, even after priority species have been designated, determining the nature and extent of desired landscape conditions have relatively little empirical justification (Moilanen, 1999). Even so, various methods have been employed to evaluate the appropriate landscape area and context for priority species.

A relatively simple, 6-step approach was put forward to identify the appropriate landscapes for forest bird conservation in the Mississippi Alluvial Valley: 1) establish priority species, 2) establish habitat priorities, 3) identify habitat requirements for priority species, 4) determine the extent and location of extant habitat, 5) set site-specific habitat goals, and 6) establish metapopulations goals (Mueller et al., 2000). A variation of this method was used to characterize priority of lands for forest restoration within the Mississippi Alluvial Valley and thereby delineate 107 discrete landscape units that were deemed capable of supporting viable populations (assumed to be ≥500 pairs) of priority silvicolous birds (Twedt et al., 2006; Fig. 4a).

Building upon this model, the Lower Mississippi Valley Joint Venture Forest Resource Conservation Working Group (Wilson et al., 2007) quantified desired forest conditions within these landscapes as:

1. >70% forest habitat, with large contiguous forest areas preferred;
2. 70-95% of forest area actively managed via silviculture intended to encourage desired stand conditions;
3. 35-50% of forest should meet desired forest conditions at any point in time, (Recognizing that habitat conditions are dynamic, silvicultural treatments should target conditions that yield extended temporal duration of desired stand conditions);
4. ≤10% of forest should be in regeneration harvests (where >80% of overstory is removed) that are >3 ha in area;
5. ~5% of area should be in shrub-scrub (thamnic woody vegetation) habitat, but this may including early seral forest stages; and
6. 5-30% of forest area should be passively managed as wilderness, natural areas, or set-aside areas that are not subjected to silvicultural manipulation.

A further refinement of this method has been used within the West Gulf Coastal Plain and Ouachita Mountains Ecoregion to characterize landscape requirements of birds reliant on open-canopy, mature pine forests (Keister et al., 2011). The first steps were again to identify priority species and define suitable habitats for priority species. However, area requirements for each of these priority species were determined by assessing the minimum territory size or home range of breeding pairs. Connectivity among territories of appropriate size was assumed to be related to the likely dispersal distance of each priority species. The minimum viable population size was estimated for each species (Hanski et al., 1996). These species-specific characteristics were used to evaluate existing landscape suitability for priority species. The sum of the total area of all habitats that were capable of supporting a breeding pair (assuming silvicultural management would yield suitable stand conditions) and that were joined by virtue of being within dispersal distance of other suitable habitat was used to evaluate landscapes. Landscapes were deemed suitable when their total areas were sufficient to support a minimum viable population of the species: Non-suitable landscapes were characterized with regard to their suitability based on the proportion of a minimum viable population that could be supported (Fig. 4b-d). Desired landscape conditions for multiple species were then jointly considered (Fig. 4, right).

3.2 Desired site conditions

Desired site conditions are dependent upon which species have been designated as priority species. Even so, as identified above, there are commonalities among stand characteristics that have been found to benefit many species and species groups. Prescribed wildlife forestry silvicultural practices can induce disturbance within forests and thereby stimulate development of desired structural conditions. Specific needs to be addressed when prescribing management actions include: 1) development and maintenance of structural (vertical and horizontal) heterogeneity, 2) achieving and retaining site appropriate tree species diversity, 3) maintaining an appropriate [sufficient yet acceptable] level of decadence, and 4) providing adequate reproduction to ensure sustainable habitat conditions and maintain future management options (Fig. 5).

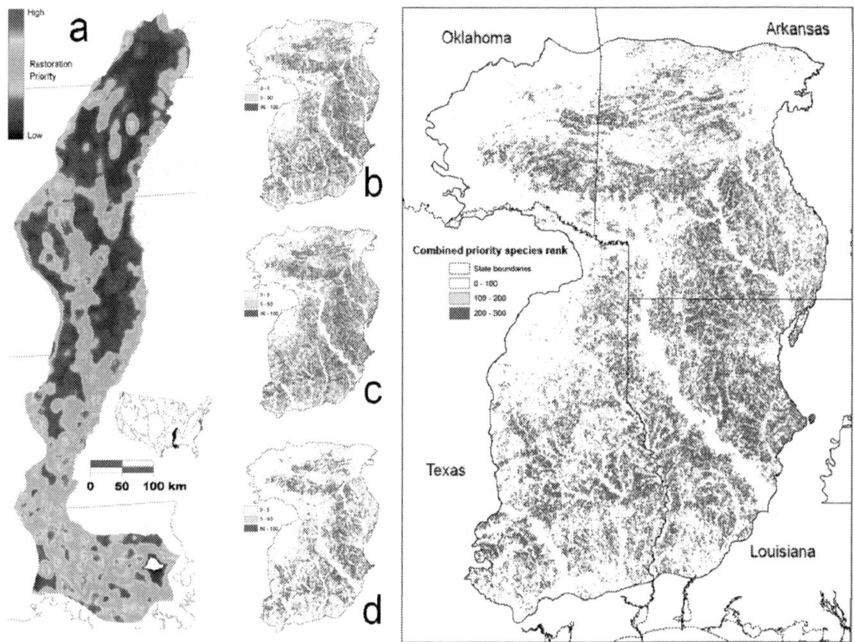

Fig. 4. Landscapes identified for priority birds based on forest restoration (a) in the Mississippi Alluvial Valley (Twedt et al. 2006) and on proportion of minimum viable population supported for Bachman's sparrow (b), brown-headed nuthatch (c), and red-cockaded woodpecker (d) within the West Gulf Coastal Plain (Keister et al. 2011).

Maintaining or enhancing diversity of species – A diversity of flora, in the canopy, mid-story, and understory, buffers fluctuations (annual, seasonal, and temporal) in phenology, productivity, susceptibility, and merchantability of forests, although this is not universally applicable to canopy species (e.g., longleaf pine stand canopies may be predominately one species). Because diversity of fruiting species affords temporal and spatial availability of fruits, and because fruit use by wildlife in temperate forests is substantial, with fruits being consumed primarily during winter (McCarty et al., 2002), maintaining and enhancing diversity of fruit bearing species should be encouraged. Moreover, a variety of fruits provide alternative foods in the event of mast crop failure of other species. Similarly, different plants host different caterpillar (larval Lepidoptera) species, some of which have species-specific relationships. Because butterfly and moth larvae are consumed by many wildlife species, especially birds, maintaining a diversity of trees and other flora within forest stands extends the temporal availability and increases the diversity of these insects (Hagar, 2004, Twedt and Best, 2004). In addition, forests with diverse species compositions are less vulnerable to catastrophic effects of disease and insects. Indeed, recent detrimental oak mortality in the Ozark Mountains Ecoregion has been attributed, in part, to the preponderance of oaks within these forests (Riggins et al. 2009). Similarly, species-specificity of many diseases and insects (e.g. chestnut blight [*Cryphonectria parasitica*] impact on American chestnut [*Castanea dentate*] or emerald ash borer [*Agrilus planipennis*] infestation of ash species [*Fraxinus* spp.]) provides argument for the maintenance of forest diversity to ameliorate the potentially catastrophic effects of regional outbreaks of disease or insect pests.

Fig. 5. Forest structure observed during summer (left) and winter (right) within bottomland hardwood forests circa 4 years after variable-retention, wildlife-forestry based silvicultural treatments (bottom) and on untreated control stands (top).

Finally, our naïve perception of the value of "undesirable species" is likely insufficient to comprehend the value of these species within the ecosystem. Therefore, preferential retention or diminution of native species may have unintended detrimental consequences (Lockhart 2004). For example, although the presence of piñon pine (*Pinus edulis*) is among the most important habitat features for many birds in piñon–juniper forests, the favored (86%) location of nests in juniper (*Juniperus osteosperma*) suggests that managers should avoid preferential thinning of juniper within these forests (Francis et al., 2011).

Increased structural heterogeneity (horizontal and vertical) – More structure generally provides a greater number of habitat niches for occupancy by different wildlife species, thereby increasing species richness, which may include priority species. Managers must, however, be cautious of excessive application of silviculture intended to promote heterogeneity as these excesses may promote what could be viewed as 'homogeneity' of this habitat structure at a more dispersed scale. As different intensity of silvicultural treatment result in different stand complexity, silvicultural decisions made at each stage of stand development affect forest structure (Table 2).

Lower Basal Area – Thinning canopies has a positive or neutral effect on diversity and abundance of most taxa, albeit intensity and biophysical setting influence wildlife response (Verschuyl et al., 2011). In general, lower basal areas are associated with increased density of understory vegetation which adds to structural diversity, provides cover for escape, thermoregulation, and nesting or bedding. Moreover patches of dense understory enhance forage opportunities for species that browse grasses, herbs, or seedlings, species that glean insects from leaves, and frugivores.

Large, old, decadent trees - These features tend to be especially characteristic of old-growth-like forest conditions. As such, these characteristics are difficult to achieve given historical forest management where short rotation lengthen and harvest of economically mature trees limit availability of these features. Large cavity trees are used as den sites by bears, as well as other meso-mammalian (e.g., martens, fishers, raccoons, porcupines, weasels) and larger avian species (e.g., woodpeckers, owls, mergansers, and ducks). Similarly, dominant emergent trees provide perches and nesting platforms for large raptors (e.g., eagles, hawks, and kites). Although avian excavators (e.g., woodpeckers) are the primary cavity producers in North America, producing 77% of nesting cavities, this is not the case globally ($\leq$26%) where most cavity formation results from damage and decay (Cockle et al., 2011). Thus, cavity-using communities are highly dependent upon maintaining decadent trees within forest stands. As such, the presence of large, decadent trees may be essential for occupancy of stands by many of these large fauna.

Abundant standing snags and downed woody debris – Diversity and abundance of birds and biomass of invertebrates are positively associated with the volume of downed coarse woody debris and snags (Riffell et al., 2011). In unmanaged oak and oak-beech forests, 25% of dead wood was standing, whereas 75% was downed (Vanderkerkhove et al., 2009). Snags (and other decadent trees, such as those with heart rot), are characteristically exploited by primary-cavity excavators (e.g., woodpeckers) as well as those species that forage on saprophytic insects. Excavated cavities are subsequently used by myriad secondary-cavity using species. Lack of snags or other decadent trees may restrict this entire suite of cavity using species (Cockle et al. 2011).

Providing sufficient and suitable regeneration – As successional changes in forest species composition are often beyond human life-spans, the need for sufficient and suitable regeneration of canopy trees is often neglected. However, within the concept of wildlife forestry, regeneration of canopy species is ongoing (generally not a single event – as in even-aged forest management). Thus, ensuring sufficient regeneration of marketable trees, and thereby enhancing the long-term merchantability of forest stands, is imperative if desired habitat conditions for priority wildlife species are to be sustained through repeated, prescribed, silvicultural treatments. Notably, where maintaining shade-intolerant species within future generations is deemed desirable for anticipated management options, silviculturally induced gaps (patch cuts or group selections) should be of sufficient area to allow regeneration and development of these shade-intolerant species.

Treatment	Increasing forest diversity, structure, and complexity →		
Regeneration harvest	Clearcut	Legacy retention; dispersed or aggregated	Wildlife forestry; variable retention with canopy gaps
Restoration planting	Monoculture	Monoculture with supplemental natural reproduction	Mixture of species; hard and soft mast; shade tolerant and intolerant
Stand modification (TSI)	Herbicide removal of understory and midstory	Partial removal of competitors	Heterogeneous retention and removal of understory competitors
Thinning	Systematic single species retention	Multi-species retention	Heterogeneous retention and thinning; underrepresented species preferably retained

Table 2. Effects of wildlife forestry and other silviculture on stand structure (expanded from Carey, 2003b).

Although not specially referring to wildlife forestry, Carey (2006) succinctly encapsulated the holistic approach of wildlife forestry as - *"Management ... of forest development and landscape dynamics is more likely to be successful in maintaining ecosystem and landscape function than just providing select structural elements in stands and select structural stages in landscapes, as is often suggested for conservation."*

3.3 Regional examples of desired stand conditions

Bottomland hardwood forests - A preponderance of closed canopy, second-growth forests with relatively homogeneous structure prompted recommendations for increased area of open canopied forests with greater structural diversity and more understory vegetation development (Wilson et al., 2007). Basal area should be reduced to 14-16 m^2 ha^{-1} with ≥25% of the basal area approaching biological maturity (i.e., in older age classes). Forest canopy cover should be 60%-70% with at least 5 stems ha^{-1} being canopy dominant trees - emergent canopies preferred. Understory vegetation should be robust yet patchily distributed with 25%-40% cover. Density of snags ≥25 cm diameter at breast height (dbh) should be >15 snags ha^{-1} or have a basal area >0.9 m^2 ha^{-1}. Volume of coarse woody debris should exceed 14 m^3 ha^{-1} and advanced reproduction should be present on 30%-40% of the stand (Table 3).

Northwestern U.S. coniferous forests - Bird species richness and abundance are positively related to proportion of broadleaf forest, from 1% to 25% canopy cover, within coniferous dominated landscapes. Thus, it is import to maintain or restore broadleaf vegetation in these coniferous dominated stands (Betts et al., 2010). Desired stand conditions reflect a structure similar to that of old-growth forest conditions (Cissel et al., 2006). Multiple tree species, representing multiple age cohorts, should occupy stands and this mix should include at least 1 shade tolerant species. Large diameter (>125 cm dbh) conifers should be present at densities of 5-15 stems ha^{-1}. At least half of the 20-30 snags ha^{-1} should be >60 cm dbh and a third of coarse woody debris should be of large (>60 cm) diameter (Table 3).

Southeastern U.S. pines - Historically open-canopied pine forests were found throughout the southeastern United States. These open forests were maintained by frequent fires. Prescribed burning, complimented with thinning of pine canopies and herbicide treatments of encroaching hardwoods can restore these historic conditions. Open canopy conditions with an abundant and floristically diverse understory increase species richness of priority open pine associated amphibians (Guyer, and Bailey, 1993, Steen et al., 2010). Similarly, abundance of endangered gopher tortoises (*Gopherus polyphemus*) and red-cockaded woodpecker (*Picoides borealis*) respond positively to prescribed fire within open pine stands. Desired stand conditions include open (40-60%) canopies that are dominated by large (≥35 cm dbh) pines. These large pines should account for ≥ 4.5 m^2 ha^{-1} of a total basal area of 11 – 16 m^2 ha^{-1}. Mid-story should be sparse (<15%) with very limited hardwoods (<5%). Ground cover and understory should be lush (>80% cover) and comprised predominately of grasses and forbs. At least 7 snags ha^{-1} should be present. Relatively frequent fire (≤3 year interval) is likely required to maintain these desired conditions (Table 3).

4. Economic considerations

Wildlife forestry's primary intent is to provide suitable habitat conditions for priority wildlife. Even so, silvicultural practices that require long-term commitments or repeated expenditures may not be undertaken and thus, may not provide desired results. Indeed, silvicultural prescriptions intended to enhance wildlife habitat must be cognizant of their expense and their long-term effect on future merchantability of the stand. With that caveat, wildlife forestry silviculture may have both positive and negative economic effects.

A positive impact of most wildlife forestry practices is a reduction in wildfire hazards. Indeed, managing tree density and species composition with prophylactic silvicultural treatments that include thinning, surface fuel treatments, and prescribed fire reduce the risk of wildfire (Graham et al., 1999).

Negative economic factors include wounding of potentially merchantable trees that remain within stands after silvicultural treatments, as these wounds may result in future economic loss due to mortality or reduction in quality of timber (Hennon and DeMars, 1997). Thus, harvest prescriptions should account for this possibility and provide recommendations to reduce unintended damage to residual trees. On the other hand, from a wildlife use perspective, basal wounds to large retained trees have the potential to increase the number of decant large trees within a stand.

Another negative economic impact of wildlife silviculture may be successional changes in species composition. As noted for ensuring adequate tree regeneration within stands, managers should be aware of unintended shifts in species composition as a consequence of prescribed wildlife silviculture, usually due to an increased presence of shade-tolerant species. Therefore, if shade-intolerant species are intended to be a part of future forest canopies, silvicultural treatments must ensure their regeneration and development (Twedt and Somershoe, in press).

All forests are subject to windthrow (toppling of live standing trees due to wind), although its severity is influenced by wind speed, climate, topography, hydrology, and soils (Ruel, 1995): Toppled trees are present in mature unharvested forests as well as along the edges of clearcuts. Even so, windthrow may increase after partial harvest silvicultural treatments. More retention within a stand, however, tends to lessen the severity of windthrow (Franklin et al., 1997). Windthrow can also be mitigated through judicious silviculture, as some species are more susceptible to windthrow, possibly due to differences in form. Trees with more extensive crown structure or shallow, poorly developed roots are more likely to be toppled by wind (Beese, 2001). Retained, sound, dominant trees are less subject to windthrow than decadent or subdominant trees. The dispersion, orientation, and shape of retention also influences susceptibility to windthrow, with tear-drop shaped, aggregated clumps, oriented with prevailing winds better able to withstand wind damage (Franklin et al., 1997). Even so, small (1.5 ha) patch cuts may be least vulnerable to wind (Beese, 2001).

Stand feature	Bottomland hardwood (Wilson et al., 2007)	Longleaf pine (Bragg, 2004; East Gulf Coastal Plain Joint Venture, unpublished)	Pacific northwest conifers (Cissel et al. 2006)	Young conifers in northern pacific rainforest Altman and Hagar, 2007)
Basal area	14 – 16 m² ha⁻¹	11 – 16 m² ha⁻¹	9 – 27 m² ha⁻¹	
Size class	≥25% of BA approaching biological maturity	>4 m² ha⁻¹, ≥35 cm dbh	several tree species of varying size and age; large emergent trees 5 – 15 ha⁻¹, >125 cm dbh conifers;	large conifer trees
Density	>5 dominant ha⁻¹	12 – 30 pines ha⁻¹, >76 cm dbh	60 – 85 ha⁻¹, 38-76 cm dbh; 250 – 500 ha⁻¹ <38 cm dbh	<500 ha⁻¹; thin to promote growth
Canopy cover	60 – 70%	40 – 60% (≤5% hardwood) Patchy	multi-layered canopies	variation in overstory achieved via silviculture, low % canopy cover to encourage understory
Midstory	25 – 40%	≤15% (≤5% hardwood) Open	open space among the lower branches of canopy	retain and protect old shrubs and shrub patches
Ground cover	25 – 40%	>80%		variation in understory associated with variable canopy
Snags & stressed trees	>15 ha⁻¹, ≥25 cm dbh; or ≥5 ha⁻¹, ≥ 51 cm dbh; or >0.9 m2 ha⁻¹, >25 cm dbh	≥7 (12-25) ha⁻¹, red heart; 10 – 50% cull in retained trees	20 – 30 ha⁻¹ >25 cm dbh; 50% > 60 cm dbh; cavities in standing trees	large
Coarse woody debris	≥14 m3 ha⁻¹	21 – 54 m3 ha⁻¹ (includes snags)	275 linear m >25 cm diameter; 33% > 60 cm dbh	retain, protect, and recruit large diameter CWD
Reproduction	30 – 40% of area with advanced reproduction	10% of area with advanced reproduction		encourage regeneration and retention of a diversity of species
Other	>10 visible cavities ha⁻¹; 1 den tree (4 ha)⁻¹	fire return ≤3 year	≥2 species (≥1 shade-tolerant); ≥2 age cohorts; >150 years old	retain and promote deciduous trees and shrubs

Table 3. Described and quantified desired site (stand) conditions that are deemed to provide enhanced habitat for priority wildlife species in different forest types and seral stages.

There is little basis for economic comparisons of wildlife forestry and other, traditional, silvicultural systems. Wildlife forestry may provide sustained revenues from exploitive use of wildlife, as well as from periodic harvest of forest products. But even without income from wildlife, little economic loss was reported from variable retention silviculture compared to even-aged, shelterwood cuts in *Nothofagus pumilio* forests (Pastur et al., 2009). Similarly, simulation models spanning 300 years that compared economic returns and ecological benefits within Pacific northwest forests being managed for biodiversity with forests being managed for timber production, and incorporating narrow riparian reserves (i.e., set asides), found significantly greater ecological benefits associated with biodiversity management (Carey et al.,1999). Concurrently, the economic cost of achieving those benefits was relatively modest, with 18% loss in net present value (NPV), only 4% loss if riparian protection was similar to that afforded within the biodiversity management model, and fully a 13% gain in NPV when newly enacted state laws for riparian protection were enforced (Carey, 2003a). More notably, under management for biodiversity, tree quality improved, decadal revenues increased by 150%, forest-based employment quadrupled, and manufacturing diversified and became more reliant on high quality products and value added manufacturing (Carey, 2003a, Carey et al., 1999, Lippke et al., 1996).

5. Restoration

If successfully implemented, wildlife forestry is self-sustaining, such that reforestation (e.g., artificial regeneration or stand regeneration) is rare. Even so, appropriate practices are increasingly being sought in conjunction with forest restoration that better enable attainment of desired forest conditions. For example, decades of bottomland afforestation, wherein agricultural lands are being restored to forested wetlands, has provided valuable insight regarding appropriate species and planting densities (Wilson et al., 2007). Management practices similar to those employed for wildlife forestry have also been used to restore longleaf pine forests on sites converted to forests dominated by other species. Similarly, within production-oriented plantation forests management alternatives are sought that will enhance biodiversity and improve wildlife habitat, yet minimally compromise product output.

Recommendations for restoration and mid-rotation management have included planting or maintaining species mixtures to ensure stand diversity and improve forest product quality (Twedt and Best, 2004, Lockhart et al. 2008). Mixed species planting of native species are favored over monocultures of exotic species. Some legacy trees (live or snags) and understory vegetation should be retained at harvest through a second rotation. Site-preparation should reflect natural disturbances and conserve coarse woody debris. Thinning earlier in the rotation and increasing rotation length also increase structural diversity (Hartley, 2002; Kerr, 1999). Finally, for some forest types, maintaining herbaceous or early succession habitat for a longer period provides benefits to priority bird species (Altman and Hagar, 2007).

6. Evaluating results of management actions

Quantitatively defined landscape conditions and desired stand conditions have been identified for only a few ecoregional landscapes and only a few forest types within these landscapes. Wildlife biologists, foresters, and conservation planners must develop a shared vision regarding what each perceives as the desired habitat condition within forests. In the landscape context, these conditions need not be the same for all forests, nor will they be identical for all priority species. Research and monitoring will be required to evaluate species responses to different ranges of habitat conditions perceived as desirable. Judgments may be needed when priority species exhibit conflicting responses to what are perceived as desired conditions.

At present we have little empirical data regarding the range of economic returns from wildlife silviculture relative to traditional production-oriented silviculture. Appropriate measures of economic returns, including an evaluation of the long-term sustainably of different management options, are needed. In addition, we have assessed the faunal responses of relatively few species to wildlife forestry silviculture. Future efforts must assess comparable responses of less-studied priority species and species groups as well as evaluate appropriate demographic responses of all species.

To determine if prescribed management actions elicit intended responses from priority species, monitoring is required. Monitoring should target specific management actions and evaluate spatial and temporal responses. Silvicultural prescriptions should consider incorporation of the principles of adaptive management during their development, implementation, and evaluation (Fig. 1). Feedback from such efforts can provide an assessment of the adequacy and sustainability of wildlife forestry silviculture. Science based knowledge of the results of these management practices is crucial to their long-term success, as *"Conservation is paved with good intentions which prove to be futile, or even dangerous, because they are devoid of critical understanding either of the land, or of economic land use"* (Leopold, 1953).

7. Acknowledgement

I am grateful to the members of the Lower Mississippi Valley Joint Venture Forest Resource Conservation Working Group for thoughtful discussions and arguments regarding the merits of wildlife forestry. Similarly, members of the West Gulf Coastal Plain and Ouachita Mountains Landbird working group provided insight and understanding regarding application of wildlife forestry principles within southern pine systems. Finally, I am grateful to the many pioneers and advocates of alternative silviculture from which most of our understanding of wildlife forestry was derived.

8. References

Altman, B. & Hagar, J. (2007). *Rainforest birds- A land manager's guide to breeding bird habitat in young conifer forests in the Pacific Northwest,* U.S. Geological Survey Scientific Investigations Report 2006-5304, Reston, Virginia

Askins, R. (2001). Sustaining biological diversity in early succession communities: the challenge of managing unpopular habitats. *Wildlife Society Bulletin,* Vol.29, No.2, pp. 407-412

Aubry, K.; Halpern, C. & Maguire, D. (2004). Ecological effects of variable-retention harvests in the northwestern United States: the DEMO study. *Forest Snow and Landscape Research,* Vol.78, No.1-2, pp. 119-134

Aukema, J. & Carey, A. (2008). *Effects of variable-density thinning on understory diversity and heterogeneity in young Douglas-fir forests.* Research Paper PNW-RP-575. U.S. Department of Agriculture, Forest Service, Pacific Northwest Research Station, Portland, Oregon

Bailey, M.; Holmes, J.; Buhlmann, K. & Mitchell, J. (2006). *Habitat Management Guidelines for Amphibians and Reptiles of the Southeastern United States.* Partners in Amphibian and Reptile Conservation Technical publication HMG-2, Montgomery, Alabama

Barclay, R. & Kurta, A. (2007). Ecology and behavior of bats roosting in tree cavities and under bark, In: *Bats in forests: conservation and management,* M.J. Lacki; J.P. Hayes & A. Kurta, (Eds.), pp. 17–59, Johns Hopkins University Press, Baltimore, Maryland

Barneycastle, C. (2001). The Sustainable Forestry Initiative of the American Forest & Paper Association, In: *Proceedings of the symposium on Arkansas forests: a conference on the results of the recent forest survey of Arkansas,* J.M. Guldin, (Ed.), 110-112, U.S. Department of Agriculture, Forest Service, Southern Research Station, General Technical Report SRS 41, Asheville, North Carolina

Bauhus, J.; Puettmann, K. & Messier, C. (2009). Silviculture for old-growth attributes. *Forest Ecology and Management* Vol.258, No.4, pp. 525-537

Beese, W. (2001). Windthrow monitoring of alternative silvicultural systems in montane coastal forests, In: *Proceedings of the Windthrow Researchers Workshop: Windthrow Assessment and Management in British Columbia,* S.J. Mitchell & J. Rodney, (Compilers), pp. 2-11, Richmond, British Columbia

Beggs, L. (2004). *Vegetation response following thinning in young Douglas-fir forests of western Oregon: Can thinning accelerate development of late-successional structure and composition?* M.S. thesis, Oregon State University, Corvallis, Oregon

Betts, M.; Hagar, J.; Rivers, J.; Alexander, J.; McGarigal, K. & McComb, B. (2010). Thresholds in forest bird occurrence as a function of the amount of early-seral broadleaf forest at landscape scales. *Ecological Applications* Vol.20, No.8, pp. 2116-2130

Bonnicksen, T.M. (2000). *America's ancient forests: from the ice age to the age of discovery.* John Wiley, New York.

Bragg, D. (2004). A prescription for old-growth-like characteristics in southern pines, In: *Silviculture in special places: Proceedings of the National Silviculture Workshop,* W.D. Sheppard & L.G. Eskew, (Compilers), pp. 80-92, U.S. Department of Agriculture, Forest Service, Rocky Mountain Research Station, Proceedings RMRS-P-34, Fort Collins, Colorado

Brown, S.; Hickey, C.; Harrington, B. & Gill, R., (Eds.). (2001). *United States Shorebird Conservation Plan.* Manomet Center for Conservation Sciences, Manomet, Massachusetts

Butcher, G.S. (2007). Common birds in decline, a state of the birds report. *Audubon,* Vol.109, No.4, pp. 58-62

Calhoun, A. & deMaynadier, P. (2004). *Forestry habitat management guidelines for vernal pool wildlife.* MCA Technical Paper No. 6, Metropolitan Conservation Alliance, Wildlife Conservation Society, Bronx, New York

Carey, A. (2000). Effects of new forest management strategies on squirrel populations. *Ecological Applications,* Vol.10, No.1, pp. 248-257

Carey, A. (2003a). Restoration of landscape function: reserve or active management? *Forestry*, Vol.76, No.2, pp. 221-230.

Carey, A. (2003b). Managing for wildlife: A key component for social acceptance of compatible forest management, In: Compatible Forest Management, A. Monserud, R.W. Haynes, & A. C. Johnson, (Eds.), Kluwer Academic Publishers, Dordrecht, The Netherlands

Carey, A. (2006). Active and passive forest management for multiple values. *Northwestern Naturalist*, Vol.87, No.1, pp. 18-30

Carey, A., Lippke, B. & Sessions, J. (1999). Intentional systems management: managing forests for biodiversity. *Journal of Sustainable Forestry*, Vol.9, No.3-4, pp. 83-125

Cissel, J.; Anderson, P.; Olson, D.; Puettmann, K.; Berryman, S.; Chan, S. & Thompson, C. (2006), *BLM Density Management and Riparian Buffer Study: Establishment Report and Study Plan*, U.S. Geological Survey, Scientific Investigations Report 2006-5087, Reston, Virginia, USA

Cissel, J.; Swanson, F. & Weisberg, P. 1999. Landscape management using historical fire regimes: Blue River, Oregon. *Ecological Applications*, Vol.9, No.4, pp. 1217–1231

Cockle, K.; Martin, K. & Wesołowski, T. (2011). Woodpeckers, decay, and the future of cavity-nesting vertebrate communities worldwide. *Frontiers in Ecology and the Environment*, Vol.9, No.7, pp. 377–382

Costello, C. & Sage Jr., R. (1994). Predicting black bear habitat selection from food abundance under 3 forest management systems, *Bears: Their Biology and Management: A Selection of Papers from the Ninth International Conference on Bear Research and Management*, pp. 375-387, , Missoula, Montana, USA, February 23-28, 1992

Egan, A. & Luloff, A. (2000). The exurbanization of America's forests: Research in rural social science. *Journal of Forestry*, Vol.98, No.3, pp. 26-30

Ford, W.; Russell, K.; Moorman, C. (Eds.). (2002). *Proceedings: the role of fire for nongame wildlife management and community restoration: traditional uses and new directions*, Nashville, Tennessee, September 15, 2000. U.S. Department of Agriculture, Forest Service, Northeastern Research Station, General Technical Report NE-288, Newtown Square, Pennsylvania, USA

Francis, C.; Ortega, C. & Hansen, J. (2011). Importance of juniper to birds nesting in piñon-juniper woodlands in northwest New Mexico. *Journal of Wildlife Management*, Vol.75, No.7, pp. 1574-1580

Franklin, J. (1989). Toward a new forestry. *American Forests*, (November/December 1989), pp. 37-44

Franklin, J.; Berg, D.; Thornburgh, D. & Tappeiner, J. (1997). Alternative silvicultural approaches to timber harvesting: Variable retention harvest systems, In: *Creating a Forestry for the 21st Century: The Science of Ecosystem Management*, K.A. Kohn & J.F. Franklin (Eds.), 111–139, Island Press, Washington, DC, USA

Franklin, J.; Forman, R. (1987). Creating landscape patterns by forest cutting: Ecological consequences and principles. *Landscape Ecology*, Vol.1, No.1, pp. 5-18

Franklin, J.; Mitchell, R. & Palik, B. (2007). *Natural disturbance and stand development principles for ecological forestry*, General Technical Report NRS-19, U.S. Department of

Agriculture, Forest Service, Northern Research Station, Newtown Square, Pennsylvania, USA

Garman, S.; Cissel, J. & Mayo, J. (2003). *Accelerating development of late-successional conditions in young managed Douglas-fir stands: a simulation study.* General Technical Report PNW-GTR-557, U.S. Department of Agriculture, Forest Service, Pacific Northwest Research Station. Portland, Oregon, USA

Graham, E. (1947). *The land and wildlife,* Oxford Press, New York, New York, USA

Graham, R.; Harvey, A.; Jain, T. & Tonn, J. (1999). *The effects of thinning and similar stand treatments on fire behavior in Western forests,* General Technical Report PNW-GTR-463, U.S. Department of Agriculture, Forest Service, Pacific Northwest Research Station, Portland, Oregon, USA

Greenwald, N.; Suckling, K. & Taylor, M. (2006). Factors affecting the rate and taxonomy of species listings under the U.S. Endangered Species Act, In: *The Endangered Species Act at 30: Renewing the Conservation Promise,* D.D. Goble, J.M. Scott & F.W. Davis (Eds.), 50-67, Island Press, Washington, DC, USA

Guyer, C. & Bailey, M. (1993). Amphibians and reptiles of longleaf pine communities, In: *Proceedings of the Tall Timbers Fire Ecology Conference Number 18, The Longleaf Pine Ecosystem; Ecology, Restoration and Management,* 139-158, Tallahassee, Florida, USA

Hagar, J.; Howlin, S. & Ganio, L. (2004). Short-term response of songbirds to experimental thinning of young Douglas-fir forests in the Oregon Cascades. *Forest Ecology and Management,* Vol.199, No.2-3, pp. 333-347

Hagar, J. (2004). *Functional relationships among songbirds, arthropods, and understory vegetation in Douglas-fir forests, western Oregon,* Ph.D. dissertation, Oregon State University, Corvallis, Oregon, USA

Hagar, J. & Friesen, C. (2009). *Young stand thinning and diversity study- response of songbird community one decade post-treatment,* U.S. Geological Survey Open-File Report 2009-1253, Reston, Virginia, USA

Hanski, I. (1998). Metapopulation dynamics. *Nature* Vol.396, pp. 41-49

Hanski, I.; Moilanen, A. & Gyllenberg, M. (1996). Minimum viable metapopulation size. *American Naturalist* Vol.147, No.4, pp. 527-541

Hartley, M. (2002). Rationale and methods for conserving biodiversity in plantation forests. *Forest Ecology and Management,* Vol.155, No.1-3, pp. 81-95

Hayward, G. (1991). Using population biology to define old-growth forests. *Wildlife Society Bulletin,* Vol.19, No.1, pp. 111-116.

Heithecker, T. & Halpern, C. (2006). Variation in microclimate associated with dispersed-retention harvests in coniferous forests of western Washington. *Forest Ecology and Management,* Vol.226, No.1-3, pp. 60-71

Hennon, P. & Demars, D. 1997. Development of wood decay in wounded western hemlock and Sitka spruce in southeast Alaska. *Canadian Journal of Forest Research,* Vol.27, pp. 1971-1978

Holmes, S. & Pitt, D. (2007). Response of bird communities to selection harvesting in a northern tolerant hardwood forest. *Forest Ecology and Management,* Vol.238, No.1-3, pp. 280-292

Hunter, W.; Buehler, D.; Canterbury, R.; Confer, J. & Hamel, P. (2001). Conservation of disturbance-dependent birds in eastern North America. *Wildlife Society Bulletin*, Vo.29, No.2, pp. 440-455

Jones, P.; Hanberry, B. & Demarais, S. (2009). Stand-level wildlife habitat features and biodiversity in southern pine forests: A review. *Journal of Forestry*, Vol.107, No.8, pp. 398-404

Kaetzel, B.; Hodges, D. & Fly, J. (2011). Landowner motivations for owning woodland on the Tennessee Northern Cumberland Plateau. *Southern Journal of Applied Forestry*, Vol.35, No.1, pp. 39-43

Kalcounis-Rüppell, M.; Psyllakis, J. & Brigham, R. (2005). Tree roost selection by bats: An empirical synthesis using meta-analysis. *Wildlife Society Bulletin*, Vol.33, No.3, pp.1123-1132

Karau, E. & Keane, R. (2007). Determining landscape extent for succession and disturbance simulation modeling. *Landscape Ecology* Vol.22, No.7, pp. 993-1006

Kaufmann, M.; Huckaby, L.; Fornwalt, P.; Stoker, J. & Romme, W. (2003). Using tree recruitment patterns and fire history to guide restoration of an unlogged ponderosa pine/Douglas-fir landscape in the southern Rocky Mountains after a century of fire suppression. *Forestry*, Vol.76, No.2, pp. 231-241

Keeton, W. (2006). Managing for late-successional/old-growth characteristics in northern hardwood-conifer forests. *Forest Ecology and Management* Vol.235, No.1-3, pp. 129–142

Keister, A.; Twedt, D.; McKnight, S. & Tirpak, J. (2011). Prioritization of open pine management for landbirds of the West Gulf Coastal Plain and Ouachita Mountains, In: *Big Thicket Science Conference*, Stephen F. Austin State University, Nacogdoches, Texas, USA, April 8-10, 2011, Available from http://atcofa.sfasu.edu/images/stories/pdf/(2011)_WCGPBTSC_Program_Book_3_30_2011_Final.pdf

Kerr, G. (1999). The use of silvicultural systems to enhance the biological diversity of plantation forests in Britain. *Forestry*, Vol.72, No.3, pp. 191-205

Kingsbury, B. & Gibson, J. (2002). *Habitat management guidelines for amphibians and reptiles of the Midwest.* Partners in Amphibian and Reptile Conservation Technical Publication HMG-1, Fort Wayne, Indiana, USA, Available from http://herpcenter.ipfw.edu/downloadables/MWHabitatGuide/Index.htm.

Kneeshaw, D. & Gauthier, S. (2003). Old growth in the boreal forest: A dynamic perspective at the stand and landscape level. *Environmental Reviews*, Vol.11, No.1 (supplement), pp. 99-114

Knight, R. 1999. Private lands: The neglected geography. *Conservation Biology*, Vol.13, No.2, pp. 223–224

Lefort, P. & Grove, S. (2009). Early responses of birds to clearfelling and its alternatives in lowland wet eucalypt forest in Tasmania, Australia. *Forest Ecology and Management*, *Vol.*258, No.4, pp. 460-471

Lencinas, M.; Pastur, G.; Gallo, E. & Cellini, J. (2009). Alternative silvicultural practices with variable retention improve bird conservation in managed South Patagonian forests. *Forest Ecology and Management*, Vol.258, No.4, pp. 472-480

Leopold, A. (1949). *A Sand County Almanac and Sketches Here and There*, Oxford University Press, New York, New York, USA.

Leopold, A. (1953). *The Land Ethic Round River*, Oxford University Press, New York, New York, USA.

Lindenmayer, D. & Franklin, J. (2002). *Conserving Forest Biodiversity: A Comprehensive Multiscaled Approach*, Island Press, Washington, DC, USA

Lindenmayer, D. & Hobbs, R. (2007). *Managing and designing landscapes for conservation: Moving from perspectives to principles*, Blackwell Publishing, Malden , Massachusetts, USA

Lindenmayer, D.; MacCarthy, M. & Donnelly, G. (1999). Factors affecting stand structure in forest – are there climatic and topographic determinants. *Forest Ecology and Management* Vol.123, No.1, pp. 55-63

Lindenmayer, D.; Manning, A.; Smith, P.; Possingham, H.; Fischer, J.; Oliver, I. & McCarthy, M. (2002). The focal-species approach and landscape restoration: a critique. *Conservation Biology*, Vol.16, No.2, pp. 338-345

Lippke, B.; Sessions, J. & Carey, A. (1996). *Economic analysis of forest landscape management alternatives*. CINTRAFOR Special Paper 21, College of Forest Resources, University of Washington, Seattle, Washington, USA

Lockhart, B. (2004). All species have value. *Journal of Forestry*, Vol.102, No.1, (January/February 2004), pp. 60

Lockhart, B.; Gardiner, E.; Leininger, T. & Stanturf, J. (2008). A stand-development approach to oak afforestation in the lower Mississippi Alluvial Valley. *Southern Journal of Applied Forestry*, Vol.32, No.3, pp. 120-129

Loeb, S. & O'Keefe, J. (2006). Habitat use by forest bats in South Carolina in relation to local, stand, and landscape characteristics. *Journal of Wildlife Management*, Vol.70, No.5, pp. 1210–1218

MacArthur, R. & Wilson, E. (1967). *The Theory of Island Biogeography*, Princeton University Press, Princeton, New Jersey, USA

Maguire, D.; Halpern, C. & Phillips, D. (2007). Changes in forest structure following variable-retention harvests in Douglas-fir dominated forests. *Forest Ecology and Management*, Vol.242, No.2-3, pp. 708-726

Maguire, C.; Manning, T; West, S & Gitzen, R. (2005). Green-tree retention in managed forests: post-harvest responses of salamanders, In: Balancing ecosystem values: innovative experiments for sustainable forestry — proceedings of a conference, C.E. Peterson & D.A. Maguire (Eds.), 265-270, General Technical Report PNW-GTR-635. U.S. Department of Agriculture, Forest Service, Pacific Northwest Research Station, Portland, Oregon, USA

Mayer, A. & Cameron, G. (2003). Consideration of grain and extent in landscape studies of terrestrial vertebrate ecology. *Landscape and Urban Planning*, Vol.65, No.4, pp. 201-217

McCarty, J.; Levey, D.; Greenberg, C. & Sargent, S. (2002). Spatial and temporal variation in fruit use by wildlife in a forested landscape. *Forest Ecology and Management*, Vol.164, No.1-3, pp. 277-291

McClellan, M. (2004). Development of silvicultural systems for maintaining old-growth conditions in the temperate rainforest of southeast Alaska. *Forest Snow and Landscape Research*, Vol.78, No.1-2, pp. 173–190

Mehlman, D.; Rosenberg, K; Wells, J. & Robertson, B. (2004). A comparison of North American avian conservation priority ranking systems. *Biological Conservation*, Vol.120, No.3, pp. 383-390

Menzel, J.; Menzel Jr, M.; Kilgo, J.; Ford, W.; Edwards, J. & McCracken, G. (2005). Effect of habitat and foraging height on bat activity in the coastal plain of South Carolina. *Journal of Wildlife Management*, Vol.69, No.1, pp. 235–245

Mitchell, R.; Hiers, J.; O'Brien, J.; Jack, S. & Engstrom, R. (2006). Silviculture that sustains: the nexus between silviculture, frequent prescribed fire, and conservation of biodiversity in longleaf pine forests of the southeastern United States. *Canadian Journal of Forest Research*, Vol.36, No.11, pp. 2724-2736

Mitchell, R.; Palik, B. & Hunter, M. (2002). Natural disturbance as a guide to silviculture. *Forest Ecology and Management*, Vol.155, No.1-3, pp. 315–317

Mladenoff, D.; White, M; Pastor, J. & Crow, T. (1993). Comparing spatial pattern in unaltered old-growth and disturbed forest landscapes. *Ecological Applications*, Vol.3, No.2, pp. 294-306

Moilanen, A. (1999). Patch occupancy models of metapopulation dynamics: efficient parameter estimation using implicit statistical inference. *Ecology*, Vol.80, No.3, pp. 1031-1043

Mueller, A.; Twedt, D. & Loesch, C. (2000). Development of management objectives for breeding birds in the Mississippi Alluvial Valley, In: *Strategies for bird conservation: the partners in flight planning process. Proceedings of the third Partners in Flight workshop*, 1-5 October 1995, Cape May, New Jersey, R. Bonney, D.N. Pashley, R. Cooper, & L. Niles, (Eds.), 12-17, Proceedings RMRS-P-16, U. S. Department of Agriculture, Forest Service, Rocky Mountain Research Station, Ogden, Utah, USA

Munks, S.; Koch, A. & Wapstra, M. (2009). From guiding principles for the conservation of forest biodiversity to on-ground practice: Lessons from tree hollow management in Tasmania. *Forest Ecology and Management*, Vol.258, No.4, pp. 516-524

Nelson, J.; Cottam, D.; Holman, E.; Lancaster, D.; McCorquodale, S. & Person, D. (2008). *Habitat guidelines for black-tailed deer: Coastal Rainforest Ecoregion*. Mule Deer Working Group, Western Association of Fish and Wildlife Agencies. Available from http://www.muledeerworkinggroup.com/Docs/Coastal_Rainforest_BTD_Habitat _Guidelines.pdf

North American Wildlife Management Plan. (1986). *North American Wildlife Management Plan: A Strategy for Cooperation*. Available from http://www.fws.gov/birdhabitat/NAWMP/files/NAWMP.pdf

North, M. & Keeton, W. (2008). Emulating natural disturbance regimes: an emerging approach for sustainable forest management, In: *Patterns and processes in forest landscapes – multiple use and sustainable management*, R. Lafortezza, J. Chen, G. Sanesi, & T.R. Crow, (Eds.), 341-372, Springer, The Netherlands

Oliver, C. & Larson, B. (1996). *Forest Stand Dynamics*. John Wiley and Sons, New York, New York, USA

Palik, B.; Mitchell, R. & Hiers, J. (2002). Modeling silviculture after natural disturbance to sustain biodiversity in the longleaf pine (*Pinus palustris*) ecosystem: balancing complexity and implementation. *Forest Ecology and Management*, Vol.155, No.1-3, pp. 347-356

Panjabi, A.; Dunn, E.; Blancher, P.; Hunter, W.; Altman, B.; Bart, J.; Beardmore, C.; Berlanga, H; Butcher, G; Davis, S.; Demarest, D; Dettmers, R.; Easton, W.; de Silva Garza, H.; Iñigo-Elias, E.; Pashley, D.; Ralph, C.; Rich, T.; Rosenberg, K.; Rustay, C.; Ruth, J.; Wendt, J. & Will, T. (2005). *The Partners in Flight handbook on species assessment. Version 2005. Partners in Flight Technical Series Number 3.* Available from http://www.rmbo.org/pubs/downloads/Handbook2005.pdf

Pastur, G.; Lencinasa, M.; Cellini, J.; Peric, P. & Estebana, R. (2009). Timber management with variable retention in Nothofagus pumilio forests of Southern Patagonia. *Forest Ecology and Management*, Vol.258, No.4, pp. 436-443

Patriquin, K. & Barclay, R. (2003). Foraging by bats in cleared, thinned and unharvested boreal forest. *Journal of Applied Ecology*, Vol.40, No.4, pp. 646-657

Perry, R.; Thill, R. & Leslie, D. (2007). Selection of roosting habitat by forest bats in a diverse forested landscape. *Forest Ecology and Management*, Vol.238, No.1-3, pp. 156-166

Poudyal, N.; Hodges, D.; Fenderson, J. & Tarkington, W. (2010). Realizing the economic value of a forested landscape in a viewshed: an assessment of the Scottsboro-Bells Bend forest area near Nashville, Tennessee. *Southern Journal of Applied Forestry*, Vol.34, No.2, pp. 72-28

Rich, T.; Beardmore, C.; Berlanga, H.; Blancher, P.; Bradstreet, M.; Butcher, G.; Demarest, D.; Dunn, E.; Hunter, W.; Iñigo-Elias, E.; Kennedy, J.; Martell, A.; Panjabi, A.; Pashley, D.; Rosenberg, K.; Rustay, C.; Wendt, J. & Will, T. (2004). *Partners in Flight North American Landbird Conservation Plan*, Cornell Lab of Ornithology, Ithaca, New York, USA, Available from http://www.partnersinflight.org/cont_plan/

Riffell, S.; Verschuyl, J.; Miller, D. & Wigley, T. (2011). Biofuel harvests, coarse woody debris, and biodiversity - A meta-analysis. *Forest Ecology and Management*, Vol.261, No.4, pp. 878-887

Riggins, J.; Galligan, L. & Stephen, F. (2009). Rise and fall of red oak borer in the Ozark Mountains of Arkansas, USA. *The Florida Entomologist, Vol.92*, No.3 (September 2009), pp. 426-433

Ruel, J. (1995). Understanding windthrow: Silvicultural implications. *Forestry Chronicle*, Vol.71, No.4, pp. 434–444

Russell, K.; Wigley, T.; Baughman, W.; Hanlin, H. & Ford, W. (2004). Responses of Southeastern amphibians and reptiles to forest management: A review, In: *Southern forest science: past, present, and future*, H.M. Rauscher & K. Johnsen, (Eds.), 319-334, General Technical Report SRS-75, U.S. Department of Agriculture, Forest Service, Southern Research Station, Asheville, North Carolina, USA

Schieck, J.; Stuart-Smith, K. & Norton, M. (2000). Bird communities are affected by amount and dispersion of vegetation retained in mixedwood boreal forest harvest areas. *Forest Ecology and Management*, Vol.126, No.2, pp. 239-254

Seymour, R. & Hunter, M. (1999). Principles of ecological forestry, In: Maintaining Biodiversity in Forest Ecosystems, M.L. Hunter, (Ed.), 22-62, Cambridge University Press, Cambridge, Massachusetts, USA

Silva Ecosystem Consultants. (1992). *Landscape ecology literature review*. Available from http://www.silvafor.org/assets/silva/PDF/Literature/LandscapeEcologyOver.pdf

Steen, D.; Rall McGee, A; Hermann, S.; Stiles, J.; Stiles, S & Guyer, C. (2010). Effects of forest management on amphibians and reptiles: generalist species obscure trends among native forest associates. *Open Environmental Sciences*, Vol.4, pp. 24-30

Streby, H. & Miles, D. (2010). Assessing ecosystem restoration alternatives in eastern deciduous hardwood forests using avian nest survival. *Open Environmental Sciences*, Vol.4, pp. 39-48

St Laurent, M.; Ferron, J.; Haché, S. & Gagnon, R. (2008). Planning timber harvest of residual forest stands without compromising bird and small mammal communities in boreal landscapes. *Forest Ecology and Management*, Vol.254, No.2, pp. 261-275

Swanson, M.; Franklin, J.; Beschta, R.; Crisafulli, C.; DellaSala, D.; Hutto, R.; Lindenmayer, D. & Swanson, F. (2011). The forgotten stage of forest succession: early-successional ecosystems on forest sites. *Frontiers in Ecology and the Environment*, Vol.9, No.2, pp. 117-125

Tubby, I. & Armstrong, A. (2002). *The establishment and management of short rotation coppice – a practitioner's guide*. Forestry Commission Practice Note. Forestry Commission, Edinburgh. Available from http://www.forestry.gov.uk/pdf/fcpn7.pdf/$FILE/fcpn7.pdf

Twedt, D. & Best, C. (2004). Restoration of floodplain forests for the conservation of migratory landbirds. *Ecological Restoration*, Vol.22, No.3, pp. 194-203

Twedt, D. & Somershoe, S. (2009). Bird response to prescribed silvicultural treatments in bottomland hardwood forests. *Journal of Wildlife Management*, Vol.73, No.7, pp. 1140-1150

Twedt, D. & Somershoe, S. (In press). Regeneration in bottomland forest canopy gaps six years after variable retention harvests to enhance wildlife habitat, *Proceedings of the Fifteenth Biennial Southern Silvicultural Research Conference*, J. Guldin, (Ed.), Hot Springs, Arkansas, USA, November 17-20, 2008

Twedt, D.; Uihlein III, W. & Elliott, A. (2006). Habitat restoration for forest bird conservation: A spatially explicit decision support model. *Conservation Biology*, Vol.20, No.1, pp. 100-110

Twedt, D. & Wilson, R. (2007). Management of Bottomland Forests for Birds, In: *Proceeding of Louisiana Natural Resources Symposium*, T.F. Shupe, (Ed.), 49-64, Louisiana State University AgCenter, Baton Rouge, Louisiana, USA, August 13-14, 2007

Vandekerkhove,K.; De Keersmaeker, L.; Menke, N.; Meyer, P. & Verschelde, P. (2009). When nature takes over from man: Dead wood accumulation in previously managed oak and beech woodlands in north-western and central Europe. *Forest Ecology and Management*, Vol.258, No.4, pp. 425-435

Van Wilgenburg, S. & Hobson, K. (2008). Landscape-scale disturbance and boreal forest birds: Can large single-pass harvest approximate fires? *Forest Ecology and Management*, Vol.256, No.1-2, pp. 136-146

Verschuyl, J.; Riffell, S.; Miller, D. & Wigley, T. (2011) Biodiversity response to intensive biomass production from forest thinning in North American forests - A meta-analysis. *Forest Ecology and Management*, Vol.261, No.2, pp. 221-232

Whitman, A. & Hagan, J. (2003). *Legacy Retention: A tool for retaining biodiversity in managed forests*. Mosaic Science Notes #2003-1, Manomet Center for Conservation Sciences, Brunswick, Maine, USA

Wilson, R.; Ribbeck, K.; King, S. & Twedt, D., (Eds.). (2007). *Restoration, Management, and Monitoring of Forest Resources in the Mississippi Alluvial Valley: Recommendations for Enhancing Wildlife Habitat*. LMVJV Forest Resource Conservation Working Group. Lower Mississippi Valley Joint Venture, Vicksburg, Mississippi, USA

Zenner, E. (2000). Do residual trees increase structural complexity in Pacific Northwest coniferous forests? *Ecological Applications*, Vol.10, No.3, pp. 800–810

11

Fire Ignition Trends in Durango, México

Marín Pompa-García and Eusebio Montiel-Antuna
Universidad Juárez del Estado de Durango, Col. Valle del Sur, Durango
Faculty of Forest Sciences, UJED
México

1. Introduction

The spatial analysis of wildfire occurrence is a key factor in understanding forest fire incidences in forest ecosystems. Most of the applications are used in Wildfire Threat Rating Systems, and many of them have been completed or are in progress around the world (Lin, 2000; San-Miguel-Ayanz *et al.*, 2003). A major issue today is how to undertake the analysis more accurately and efficiently for the purposes of planning and to design suitable strategies and actions for forest fire management (Gollberg *et al.*, 2001).

In Mexico, most hazard systems use remote sensing techniques that measure weather and climate assessments, topography and fire interaction. (Flasse and Ceccato, 1996; Giglio *et al.*, 2003; Sepulveda *et al.*, 2001). In Durango State for instance, Drury and Veblen (2008) applied the spatial pattern of forest fire occurrence and showed that the rate of forest fires is correlated to years when extreme variations in weather occur. Renteria (2004) developed models for forest fire risk, which are now applied as support tools for decision-making in fuel management. Furthermore, Avila *et al.* (2010a) analysed the spatial patterns of forest fire occurrence, showing that the ignition locations did not exhibit a random distribution. In a subsequent spatial analysis using geographical weighted regression, Avila *et al.* (2010b) explicitly identified that human activities are the main factors determining the occurrence of forest fires, followed by vegetation and lack of precipitation.

All of these studies were concerned with the analysis of spatial patterns of forest fires, but the majority did not examine fire ignition history. Although electronic data is not available, a Geographic Information System (GIS) layer can be generated from local fire reports or associated records, including attribute data such as the number, date, size, cause, and other relevant information relating to the fires. Creating density points of fire locations helps to identify high fire frequency areas and to find any noticeable trends. It is known from experience that forest fires in Durango have tended to occur in patches, but no study has taken quantitative measurements. This information should give managers and researchers an adequate perspective as to where efforts should be focused.

Recently interest in several statistical techniques that focus on local measurements of spatial dependence have been growing. These techniques make it possible to have a discussion about tests for the detection of clusters without any preconceptions about their locations or spatial trends. The Getis-Ord G-statistic is used for the detection of clusters and is especially useful in cases where global traditional statistics, such as kernel estimation, k-function analysis, Moran's I index and the semi-variogram, did not display any global spatial pattern. However, in these cases there may still be significant points of clustering (Ord and Getis, 1995). Therefore, wildfire clustering metrics would be useful for providing knowledge to build more sophisticated modelling methods, including fire risk systems, optimizing fuel treatments, and prevention planning.

In this study, the objective was to assess wildfire ignition history in order to develop clustering maps that would be easy for managers to understand in setting up a strategy for fighting forest fires. It was hypothesized that the locations of ignition would have a clustered distribution. However, this study did not attempt to model ignition probability or historical fire regimes.

2. Methods

2.1 Study area

The state of Durango is located in the northwest of Mexico (22°16'–26°53'N, 102°29'–107°16'W) and covers a surface of 123,181 km^2 with a high diversity of ecosystems (Rodríguez *et al.*, 2010). It is divided into four ecozones, which are large ecological units containing landscapes of similar climate, topography, and vegetation (Figure 1). The coniferous and oak forest ecozone, situated on the plateau of the Sierra Madre Occidental, contains several species of pine and oak for commercial activities. The deciduous tropical forest is located in one of the numerous large canyons that cut through the irregular terrain and is mainly composed of tropical and subtropical species. The xeric shrub-lands have species with the ability to grow in dry and saline flat areas. Additionally, the grasslands are big areas with several species of grass, sometimes mixed with scrublands, in rolling terrains, which are mainly used for cattle grazing. Most of the ecosystems are owned communally, but a sprinkling of private owners exists.

2.2 Data sources

In order to measure the clustering of forest fires, the study area was further subdivided into 1,343 physiographical units (also known as terrestrial systems), digitized on a 1:250,000 scale (SEMARNAT and PGC, 2001). These spatial units provide information regarding the location, slope, land use and main vegetation (Pompa *et al.*, 2011).

A forest fire database, that tracked the dating information with the location of ignition within the area of study, was obtained from a daily report collected in the field by the National Commission for Forestry, for the period 2001 to 2010. This database layer was overlaid with the terrestrial systems data to provide a complete dataset for each system, including the physical properties of the system and the frequency of forest fires.

2.3 Statistical analysis

Considering that the study area was subdivided into 1,343 polygons whose Cartesian coordinates are known and where each feature has a respective number of fire ignitions, the G-statistic introduced by Getis and Ord (1992) can detect whether features with high values or features with low values are more likely to cluster.

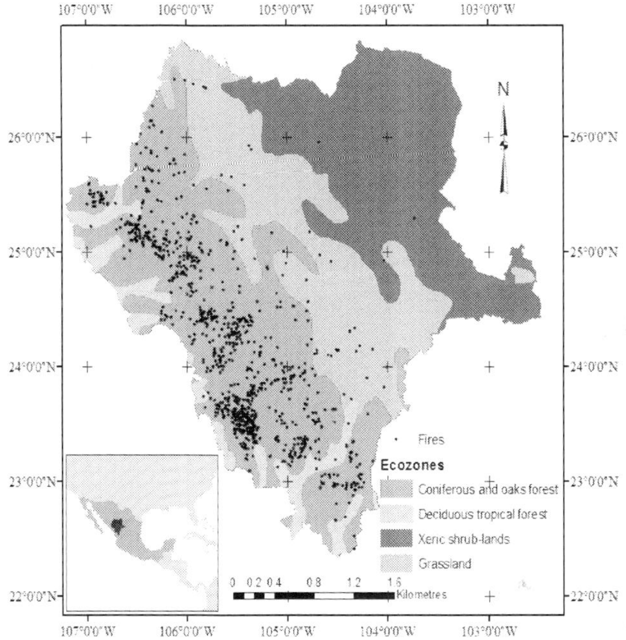

Fig. 1. Location of the Durango State in Mexico, indicating ecozones and fire ignitions from 2001 to 2010.

The G-statistic is defined as:

$$G = \frac{\sum\limits_{i=1}^{N} \sum\limits_{j=1}^{N} w_{ij}(d) x_i x_j}{\sum\limits_{i=1}^{N} \sum\limits_{j=1}^{N} x_i x_j} \quad i \neq j \tag{1}$$

Where $x_i x_j$ is the measured attribute for features i and j, respectively; and $w_{ij}(d)$ is a symmetric one/zero spatial weight matrix for detecting the vicinity between i and j at the distance given by d.

In order to indicate how the observed G-statistic is significantly different from the expected value (and hence significantly different from a random distribution), the following formula was applied:

$$Z - score = \frac{G(d) - G_E(d)}{Std\ Dev} \tag{2}$$

The test procedure assumes no global autocorrelation, but when global autocorrelation exists it has a significant impact on the expected value of G. This makes any proposed inferences on the presence of local clusters misleading (Haining, 2003). To remedy this problem, the study area was divided into small polygons.

The G-statistic was computed by physiographical units, using the hot spot analysis available in ESRI ArcMap™ version 9.3. G is calculated by looking at each feature within the context of neighbouring features. If a feature's value is high, and the values for all of its neighbouring features are also high, it is part of a hot spot. The local sum of a feature and its neighbours are compared proportionally to the sum of all features. When the local sum is much different than the expected local sum, and that difference is too large to be the result of random chance, a statistically significant Z score is the result (ESRI, 2008).

Due to the fixed spatial scale of the dataset and its reflection of good spatial autocorrelation, the level of clustering among fires was assessed using the fixed distance band option and calculated using the Euclidian distance method. Therefore, the Z-scores were reliable.

3. Results

In order to test whether or not the observations are spatially dependent, Figure 2 shows the clustering found for forest fires in terms of probabilities. The red grid cells indicate a cluster with high attribute values or clustering (hot spots) and are where fires are most likely to occur; while green tones imply low clustering (cold spots) and represent locations where the fires are less likely to occur. Finally, yellow tones denote areas where there is no concentration of either high or low values surrounding the target feature. This occurs when the surrounding values are all near the mean or when the target feature is surrounded by a mix of high and low values.

As a statistical test on the validity of the clustering, p-values test was carried out. Figure 3 shows the accuracy of the G-statistics. The best results are represented in red, and the worst are in green. A more conservative significance level was used (α=0.0001) in order to compensate for the effect of the large sample size (N=1,343), which may detect local pockets. Figure 2 reflects the clustering degree of occurrence in the ignition plots and is evidently varied among ecozones, which appeared to be supported by the spatial patterns.

The G-statistic estimates of spatial clustering allowed us to generate maps of annual clustering for forest fires from 2001 to 2010 (Figs. 4 - 13). For comparative purposes, the estimated values in all of these figures were subjected to the same scaling.

The fires exhibited most clustering in the ecozones of the coniferous and oak forests, as a result of fire progressively aggregating in this zone over the considered time period. This suggests that the increased incidences are following a rather stable pattern over time. These areas neighbour the deciduous tropical forest, which have rugged topography and are difficult to access. The ecozone that had less fire activity was the xeric shrub-lands, characterized by sandy areas and an arid climate. Although this study did not attempt to assess intra-ecozone variability and causes of such interaction, the results in Figure 2 lead us to believe that some features (physical, anthropogenic factors, etc) vary considerably among ecozones, and even within each, individual ecozone. This type of analysis requires more detailed data pertaining to the factors responsible for the ignition and spread of fires (Flannigan *et al.*, 2005).

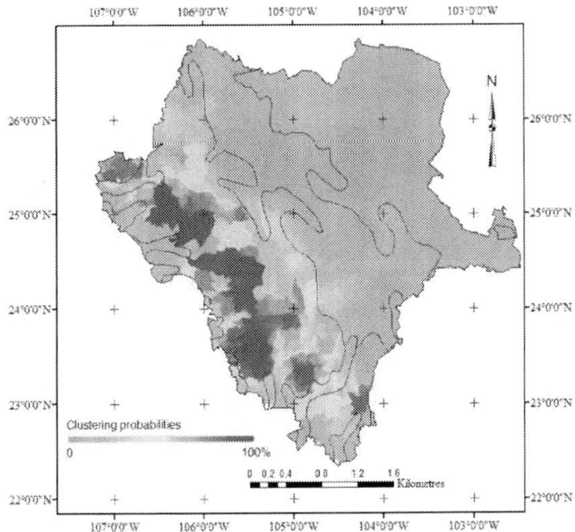

Fig. 2. Estimated spatial intensity of the occurrence of forest fires in Durango for the whole period 2001-2010.

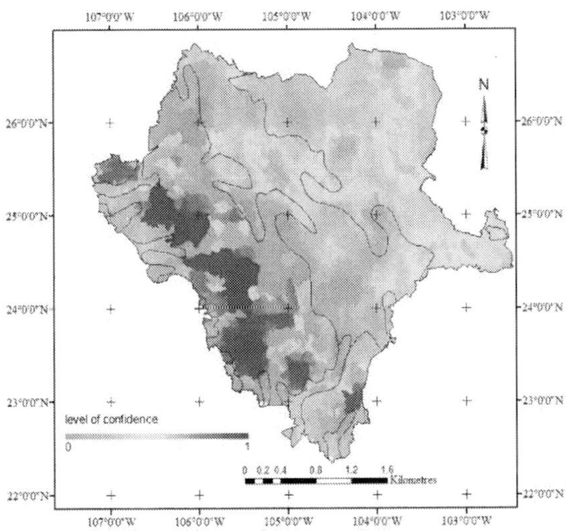

Fig. 3. Level of confidence of the clustering of forest fires in Durango for the whole period 2001-2010.

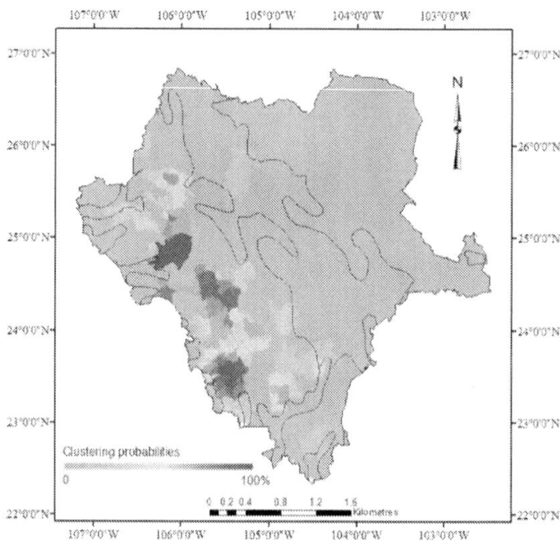

Fig. 4. Estimated spatial intensity of the occurrence of forest fires in Durango for the 2001 year.

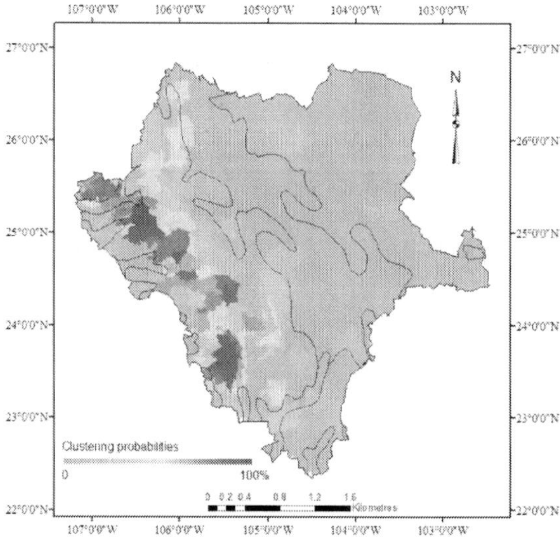

Fig. 5. Estimated spatial intensity of the occurrence of forest fires in Durango for the 2002 year.

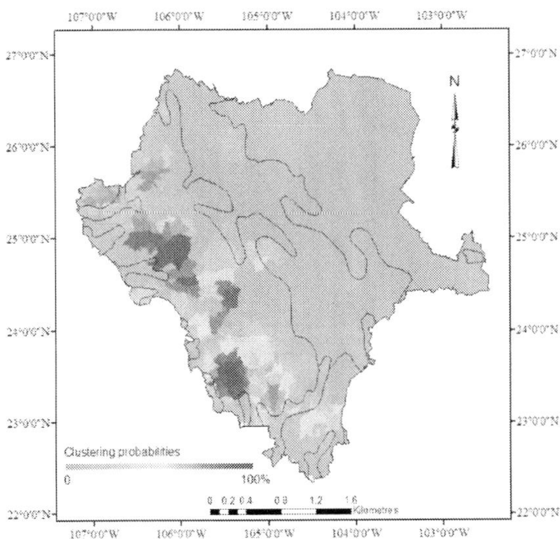

Fig. 6. Estimated spatial intensity of the occurrence of forest fires in Durango for the 2003 year.

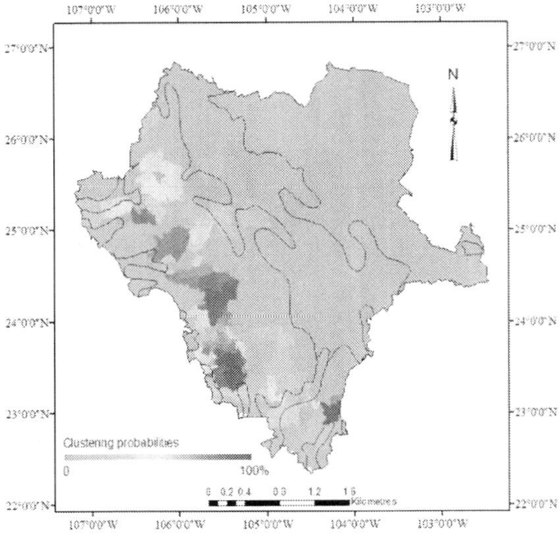

Fig. 7. Estimated spatial intensity of the occurrence of forest fires in Durango for the 2004 year.

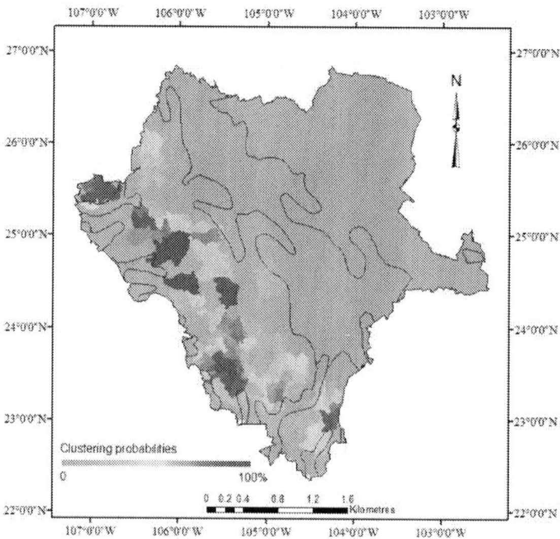

Fig. 8. Estimated spatial intensity of the occurrence of forest fires in Durango for the 2005 year.

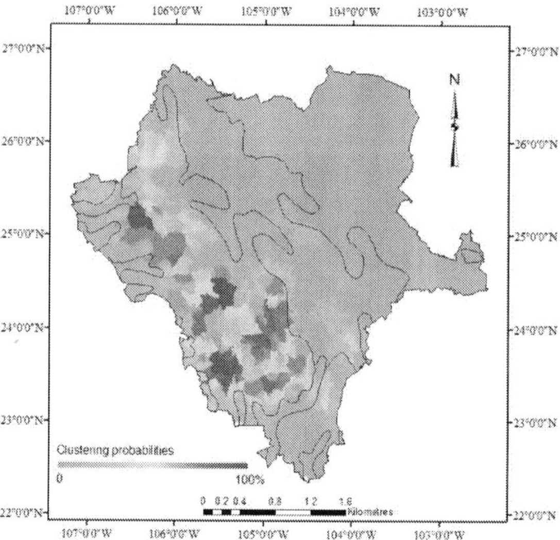

Fig. 9. Estimated spatial intensity of the occurrence of forest fires in Durango for the 2006 year.

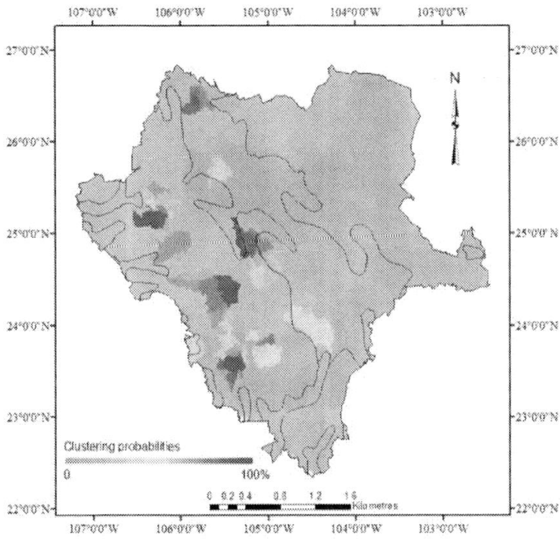

Fig. 10. Estimated spatial intensity of the occurrence of forest fires in Durango for the 2007 year.

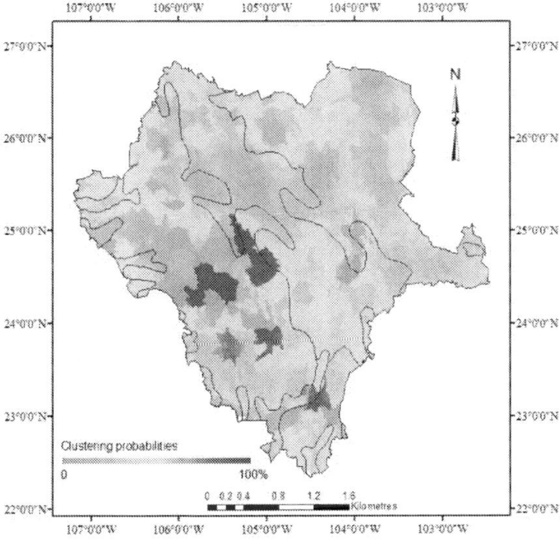

Fig. 11. Estimated spatial intensity of the occurrence of forest fires in Durango for the 2008 year.

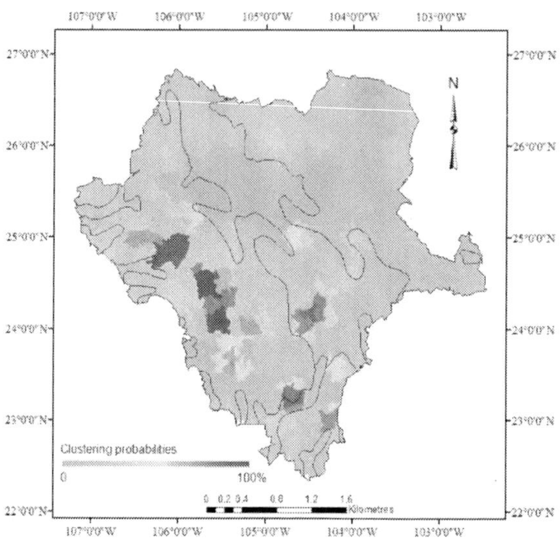

Fig. 12. Estimated spatial intensity of the occurrence of forest fires in Durango for the 2009 year.

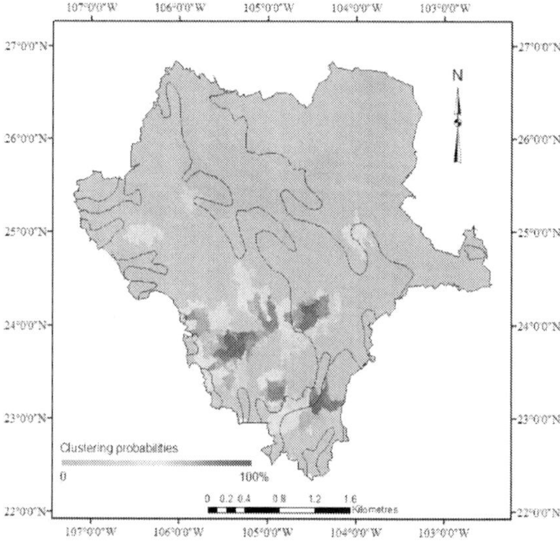

Fig. 13. Estimated spatial intensity of the occurrence of forest fires in Durango for the 2010 year.

4. Discussion

Many studies of forest fires have been conducted in Mexico over the past few years, but no effort has been made to quantitatively compare the spatial characteristics of clustering. This spatial analysis allows us to answer the hypothesis that had been posed earlier. Several clusters of fires had been formed in Durango between 2001 and 2010, as was confirmed by our results. The G-statistic was satisfactory as a quantitative tool for such estimations.

In spite of the difficulties in interpreting their causes, the clustering metrics that were used show the relative differences in clustering among terrestrial units. This approach is consistent with the results of Parisien *et al.* (2006) which showed that several clusters of large fires were formed in Canada from 1980 to 1999. They hold that the clustering of fires is a function of cumulative year intervals, which supports an analysis made a few years before by Vasquez and Moreno (2001). They showed that fires aggregate spatially and over time, producing larger interconnected burned patches. Wang and Anderson (2010) have demonstrated that the interaction is dynamic; it changes from year to year.

However, clustering depends on a dataset that has been complied over a long period of time, and therefore, it is likely that it would have shown different patterns in clustering. Many studies have reported from simulations that the clustering varies with vegetation cover (Keane *et al.*, 2002; Miller, *et al.,* 2008). According to Cumming (2001), and Duncan and Schmalzer (2004), the extent and configuration of flammable vegetation and non-flammable landscape features clearly influence patterns of fire ignition. They also speculate that these patterns are likely to respond differently to changes in spatial scale. Indeed, Parisien *et al.* (2006) show that splitting the clusters, and therefore reducing study units, will avoid some bias in the estimates and surely dilute the effect of clustering. In this approach, the choice of terrestrial systems as study units was not entirely arbitrary, as it was based on units for which fire activity is known to vary throughout Durango.

Flannigan *et al.* (2005) showed that smaller study units nested in ecozones performed more poorly than the terrestrial units used here. To this end, these units provided a useful and objective means to spatially stratify the ecozones in order to study fire. Wang and Anderson (2010) highlighted the importance and challenges of analysing different spatial data sets, assessing and identifying the spatial scales that are the most relevant for the study of spatial fire patterns. Podur *et al.* (2003) explicitly identified the spatial scales at which fires exhibit clustered, random, or inhibited distribution.

In addition, our results reaffirm the importance of recording historical ignitions. They are thus consistent with those documented by Van Wagner (1988), Rollins *et al.* (2002) and Podur *et al.* (2003) who noted that historical fire records have been useful in understanding the spatial distribution of fires and in helping managers to conceptualize where ignitions could occur, due the assessment of fire clustering can provide useful new information in predicting potential fire dangers across the ecosystems. They will also enable managers to understand the level of the fire threat. However, creating an ignition density from fire locations only indicates where fires have started, not where or under which conditions the fires will spread or the impact which will occur. Thus, fire occurrence data alone are of limited value to risk assessment (Stratton, 2006).

Historical recording was attractive in terms of expediency, simplicity, low costs, and transparency. Although it is not possible to ensure that this database contains all fire ignitions, it does represent the vast majority of them. Therefore, the lack of more available temporal explicit data has made it difficult to undertake a detailed study of spatial fire patterns over a considerable period of time (e.g. decades).

As noted earlier, Avila *et al.* (2010a) have noted the clustering of forest fires; however, they used the Moran I index. The G-statistic focuses on the clustering around each terrestrial system, since it does not take into account the rate of forest fires in the area of study itself. This approach helps to monitor local behaviour due to G being more sensitive to high clusters than low ones. On the other hand, Moran's index is mainly affected by the scale of the clusters (Zhang, 2007). The question of the interaction between local and global coefficients is an important one and much more remains to be done in this context (Ord and Getis, 1995).

5. Conclusion

In this study, the history of fire ignitions over an ecologically heterogeneous landscape during the period of a decade was compared. The fires of all ecozones from 2001 to 2010 exhibited a clustered spatial distribution remarkable in oak forest ecosystems. However, it was beyond the scope of the present study to thoroughly analyse the underlying mechanisms responsible for the spatially clustered fires. Regardless of the factors that may have contributed to any increase in wildfire frequency, the assessment of fire clustering can provide useful new information to researchers in predicting potential fire dangers and behaviours across the landscape. Also, it is important to take into consideration the importance of the updated geographic data from preceding data fields. Results such as these will hopefully encourage more detailed analyses that investigate the relative roles of weather, fuels, landscape features, and others factors on the progression of fire ignition. The same methods could be applied to smaller and ecologically homogeneous landscape units, such as stand management units, over a finer timescale (e.g. months or even days). Finally, it should also be possible to use these methods to study clustering fires at other levels where ignition data are available.

6. Acknowledgement

This work was conducted within the framework of Academic Mobility supported by the UJED (University Juarez of Durango State). We are grateful to CONAFOR (Mexican National Forestry Commission) in Durango for sharing their data and thanks to COCYTED (Science & Technology Council in Durango State) and SRNYMA (Ministry of Environment and Natural Resources in Durango State). Also, we would like to extend our thanks to David Mair and Peter Englefield for providing constructive comments on the manuscript.

7. References

Avila F D, Pompa G M, and Vargas P E (2010a). Spatial analysis of fire occurrence in the Durango State. *Revista Chapingo. Serie Ciencias Forestales* 16(2): 253–260, ISSN 0186-3231,State of Mexico, Mexico.

Avila F D, Pompa G M, Antonio N X, Rodríguez T D, Vargas P E and Santillán P J (2010b). Driving factors for forest fire occurrence in Durango State of Mexico: A geospatial perspective. *Chinese Geographical Science* 20:6, 491, ISSN: 1002-0063 (print version) ISSN: 1993-064X (electronic version), Mainland, China.

Cumming S G (2001). Forest type and fire in the Alberta Boreal Mixedwood: What do fires burn? *Ecological Application* 11, 97–110. doi:10.2307/3061059, ISSN 1051-0761.

Drury S A and Veblen T T (2008). Spatial and temporal variability in fire cccurrence within the Las Bayas Forestry Reserve, Durango, Mexico. *Plant Ecology*, 197(2): 299–316. DOI: 10.10- 07/s11258-007-9379-5, ISSN: 1385-0237 (print version), ISSN: 1573-5052 (electronic version).

Duncan B W and Schmalzer P A (2004). Anthropogenic influences on potential fire spread in a pyrogenic ecosystem of Florida, USA. *Landscape Ecology* 19, 153–165. doi:10.1023/B:LAND.0000021714. 97148.AC, ISSN: 0921-2973 (print version), ISSN: 1572-9761 (electronic version).

ESRI (2008). ArcMap 9.3. *Environmental Systems Research Institute*: Redlands, CA.

Flannigan M D, Logan K A, Amiro B D, Skinner W R and Stocks B J (2005). Future area burned in Canada. *Climatic Change* 72, 1–16. doi:10.1007/S10584-005-5935-Y . ISSN: 0165-0009 (print version), ISSN: 1573-1480 (electronic version).

Flasse S P and Ceccato P (1996). A contextual algorithm for AVHRR fire detection, *Int. J. Rem. Sens.*, 17:419-424, ISSN: 0143-1161(print version), ISSN: 1366-5901 (electronic version).

Getis A and J K Ord. (1992). The analysis of spatial association by use of distance sStatistics. *Geographical Analysis* 24 (July), 189–206, ISI-JCR 14/61 (2009), ISSN: 1538-4632 (electronic version), Ohio, USA.

Giglio L, JDescloitres J, Justice C O and Kaufman Y J (2003). An enhanced contextual fire detection algorithm for MODIS, *Remote Sens. Environ*, 87, 273–282, ISSN: 0034-4257.

Gollberg G E, Neuenschwander L F and Ryan K C.. (2001). Introduction: Integrating spatial technologies and ecological principles for a new age in fire management. *International Journal of Wildland Fire* 10:263–265, ISSN: 1049-8001, ISSN: 1448-5516, Australia.

Haining R P (2003). *Spatial Data Analysis: Theory and Practice*. Cambridge, UK: Cambridge University Press. 432 p. ISBN 0521-77437-3

Keane R E, Parsons R A and Hessburg P F (2002). Estimating historical range and variation of landscape patch dynamics: Limitations of the simulation approach. *Ecological Modelling* 151, 29–49. doi:10.1016/S0304-3800(01)00470-7, ISSN: 0304-3800, USA.

Lin C-C (2000). The development, systems, and evaluation of forest fire danger rating: a review. *Taiwan Journal of Forest Science* 15, 507–520, ISSN: 1026-4469. Taiwan.

Miller C, Parisien M-A, Ager A A and Finney M A (2008). Evaluating spatially explicit burn probabilities for strategic fire management planning. *Transactions on Ecology and the Environment* 19: 245–252, ISSN: 1743-3541, UK.

Ord J K and Getis A (1995). Local spatial autocorrelation statistics: Distributional issues and an application. *Geographical Analysis* 27: 286–306, ISI-JCR 14/61 (2009), ISSN: 1538-4632 (electronic version), Ohio, USA.

Podur J, Martell D L and Csillag F (2003). Spatial patterns of lightning-caused forest fires in Ontario, 1976–1998. *Ecological Modelling* 164, 1–20. doi:10.1016/S0304-3800(02)00386-1, ISSN: 0304-3800, USA.

Pompa GM, Antonio NX, Carrasco MJ, Mendoza BM (2011). Spatial patterns of soil degradation in Mexico. African Journal of Agricultural Research 6(5), 1109-1113, ISSN 1991-637X, Nairobi.

Parisien M-A, Peters VS, Wang Y, Little J M, Bosch E M and Stocks B J (2006). Spatial patterns of forest fires in Canada, 1980–1999. *International Journal of Wildland Fire* 15(3): 361–374, ISSN: 1049-8001, ISSN: 1448-5516, Australia.

Renteria A J B (2004). Development of models for fuel control in the management of forest ecosystems in Durango. Mexico: University of Nuevo Leon, 125.

Rodríguez FJ, Pompa GM, Hernández DC, Juárez RA. 2010. Patrón de distribución espacial de la pérdida, degradación y recuperación vegetal en Durango, México. Avances en Investigación Agropecuaria 14 (1), 53-65. ISSN: 0188789-0, México.

Rollins M G, Morgan P and Swetnam T (2002). Landscape-scale controls over 20th century fire occurrence in two large Rocky Mountain (USA) wilderness areas. *Landscape Ecology* 17, 539–557. doi:10.1023/A:1021584519109, ISSN: 0921-2973 (print version), ISSN: 1572-9761 (electronic version).

San-Miguel-yanz J, Carlson J D, Alexander M, Tolhurst K, Morgan G, Sneeuwjagt Rand Dudfield M (2003). Current methods to assess fire danger potential. In 'Wildland fire danger estimation and mapping – the role of remote sensing data'. Series in *Remote Sensing* Volume 4. (Ed. E Chuvieco) pp. 21–61. (World Scientific Publishing: Singapore)

SEMARNAT (Ministry of Environment and Natural Resources) and PGC (Post Graduate College), 2001. Evaluation of soil degradation caused by man in Mexico (1:250,000 scale). National Memory.

Sepúlveda B J R, Meza S W, Zúñiga C G, Solís G y M and Olguín E (2001). SIG para determinar riesgo de incendios forestales en el noroeste de México. *Publ. Téc. Núm. 1.* Instituto Nacional de Investigaciones Forestales, Agrícolas y Pecuarias. Campo Experimental Costa de Ensenada. Ensenada. México. 37 p.

Stratton R D (2006). Guidance on spatial wildland fire analysis: models, tools, and techniques. *Gen. Tech. Rep. RMRS-GTR-183.* Fort Collins, CO: U.S. Department of Agriculture, Forest Service, Rocky Mountain Research Station. 15 p

Van Wagner C E (1988) The historical pattern of annual burned area in Canada. *Forestry Chronicle* 64, 182–185, ISSN: 0015-7546 (print version), ISSN: 1499-9315 (electronic version), Mattawa, Canada.

Vasquez A and Moreno J M (2001) Spatial distribution of forest fires in Sierra de Gredos (Central Spain). *Forest Ecology and Management* 147, 55–65. doi:10.1016/S0378-1127(00)00436-9, ISSN: 0378-1127, Australia.

Wang Yonghe and Anderson Kerry R (2010). An evaluation of spatial and temporal patterns of lightning- and human-caused forest fires in Alberta, Canada, 1980–2007. *International Journal of Wildland Fire* 19(8). p. 1059–1072, ISSN: 1049-8001, ISSN: 1448-5516, Australia.

Zhang Song-lin (2007). Comparison between general Moran's Index and Getis-Ord general G of spatial autocorrelation. *Acta Scientiarum Naturalium Universitatis Sunyatseni;* 2007-04, ISSN: 0529-6579, Guangdong, China.

12

Remote Sensing of Forestry Studies

Kai Wang[1], Wei-Ning Xiang[1], Xulin Guo[2] and Jianjun Liu[3]
[1]Shanghai Key Laboratory of Urbanization and Ecological Restoration (KLUER)
East China Normal University
[2]Department of Geography and Planning
University of Saskatchewan
[3]College of Forestry, Northwest A&F University
[1,3]China
[2]Canada

1. Introduction

Forestry can be simply explained as the science and technology linked with tree resources or forests. According to the Global Forest Resources Assessment 2010 (FRA 2010), "the world's total forest area was just over 4 billion hectares, corresponding to 31 percent of the total land area or average of 0.6 ha per capita". Forest as one of the most important resources on this planet plays a pivotal role in the progress of human civilizations. Forestry studies have always been the hot topic since the naissance of this discipline. With the advent of satellite remote sensing, the forestry studies have made the unprecedented development. This introductory review is intended to introduce the basics of remote sensing in forestry studies, summarize the recent development, and elucidate several typical applications in this area.

1.1 Remote sensing

Aerial photos (i.e. airborne remote sensing) or satellite imagery (i.e. spaceborne remote sensing) are widely used in forest studies. If the earliest platforms such as homing pigeons, kites, and hot air balloons, which were quite uncertain and unstable platforms with relatively low altitude, are taken into account, the history will be even longer (Colwell, 1964). Up to now, hundreds of Earth Observation Satellites are in orbit, and delivering assorted remotely sensed data ranging from optical data to radar data, from multispectral imagery to panchromatic imagery, and from local scale to global scale. Remote sensing has long been identified as an effective and efficient tool in forestry studies, such as forest inventory, forest health and nutrition, forest sustainability, forest growth, and forest ecology (Kohl et al., 2006).

Remote sensing is the "noncontact recording of information from the ultraviolet, visible, infrared, and microwave regions of the electromagnetic spectrum by means of instruments such as cameras, scanners, lasers, linear arrays, and/or area arrays located on platform such as aircraft or spacecraft, and the analysis of acquired information by means of visual and digital image processing" (Jensen, 2007). Franklin (2010) emphasized that "remote sensing is both technology (sensors, platforms, transmission and storage devices, and so on) and methodology (radiometry, geometry, image analysis, data fusion, and so on)". There are numerous outstanding textbooks providing comprehensive introduction to the remote sensing science (e.g. Lillesand et al., 2007; Campbell, 2007; Jensen, 2007) and the digital image processing (e.g. Jensen, 2004; Richards and Jia, 2006; Lee and Pottiter, 2009).

1.2 Remote sensing process

Scientists generally summarized the remote sensing process into four phases: statement of the problem, data collection, data-to-information conversion, and information presentation (Jensen, 2007). Several notes derived from Jensen (2007) are provided as follows:

- A successful study or program must get started from properly stating the research question and forming the research hypothesis.
- Then in situ observation and/or remote sensing may be used for data collection. By far the most widely used remote sensing data in forest studies are optical satellite imagery, which utilizes the spectral information in visible and infrared portions (mainly ranging from 400 nm to 2500 nm) of the electromagnetic spectrum (Figure 1). The three types of optical remote sensing data are multispectral imagery (e.g. Landsat TM [Thematic Mapper] image), panchromatic imagery (e.g. SPOT [Satellite Pour l'Observation de la Terre] 5 panchromatic image), and hyperspectral imagery (e.g. Hyperion image). In addition, Radar (Radio Detection and Ranging) and Lidar (Light Detection and Ranging) are becoming research hotspots in forestry studies.
- Data-to-information conversion refers to remote sensing data analysis employing a range of image processing techniques, such as algorithms for image preprocessing, classification, feature extraction and change detection, and a variety of modeling methods.
- Information extracted from remote sensing data should be represented in a proper way in order to communicate effectively. In general, with the assistance of cartography, GIS (Geographic Information System), spatial statistics and the knowledge of research fields, the extracted information could be made into image map, thematic map, spatial database file, or graph. Poor information presentation results from the ignorance or violation of fundamental rules (e.g. cartographic theory or database topology design).
- The remote sensing process inevitably introduces errors to the generated information, especially in the phase of data-to-information conversion and information presentation, and these errors should be properly identified and reported (e.g. classification accuracy).

2. Forestry information needs and remote sensing

In practice, researchers choose one or several types of remotely sensed data according to their information needs. The information needs are converted to specific properties of remotely sensed data, such as spatial resolution, spectral resolution, temporal resolution, etc.

Tables 1 and 2 list commonly used sensors on Earth observation satellite that are still operational. These sensors provide diverse remotely sensed data with a unique configuration of image resolutions, such as spatial resolution, spectral resolution and temporal resolution (see Figure 2). Jensen (2007) defined spatial resolution as "a measure of the smallest angular or linear separation between two objects that can be resolved by the remote sensing system". In other words, the spatial resolution stands for how detailed information the remotely sensed data could provide. Temporal resolution can be defined as "how often the sensor records imagery of a particular area" (Jensen, 2007). For example, the well-known Landsat TM has 16-day temporal resolution. Another key property of a remote sensing system is spectral resolution, which is defined as "the number and dimension (size) of specific wavelength intervals (referred to as bands or channels) in the electromagnetic spectrum to which a remote sensing instrument is sensitive". Likewise, take the Landsat TM as an example. The Landsat TM imagery has 7 bands (6 optical bands plus 1 thermal band). It is noteworthy that the spectral resolution is mainly applied to describe optical imagery, and it cannot be used for Radar or Lidar remotely sensed data. In fact, the aforementioned data collection can be described as selecting the unique configuration of image resolutions (or properties), which can be used to meet certain research needs.

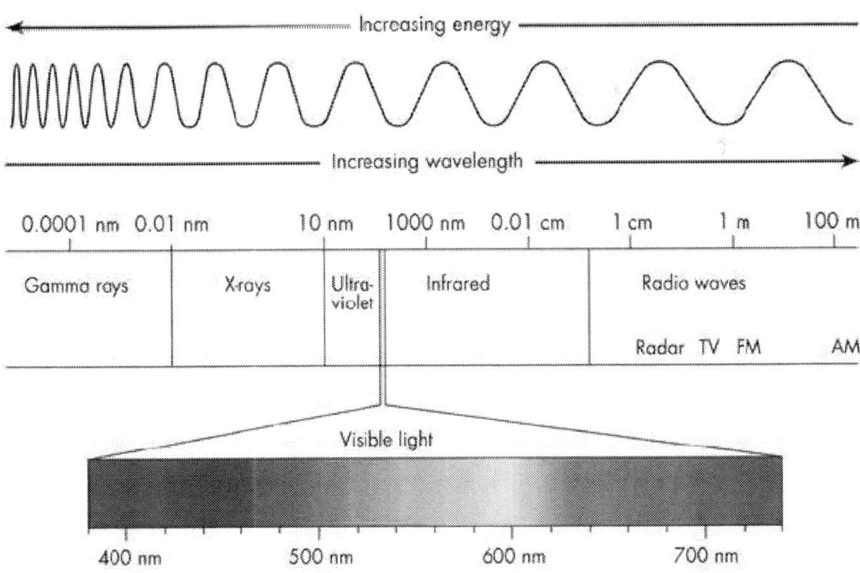

Fig. 1. Electromagnetic spectrum (Courtesy of the Antonine Education, UK) (http://www.antonine-education.co.uk/physics_gcse/Unit_1/Topic_5/topic_5_what_are_the_uses_and_ha.htm).

Wulder et al. (2009) endeavored to clearly demonstrate the relationship of information needs and the selection of appropriate data and processing methods in remote sensing for studies of vegetation condition. The issues need to be taken into account, including "the scale at which the target must be measured (e.g. landscape-level or tree level information); the attributes of interest (change, condition, spatial extent); cost; timeliness; and, repeatability" (Wulder et al., 2009).

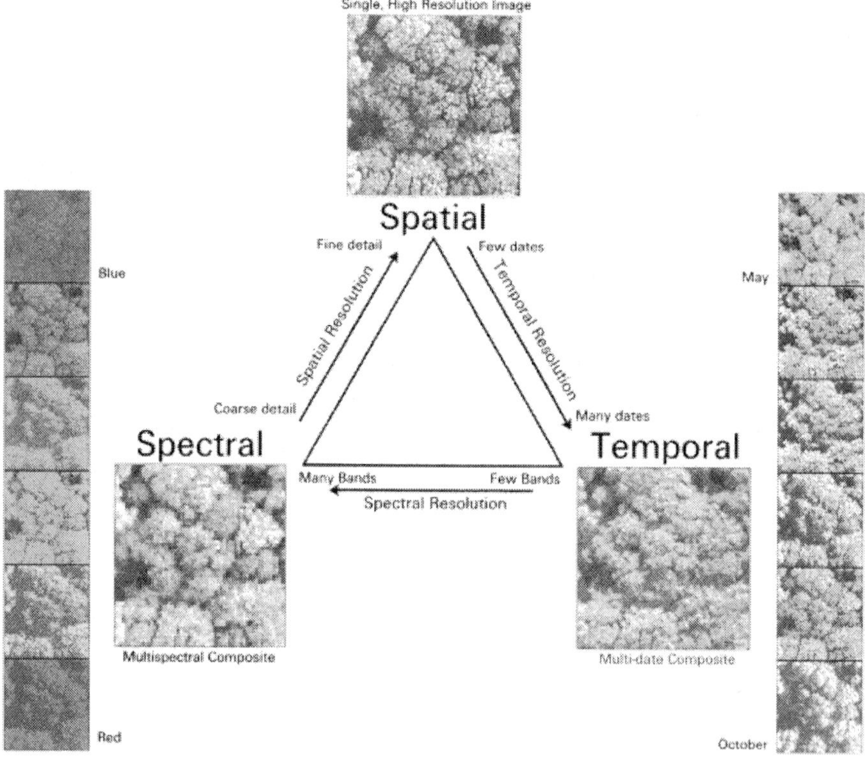

Fig. 2. Given a limited bandwidth, trade-offs have to be made between spectral, temporal, and spatial properties of the imagery acquired. For users who require high spatial resolution data, it is possible that multitemporal data can substitute for limited multispectral properties. (*With permission from* Key, T., T.A. Warner, J.B McGraw, and M.A. Fajvan. 2001. *Remote Sensing of Environment 75: 100-112*).

Satellite Program	Satellite Platform	Sensor	Data Operator
Optical Remote Sensing			
POES (Polar Orbiting Environmental Satellites)	NOAA 18	AVHRR	NOAA
EOS (Earth Observing System)	TERRA/AQUA	MODIS	NASA/USGS
Landsat	LANDSAT 5	TM	NASA/USGS
SPOT (Satellite Pour l'Observation de la Terre)	SPOT 4	HRVIR VEGETATION	Spot Image
	SPOT 5	HRG VEGETATION	Spot Image
IRS (Indian Remote Sensing Satellites)	IRS P6 (ResourceSat-1)	LISS III LISS IV AWiFS	ISRO (India Space Research Organization)
DMC (Disaster Monitoring Constellation)	Beijing-1	SLIM-6	DMC International Imaging Ltd
CBERS (China-Brazil Earth Resources Satellite)	CBERS-2B	CCD HRC IRMSS WFI	CAST (China)/INPE (Brazil)
Digital Globe Constellation	WorldView 2	WV110	DigitalGlobe Corporate
	QuickBird 2	BGIS 2000	DigitalGlobe Corporate
GeoEye	GeoEye-1	GIS MS	GeoEye Inc.
Radar Remote Sensing			
RADARSAT Constellation	RADARSAT 2	SAR	CSA/MDA
TanDEM-X	TanDEM-X	TSX-SAR	DLR/Astrium
TerraSAR	TerraSAR-X	TSX-SAR	DLR/Astrium

Table 1. The Current Commonly-used Optical and Radar Sensors

Medium resolution sensors	Spatial resolution (m)[a]	Swath (km)	Spectral range (nm) / Bands	Temporal coverage	Revisit (day)
Coarse spatial resolution optical sensors					
NOAA-18 (AVHRR)	1100	2900	Variable / 5	2005-Present	1
Terra/Aqua (MODIS)	250-1000	2330	Variable / 36	1999-Present	1-16
Moderate spatial resolution optical sensors					
Landsat (TM)	30	180	450-2350 / 6	1984-Present	16
IRS-P6 (LISS III)	23.5	141	520-1700 / 4	2003-Present	24
SPOT 4 (HRVIR)	20	60	500-1750 / 4	1998-Present	1-3
SPOT 5 (HRG)	10 (MS); 20 (SWIR)	60	500-1730 / 5	2002-Present	1-3
CBERS-2	20 (Pan and MS)	120	450-890 / 4	2003-Present	3
DMC (SLIM-6)	22/32	600	520-990 / 3	2002-Present	1
Fine spatial resolution optical sensors					
WorldView 2 (WV110)	0.46 (Pan); 1.85 (MS)	16.4	400-1040 / 8	2009-Present	1.1-3.7
QuickBird 2 (BGIS 2000)	0.65 (Pan); 2.62 (MS)	18.0	430-918 / 4	2001-Present	2.5-5.6
GeoEye-1 (GIS MS)	0.41 (Pan); 1.65 (MS)	15.2	450-920 / 4	2008-Present	2.1-8.3

[a] MS = multispectral, SWIR = shortwave infrared, Pan = panchromatic

Table 2. The Detail of the Current Commonly-used Optical Sensors

3. Underdeveloped remote sensing technologies in forestry studies

Optical sensors have been commonly used in forestry studies. However, the use of hyperspectral sensors, Radar and Lidar is still relatively underdeveloped. It is worth paying more attention to the application of hyperspectral sensors, Radar and Lidar in forestry studies.

3.1 Hyperspectral sensors

Optical sensors mentioned above, which are divided from the dimension of spatial resolution, are categorized into multispectral sensors. By contrast, there is a group of sensors called hyperspectral sensors, which accordingly generate hyperspectral data. Wang et al. (2010) stated that "hyperspectral data have the ability to collect ample spectral information across a continuous spectrum generally with 100 or more contiguous spectral bands". Shippert (2004) listed the existing hyperspectral sensors acquiring imagery from space, including the Hyperion sensor on NASA's EO-1 (National Aeronautics and Space Administration's Earth Observing-1), the CHRIS (Compact High Resolution Imaging Spectrometer) sensor on the European Space Agency's PROBA (PRoject for On-Board Autonomy) satellite, and the FTHSI (Fourier Transform Hyperspectral Imager) sensor on the U.S. Air Force Research Lab's MightySat II satellite.

3.2 Radar and lidar

Besides optical sensors, Radar and Lidar play more and more important roles in remote sensing of forest studies. Radar, the acronym of radio detection and ranging, is based on the transmission of long-wavelength microwaves (e.g., 3–25 cm) through the atmosphere and then recording the amount of energy backscattered from the terrain (Jensen, 2007). Wang et al. (2009) briefly introduced the Phased Array type L-band Synthetic Aperture Radar (PALSAR) on board Advanced Land Observing Satellite (ALOS), and RADARSAT-2 operated by the Canadian Space Agency (CSA) and MacDonald Dettwiler and Associates Ltd (MDA). Both could provide fully polarized SAR data to support PolSAR (Polarimetric SAR) technology (i.e., PolSAR decomposition), which has achieved promising results in many environmental researches (e.g., Lee et al., 2001; McNairn et al., 2009; Shimoni et al., 2009). Light detection and ranging (Lidar), also called Laser altimetry, is an active remote sensing technology that utilizes a laser to illuminate a target object and a photodiode to register the backscatter radiation (Lim et al., 2003; Hyyppa et al., 2009). It has been widely accepted that Lidar is capable of accurate (or even precise) vertical information (Wang et al., 2010). Therefore, it is believed that Lidar will bring forestry studies into an unprecedented age.

4. Case studies

Young and Giese (2003) summarized forest science and management into three categories: A. forest biology and ecology (e.g. forest biomes of the world, forest ecophysiology, forest soils, forest ecosystem ecology, landscape ecology, and forest trees: disease and insect interactions); B. forest management and multiple uses (e.g. forest management and stewardship, nonindustrial private forests, measuring and monitoring forest resources, silviculture and ecosystem management, forest-wildlife management, forest and rangeland management, forest and watershed management, forest and recreation behavior, behavior and management of forest fires, timber harvesting, wood products, and economics and the management of forests for wood and amenity values); and C. forests and society (e.g. urban forest, and social forstry: the community-based management of natural resources). As a matter of fact, remote sensing has more or less served all the three categories. Several examples in remote sensing of forestry studies are provided as follows. The selected examples were included in the papers that were either highly cited or newly published Science Citation Index (SCI) papers.

4.1 Species composition (biodiversity)

Turner et al. (2003) stated that the recent advances in remote sensing, such as the availability of remotely sensed data with high spatial and spectral resolutions, make it possible to detect key environmental parameters, which can be applied to determine the distribution and abundance of species across landscapes via ecological models. This approach, in general referred to as indirect remote sensing of biodiversity, plays a major role in this research area. For example, Defries et al. (2000) applied the 1km Advanced Very High Resolution Radiometer (AVHRR) to estimate and map percentage tree cover and associated proportions of trees with different leaf longevity (evergreen and deciduous) and leaf type (broadleaf and needleleaf).

4.2 Forest ecophysiology

Kokaly and Clark (1999) developed an approach to estimate the concentrations of nitrogen, lignin, and cellulose in dried and ground leaves using ban-depth analysis of absorption features (centered at 1.73 µm, 2.10 µm, and 2.30 µm) and stepwise multiple linear regression. As mentioned above, hyperspectral remote sensing was used to estimate the leaf pigment of sugar maple (*Acer saccharum*) in the Algoma Region, Canada, and promising results were obtained (Zarco-Tejada et al., 2001).

4.3 Forest ecosystem ecology

Jin et al. (2011) developed an algorithm based on a semi-empirical Priestley-Taylor approach to estimate continental-scale evapotranspiration (ET) using MODIS satellite observations. The seasonal variation in ET has been indicated as a key factor to the soil moisture and net ecosystem CO_2 exchange through water loss from an ecosystem. Lefsky et al. (2002) reviewed Lidar remote sensing for ecosystem studies. Lidar is capable of accurately measuring vertical information besides the horizontal dimension, such as the three-dimensional distribution of plant canopies and subcanopy topography (Lefsky et al., 2002). More specifically, lidar can provide accurate estimates vegetation height, cover, canopy structure, leaf area index (LAI), aboveground biomass, etc (Lefsky et al., 2002).

4.4 Forest trees: Disease and insect interactions

Due to the difference between dead, diseased, and healthy trees in visible and near-infrared reflectance values, Everitt et al. (1999) used color-infrared digital imagery and successfully detected oak wilt disease in live oak (*Quercus fusiformis*). The outbreak of mountain pine beetle (MPB) has resulted in a huge monetary loss in the western of the North America. It is urgent to efficiently survey the location and the extention of beetle impacts. Wulder et al. (2006) indicated the potential and limits of the detection and mapping of MPB using remotely sensed data, and suggested methods and data sources accordingly.

4.5 Measuring and monitoring forest resources

Cohen et al. (1995) stated that "remote sensing can play a major part in locating mature and old-growth forests", and applied a number of remote sensing techniques to estimate forest age and structure. Over a 1,237,482 ha area was investigated and an accuracy of 82 per cent was obtained. Maps of species richness have been recognized as a useful tool for biodiversity conservation and management due to its capability of explicitly describing information on the spatial distribution and composition of biological communities (Hernandez-Stefanoni et al., 2011). Hernandez-Stefanoni et al. (2011) tested remotely sensed data with regression kriging estimates for improving the accuracy of tree species richness maps, and concluded that this research will make a great step forward in conservation and management of highly diverse tropical forests.

4.6 Forest-wildlife management

Stoms and Estes (1993) proposed a framework to guide the application of remotely-sensed data in mapping and monitoring biodiversity. From then on, there are lots of works focusing on this field, e.g. Tuomisto et al. (1995), Nagendra (2001), Kerr and Ostrovsky (2003), Wang et al. (2009), Wang et al. (2010). From the perspective of remote sensing techniques, Franklin et al. (2001) developed an integrated decision tree approach to mapping land cover using remotely sensed data in support of grizzly bear habitat analysis.

4.7 Forest fires

Giglio et al. (2003) presented an enhanced contextual fire detection algorithm in order to identify smaller, cooler fires with a significantly lower false alarm rate, and promising results were obtained. Lentile et al. (2006) reviewed "current and potential remote sensing methods used to assess fire behavior and effects and ecological responses to fire".

4.8 Urban forest

Jensen et al. (2003) investigated the relationship between urban forest leaf area index (LAI) and household energy usage in a mid-size city, and concluded that the increase of LAI resulted in the less energy usage. Zhang et al. (2007) applied remote sensing to map the distribution, classification and ecological significance of urban forest in Jinan city.

5. Annotated bibliography of selected reference books

Franklin, S.E. 2001. *Remote Sensing for Sustainable Forest Management*. CRC Press, Boca Raton, FL, USA.

This book provided tools for "understanding and selecting remote sensing solutions to problems of forest management and sustainability". Examples of forest change detection, forest defoliation monitoring, forest classification, and forest growth modeling were provided, with highlights on methods from an operational perspective. The author underlined that "the remote sensing methods that need to be adopted and adapted to the forest science issues that are emerging through the sustainable forest management approach".

Wulder, M.A., and S.E. Frankllin, eds. 2006. *Understanding Forest Disturbance and Spatial Pattern: Remote Sensing and GIS Approaches.* CRC Press, Boca Raton, FL, USA.

This book provided the in-depth, detailed information through "the general biological or landscape ecological context of forest disturbance" to "remote sensing and GIS technological approaches and pattern description and analysis". Examples in this book allowed readers to "develop an understanding of the application of both remote sensing and GIS technologies to forest change and the impacts of fire, insect infestation, forest harvesting, and other potential change influences – such as extreme weather events".

Jensen, J.R. 2007. *Remote Sensing of the Environment: An Earth Resource Perspective, 2nd ed.* Prentice Hall, Upper Saddle River, NJ, USA.

This popular book introduced "the fundamentals of remote sensing from an earth resource (versus engineering) perspective". It covers the topics as follows: electromagnetic radiation principles (Ch. 2), photogrammetry (Ch. 6), image interpretation basics (Ch. 5), remote sensing platforms (Ch. 3: aerial platforms, Ch. 7: multispectral remote sensing systems, Ch. 8: thermal remote sensing systems, Ch. 9:active and passive microwave remote sensing, and Ch. 10: Lidar remote sensing), applications of remote sensing technology (Ch. 11 to Ch. 14), and in situ remote sensing (Ch. 15). This book has been chosen as the textbook for the course of introduction to remote sensing by many North American universities.

Jensen, J.R. 2004. *Introductory Digital Image Processing: A Remote Sensing Perspective, 3rd, ed.* Prentice Hall, Upper Saddle River, NJ, USA.

This book presented "digital image processing of aircraft- and satellite-derived, remotely sensed data for Earth management applications" with extensive illustrations. A state-of-the-art synopsis of the remote sensing processing algorithms and methods were provided.

Warner, T.A., M.D. Nellis, and G.M. Foody, eds. 2009. *The SAGE Handbook of Remote Sensing.* SAGE Publications Ltd., London, UK.

The handbook summarized the recent development of environmental remote sensing from theory to application, from data to algorithm, and from history to future direction. A professor graded this book as "a who's who of contemporary remote sensing". It is an excellent textbook for independent learning in advanced studies.

6. References

Campbell, J.B. 2007. Introduction to Remote Sensing, 4th ed. Guilford Press, New York, NY, USA.

Cohen, W.B., T.A. Spies, and M. Fiorella. 1995. Estimating the age and structure of forests in a multi-ownership landscape of western Oregon, USA. International Journal of Remote Sensing, Vol. 16, No. 4, pp. 721-746.

Colwell, R.N. 1964. Aerial photography – a valuable sensor for the scientist. American Scientist, Vol. 52, No. 1, pp. 16-49.

Defries, R.S., M.C. Hansen, J.R.G. Townshend, A.C. Janetos, and T.R. Loveland. 2000. A new global 1-km dataset of percentage tree cover derived from remote sensing. Global Change Biology, Vol. 6, pp. 247-254.

Everitt, J. H., D.E. Escobar, D.N. Appel, W.G. Riggs, and M.R. Davis. 1999. Using airborne digital imagery for detecting oak wilt disease. Plant Disease. Vol. 83, No. 6, pp 502-505.

Franklin, S.E. 2010. Remote Sensing for Biodiversity and Wildlife Management: Synthesis and Applications. McGraw-Hill Companies, Inc., USA.

Franklin, S.E., G.B. Stenhouse, M.J. Hansen, C.C. Popplewell, J.A. Dechka, and D.R. Peddle. 2001. An Integrated Decision Tree Approach (IDTA) to mapping landcover using satellite remote sensing in support of grizzly bear habitat analysis in the Alberta yellowhead ecosystem. Canadian Journal of Remote Sensing, Vol. 27, No. 6, pp. 579-592.

Giglio, L., J. Descloitres, C.O. Justice, and Y.J. Kaufman. 2003. An enhanced contextual fire detection algorithm for MODIS. Remote Sensing of Environment, Vol. 87, No. 2-3, pp. 273-282.

Hernandez-Stefanoni, J.L., J.A. Gallardo-Cruz, J.A. Meave, and J.M. Dupuy. 2011. Combining geostatistical models and remotely sensed data to improve tropical tree richness mapping. Ecological Indicators, Vol. 11, No. 5, pp. 1046-1056.

Hyyppa, J., W. Wagner, M. Hollaus, and H. Hyyppa. 2009 Airborne laser scanning. In The SAGE Handbook of Remote Sensing; Warner, T.A., Nellis, M.D., Foody, G.M., eds.; SAGE Publications Ltd., London, UK; pp. 199-211.

Kerr J.T., and M. Ostrovsky. 2003. From space to species: ecological applications for remote sensing. Trends in Ecology & Evolution, Vol. 18, No. 6, pp. 299-305.

Kohl, M., S. Magnussen, and M. Marchetti. 2006. Sampling Methods, Remote Sensing and GIS Multiresource Forest Inventory. Springer-Verlag Berlin Heidelberg, Germany.

Kokaly, R.F., and R.N. Clark. 1999. Spectroscopic determination of leaf biochemistry using band-depth analysis of absorption features and stepwise linear regression. Remote Sensing of Environment, Vol. 67, pp. 267-287.

Jensen, J.R. 2004. Introductory Digital Image Processing, 3rd ed. Prentice Hall, Upper Saddle River, NJ, USA.

Jensen, J.R. 2007. Remote Sensing of the Environment: An Earth Resource Perspective, 2nd ed. Prentice Hall, Upper Saddle River, NJ, USA.

Jensen, R.R., J.R. Boulton, and B.T. Harper. 2003. The relationship between urban leaf area and household energy usage in Terre Haute, Indiana, U.S. Journal of Arboriculture, Vol. 29, No. 4, pp. 226-230.

Jin, Y., J.T. Randerson, and M.L. Goulden. 2011. Continental-scale net radiation and evapotranspiration estimated using MODIS satellite observations. Remote Sensing of Environment, Vol. 115, No. 9, pp. 2302-2319.

Lee, J.S., and E. Pottier. 2009. Polarimetric Radar Imaging: From Basics to Applications. CRC Press, Taylor & Francis Group, Boca Raton, FL, US.

Lee, J.S., M.R. Grunes, and E. Pottier. 2001. Quantitative comparison of classification capability: fully polarimetric versus dual and single-polarisation SAR. IEEE Transactions on Geoscience and Remote Sensing, Vol. 39, No. 11, pp. 2343-2351.

Lefsky, M.A., W.B. Cohen, G.G. Parker, and D.J. Harding. 2002. Lidar remote sensing for ecosystem studies. BioScience, Vol. 52, No. 1, pp. 19-30.

Lentile, L.B., Z.A. Holden, A.M.S. Smith, M.J. Falkowski, A.T. Hudak, P. Morgan, S.A. Lewis, P.E. Gessler, and N.C. Benson. 2006. Remote sensing techniques to assess active fire characteristics and post-fire effects. International Journal of Wildland Fire, Vol. 15, No. 3, pp. 319-345.

Lillesand, T., R.W. Kieffer, and J. Chipman. 2007. Remote Sensing and Image Interpretation. 6th ed., John Wiley and Sons, New York, US.

Lim, K., P. Treitz, M. Wulder, B. St-Onge, and M. Flood. 2003. LiDAR remote sensing of forest structure. Progress in Physical Geography, Vol. 27, No.1, pp. 88-106.

McNairn, H., J. Shang, X. Jiao, and C. Champagne. 2009. IEEE Transactions on Geoscience and Remote Sensing, Vol. 47, No.12, pp. 3981-3992.

Nagendra, H. 2001. Using remote sensing to assess biodiversity. International Journal of Remote Sensing, Vol. 22, No. 12, pp. 2377-2400.

Richards, J.A., and X. Jia. 2006. Remote Sensing Digital Image Analysis: An Introduction. 4th ed., Springer-Verlag Berlin Heidelberg, Germany.

Shimoni, M., D. Borghys, R. Heremans, C. Perneel, and M. Acheroy. 2009. Fusion of PolSAR and PolInSAR data for land cover classification. International Journal of Applied Earth Observation and Geoinformation, Vol. 11, pp. 169-180.

Shippert, P. 2004. Why use hyperspectral imagery? Photogrammetric Engineering and Remote Sensing Vol. 70, pp. 377-380.

Stoms, D.M., and J.E. Estes. 1993. A remote sensing research agenda for mapping and monitoring biodiversity. International Journal of Remote Sensing, Vol. 14, No. 10, pp. 1839-1860.

Tuomisto, H., K. Ruokolainen, R. Kalliola, A. Linna, W. Danjoy, and Z. Rodriguez. 1995. Dissecting Amazonian Biodiversity. Science, Vol. 269, No. 5220, pp. 63-66.

Turner, W., S. Spector, N. Gardiner, M. Fladeland, E. Sterling, and M. Steininger. 2003. Remote sensing for biodiversity science and conservation. Trends in Ecology & Evolution, Vol. 18, No. 6, pp. 306-314.

Wang, K., S.E. Franklin, X. Guo, and M. Cattet. 2010. Remote sensing of ecology, biodiversity and conservation: a review from the perspective of remote sensing specialists. Sensors, Vol. 10, No.11, pp. 9647-9667.

Wang, K., S.E. Franklin, X. Guo, Y. He, and G.J. McDermid. 2009. Problems in remote sensing of landscapes and habitats. Progress in Physical Geography, Vol. 33, No. 6, pp. 747-768.

Wulder, M.A., C.C. Dymond, J.C. White, D.G. Leckie, and A.L. Carroll. 2006. Surveying mountain pine beetle damage of forests: A review of remote sensing opportunities. Forest Ecology and Management, Vol. 221, No. 1-3, pp. 27-41.

Wulder, M.A., J.C. White, N.C. Coops, and S. Ortlepp. 2009. Remote sensing for studies of vegetation condition: theory and application. In The SAGE Handbook of Remote Sensing; Warner, T.A., Nellis, M.D., Foody, G.M., eds.; SAGE Publications Ltd., London, UK; pp. 199-211.

Young, R.A., and R.L. Giese, eds. 2003. Introduction to Forest Ecosystem Science and Management, 3rd ed. John Wiley & Sons, Inc., Hoboken, NJ, USA.

Zarco-Tejada, P.J., J.R. Miller, T.L. Noland, G.H. Mohammed, and P.H. Sampson. 2001. Scaling-up and model inversion methods with narrowband optical indices for chlorophyll content estimation in closed forest canopies with hyperspectral data. IEEE Transactions on Geoscience and Remote Sensing, Vol. 39, No. 7, pp. 1491-1507.

Zhang, W., X. Zhang, L. Li, and Z. Zhang. 2007. Urban forest in Jinan City: distribution, classification and ecological significance. Catena, Vol.69, No.1, pp. 44-50.

13

GIS for Operative Support

Gunnar Bygdén
Olofsfors AB Nordmaling
Sweden

1. Introduction

Geographic Information System (GIS) is often used when planning the cuttings in the forest. Starting with a good geographical map and pre-planned stand marking at the office, the planning person can walk up the borders of the stand with a high precision GPS and mark /correct the stand borders on a forestry GIS map as an overlay. The planning person may also mark areas where special environmental consideration should be made. There could be special valuable trees or ancient remains that should be avoided from machine traffic or cutting. The advantage by doing this as an overlay on the geographical map is that the stand area can be well calculated and used for estimation of the expected volume together with earlier stand information (e.g. stand inventory). In the forestry map indications could also be made where to put landing, alarm coordinates and other needed information for the machine operators.

Unfortunately the planning person often has too little knowledge about forestry machines and expects the machine operators to choose where to go. Preferably the planning person should also make proposals of where to put the major temporary forestry roads, thus where to find the best ground for such roads. The machine operators on the other hand have seldom time for walking around the stand but like to start more or less directly when arriving to the stand. They start often by cutting around where to put the landing and then continue along the borders of the stand. Thus, the machines may enter areas with less good soil bearing capacity without any previous information. Landing may also be placed on less good spots from the stand point of view. The result is severe rutting around the landing area and a number of wheel ruts in the forest. In worst case the machines may get stuck in forest with expensive rescuing operations. And, a disappointed forest owner looking at all the mess in the forest and will not accept to sell the timber to this company any more.

All these problems belong actually to poor soil knowledge of all involved personal. Less good planning, too heavy machines, too much traffic together with bad weather conditions for the soil conditions are factors causing severe disturbance of the forest soils. The aim must be to adapt machines, traffic and planning to actual soil conditions.

2. Some soil mechanics

Many people believe that the soil bearing capacity depends on the soil type. Of cause, a peat land is soft and a gravelly soil is strong, but for mineral soils the bearing capacity is not so very different regardless of particle size when it is dry. However, it is with moisture content the particle size class differs. A gravelly soil does not change very much with moisture, a silt soil changes drastic behaviour and a clay soil becomes sticky and slippery with increased moisture content. In the forest we have also to consider a humus layer with high elasticity full with roots from trees and ground vegetation (Wästerlund 1989). The roots may have a significant influence on the ground bearing capacity (Fig. 1) and at moist conditions they may contribute as much as 50 % of the ground bearing strength. Normally the shear strength (τ) of a soil is described as $\tau = c + \sigma tan\varphi$, which is the cohesion (c) plus load (σ) times the friction angle ($tan\varphi$) of the soil. Both parameters will change with increasing moisture content. However, the root component (Sr) is missing and therefore it has been suggested to add a root component, such as $\tau = c + \sigma tan\varphi + Sr$ for forestry ground (Wästerlund 1990).

Fig. 1. Surface cover blown away in a spruce stand down to 20 cm depth showing all the surface roots. A farm tractor has been passing causing root damage. Photo: I. Wästerlund.

With this important influence of the root armouring of the ground, it becomes clear that deep rutting should be avoided. When the roots are broken, the bearing capacity of a wet ground may decrease to only the half. Furthermore, the rutting of the soil means that the machine is using energy to deform the ground and the wheel ruts are causing the machine to drive uphill seen from the wheel perspective. Thus, the rutting will cause increased fuel consumption and make the operation more expensive.

The bearing capacity of the ground can be measured with a cone penetrometer (Fig. 2). This tool has a cone of 21 mm in diameter with 30° angle and is pressed down into the soil with a rate of 3 cm/s (WES standard). By checking the ground before passage after a heavy rain, it can be determined if the machine can pass or not or will be making deep wheel ruts. A measured value below 0.3 MPa means that the soil is very soft, between 0.3-0.7 MPa the soil is soft, between 0.7 and 1 the soil is medium hard but may be soft after rain, and above 1 MPa the soil has a good bearing capacity.

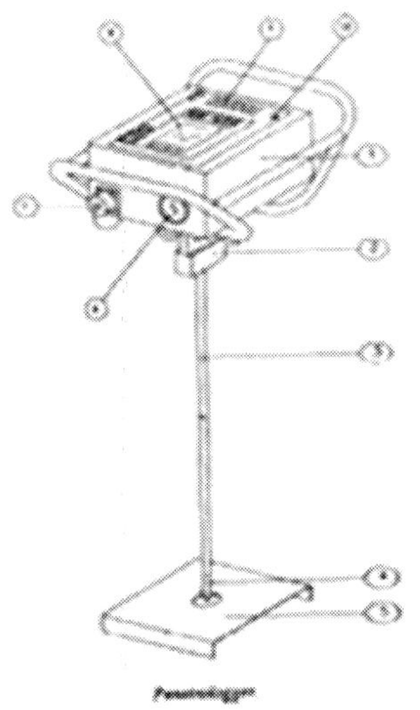

Fig. 2. A registration cone penetrometer.

Andersson (2010) made a study of forest machine rutting in the soil after operation comparing soil strength measured with a cone penetrometer, machine ground pressure and depth of ruts. The soil strength was measured with a cone penetrometer. His conclusion was that on soils below 0.7 MPa severe rutting occurred when the machines ground pressure was above 80 kPa (Fig. 3).

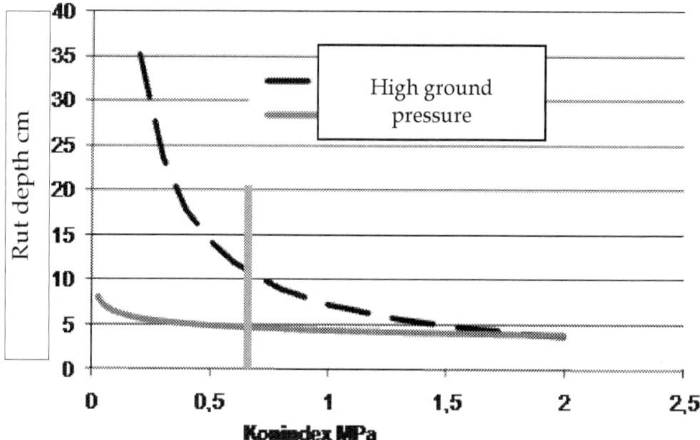

Fig. 3. With high ground pressure and low bearing capacity (below 0.7 MPa) severe rutting may occur.

Ground pressure in his case was calculated as axle load divided with number of wheels and contact area. The contact area was calculated as wheel radius times the width of tire. It is possible to increase the contact area by increasing the width of the tire, or putting tracks on the wheels. Best effects seem to be with bogey tracks. Bygdén et al. (2003) showed that bogey tracks could both reduce the rutting of the soil and decrease the rolling resistance on soft soils. With bogey tracks the area is calculated as width of the track times the contact length and often the ground pressure is reduced by 30-50 % compared to wheels when using bogey tracks, especially on forwarders. This is important for reduction of rutting on soft soils.

3. Proposal

Soil maps are coming slowly into use for forestry operations. Within a few years ahead many national geological institutions will be able to provide soil maps that can be used together with GIS. At least one forest machine manufacturer has started to prepare a system with an integrated soil map to the GIS system. With help of colours soft soils can be marked and guide the machine operators where soft soils can be expected, for instance a red colour could mark soft soils, pink medium soils and yellow colour means normal soils. Thörnewall (2007) introduced a scale from 1 to 5, where 1 was easy and 5 difficult. He put one scale for soil type, one scale for inclination and one for surface structure and added the numbers. An easy terrain with strong soils could be 1+1+1 = 3 and a difficult terrain could be 4+3+2 = 9. For road construction he put a price for each number and thereby steering the road construction towards easy terrain. A similar system could also be used for payment for forestry operations, e.g. when soils do have less good bearing capacity, the payment would increase as an incitement to use smaller machines or buy proper bogey tracks to reduce the ground pressure and decrease soil disturbance.

Thus, the colour scheme shown in the machine computer should be a warning signal for operator to avoid that area marked with red and be careful when pink colour appears. The planning person must also notice the warning colours and try to avoid operations on those areas or prescribe certain machines or protective measures to be taken such as reducing the ground pressure. The protective measures could be to prescribe use of bogey tracks to the conditions, e.g. aggressive types at snow conditions or types with good supporting construction for soft conditions. A number of different types of wheel and bogey tracks for varying conditions could be supplied through Olofsfors AB.

Both planning personal and operators should work together to calculate the ground pressures of used machines on the district. The calculation must be done per axle of the machine and it is the heaviest axle when loaded that determines the trafficability of the machine. Both categories must be well aware that ground pressures above 80 kPa may cause severe rutting on soils with strength below 0.7 MPa and a heavy rain could reduce the bearing capacity for soils up to 1 MPa. Tracks may increase both the trafficability and the stability of many machines by increasing the contact area. The ambition to have machines for forestry operations with ground pressure below 75 kPa will save both the fuel consumption on weak soils and the environment.

4. Conclusion

With a soil map as a background in the ordinary site map there will be better possibilities for the operators to avoid rutting and compaction. Route planning will also be better with the knowledge of where the sensitive parts are situated in the logging area. A scaled map with numbers and warning colours should be a useful tool to show less bearing capacity in the logging area.

To measure the machines weight and get exact ground pressure is also a way to minimize soil disturbance.

Decreasing the machine ground pressure with proper bogie tracks should be regular in sensitive areas.

5. References

Andersson, E., 2010. Increased harvesting possibilities with suitable ground pressure from harvesting machines. SLU, dept of Forest Resource Management, Umeå. Report No 276.

Bygdén, G., Eliasson, L. & Wästerlund, I., 2003. Rut depth, soil compaction and rolling resistance when using bogie tracks. J. Terramechanics, 40(3):179-190.

Suvinen, A., 2006. A GIS-based simulation model for terrain tractability. J. Terramechanics 43:427-449.

Thörnewall, D., 2007. Method for use of Geographic Information Systems in long term road planning. SLU, dept of Forest Resource Management, Umeå. Report No 176.

Wästerlund, I., 1989. Strength components in the forest floor restricting maximum tolerable machine forces. J. Terramechanics, Vol. 26 (2):177-182.

Wästerlund, I., 1990. Soil strength in forestry measured with a new type kind of test-rig. Proc. 10th International Conference of the ISTVS/Kobe, Japan/August 20-24, 1990. Vol. 1:73-82.

Section 2

Forest Utilization

14

Adding Value Prior to Pulping: Bioproducts from Hemicellulose

Lew Christopher
Center for Bioprocessing Research and Development
Department of Chemical and Biological Engineering
South Dakota School of Mines and Technology
USA

1. Introduction

The global trend for production of biofuels and bioproducts from renewable resources is currently steered by three important drivers: 1) increasing demand and prices of petroleum-derived fuel; 2) increasing food needs; and 3) increasing greenhouse gas emissions. Biomass is the single renewable resource that has the potential to supplant the use of liquid transportation fuels now and help create a more stable energy future.

As recently reported by the US Department of Energy (US DOE), nearly 1.3 billion dry tons of biomass could be available for large-scale bioenergy and biorefinery industries, enough to displace 30% or more of the nation's current consumption of liquid transportation (Perlack et al., 2005). About a third of the biomass resources in US are wood-based and two-thirds are of agricultural origin. However, the biomass share of the US energy supply in 2004 was less than 3% of the total, compared to 40% and 23% derived from petroleum and coal, respectively. In 2009, the US produced 10.75 billion gallons of ethanol (mainly corn-based), and together with Brazil, both countries accounted for 89% of the world's production. The 2007 Energy Independence and Security Act and the new US Renewable Fuels Standard of 2008 call for the production of 36 billion gallons of biofuels, mainly ethanol and biodiesel, annually by 2022, with 21 billion gallons coming from "advanced biofuels" of which 16 billion gallons is expected as biofuels derived from lignocellulosic biomass.

Breakthrough technologies are needed to make cellulosic ethanol cost-competitive with corn-based ethanol by 2012. Plant-derived biofuels as a carbon-neutral technology have to achieve at least 60% lower emissions than petroleum fuel based on lifecycle studies that include all emissions resulting from making the fuel from the field to the tank. Meeting these goals will require significant and rapid advances in biomass feedstock and conversion technologies; availability of large volumes of sustainable biomass feedstock; demonstration and deployment of large scale, integrated biofuels production facilities; and development of an adequate biofuels infrastructure. Although significant progress has been recently made towards commercialization of cellulosic ethanol, there are still economic, social and environmental challenges that need to be addressed. A minimum profitable ethanol selling price of \$2.50/gallon can compete on an energy-adjusted basis with gasoline derived from oil costing \$75-\$80/barrel. At the lower oil prices (\$45 - \$50/barrel), cellulosic technology may not be as competitive and could require policy supports and regulatory mandates to drive the market. The biofuels and bioproducts strategies need to be based on a thorough assessment of opportunities and costs associated with the upward pressure on food prices, intensified competition for land and water, and deforestation. As the feedstock costs comprise more than 20% of the production costs, it has now been recognized that biomass waste such as agricultural and forest waste can provide a cost-effective alternative to improve the economic viability of bioethanol production (Zheng et al., 2009).

Due to the increasing off-shore competition, global movement and incentives for green fuels and chemicals, the North American pulp and paper and other fiber processing industries need to create additional revenues and diversify their products and markets to remain competitive. To achieve this, the forest-based industries, and in particular the pulp and paper industry, need to evolve into integrated forest biorefineries (IFBR). IFBR is defined as a processing and conversion facility that fully integrates forest biomass and other biomass waste for simultaneous production of marketplace products, including fibers for pulp and paper products, fuels, chemicals and materials (Fig. 1).

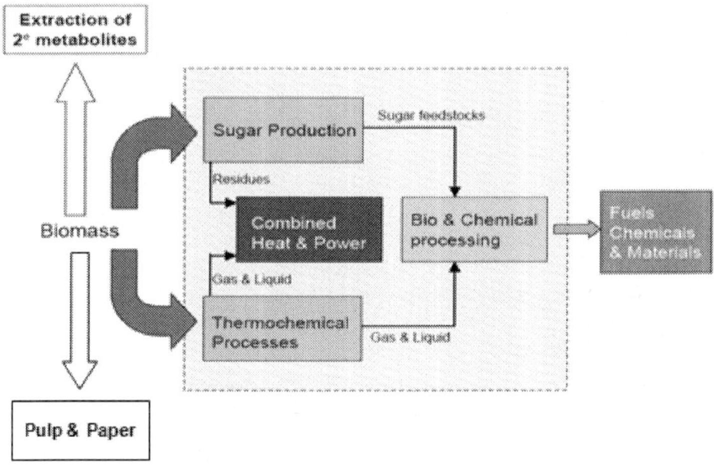

Fig. 1. The Integrated Forest Biorefinery (IFBR) concept

The biorefinery concept is analogous to today's petroleum refineries, which produce multiple fuels and products from petroleum. Co-products of biofuels production, such as corn gluten feed and meal, corn oil, glycerin, natural plastics, fibers, cosmetics, liquid detergents and other bioproducts, also increase with biofuel production. Currently, however, of the 100 million metric tons of chemicals produced annually in the US, only about 10% are biobased (National Academy Press, 2000).

Based on the billion ton vision of the US DOE, nearly 400 million tons of hemicellulose are available in US per annum for bioprocessing to fuels and chemicals. In addition, every year approximately 15 million tons of hemicellulose are produced by the pulp and paper industry alone, and according to preliminary results, this can yield in excess of 2 billion gallons of ethanol and 600 million gallons of acetic acid, with a net cash flow of $3.3 billion. Unfortunately, during the current fiber processing of woody feedstocks such as pulping and bleaching, hemicellulose is not efficiently utilized and often discarded as industrial waste with limited usage. However, a number of bioproducts such as biochemicals and biomaterials can be produced from biomass hemicellulose to enhance the value extracted from wood fiber and improve the process economics in a forest-based biorefinery. The aim of this paper was to review the potential of hemicellulose and its monomers derived from pulp and paper processing of wood for production of biofuels and value-added bioproducts.

2. Hemicellulose

The lignocellulosic biomass has an estimated annual production on earth of about 60 billion tons. Biomass is composed of three major polymers – cellulose, hemicelluloses and lignin - and their ratio, composition and structure determine biomass properties. Hemicellulose forms an interface in the cell wall matrix with binding properties mediated by covalent and non-covalent interactions with lignin, cellulose and other polysaccharides (Kato, 1981). The close association between the biopolymers in plant biomass is realized via chemical bonds, predominantly between lignin and hemicelluloses, in lignin-carbohydrate complexes (LCCs) that include benzyl-ether, benzyl-ester and phenyl-glycoside types of linkages. The cellulose and hemicelluloses form the carbohydrate composition of lignocelluloses (Fig. 2).

2.1 Hemicellulose composition of wood

Overall, hardwoods contain on average less lignin and extractives and more hemicellulose than softwoods (Table 1). In hardwoods, the major hemicellulose component is the O-acetyl-4-O-methylglucuronoxylan, whereas in the softwood species, the O-acetyl-galactoglucomannan is the predominant one (Fengel & Wegener, 1984). The building blocks of hemicelluloses are hexoses (glucose, mannose and galactose) and pentoses (xylose and arabinose) which exist in a pyranose (p) and furansose (f) forms (α and β).

Due to the lower degree of polymerization (DP), the chemical and thermal stability of hemicelluloses is lower and their alkali solubility – higher than that of cellulose. Whereas cellulose is present in all plants as a ß-1,4-glucose polymer with a non-branching structure, xylan is composed of ß-1,4-linked xylose units forming a xylan backbone with a DP of 150-200 and of side chains connected to the backbone (Table 2).

Wood components	Hardwoods	Softwoods
Cellulose	40-50	45-50
Hemicellulose	22-35	20-30
(Galacto)glucomannan	2-5	15-20
Glucuronoxylan	20-30	5-10
Lignin	20-30	25-35
Extractives	1-5	3-8

Table 1. Chemical composition of wood

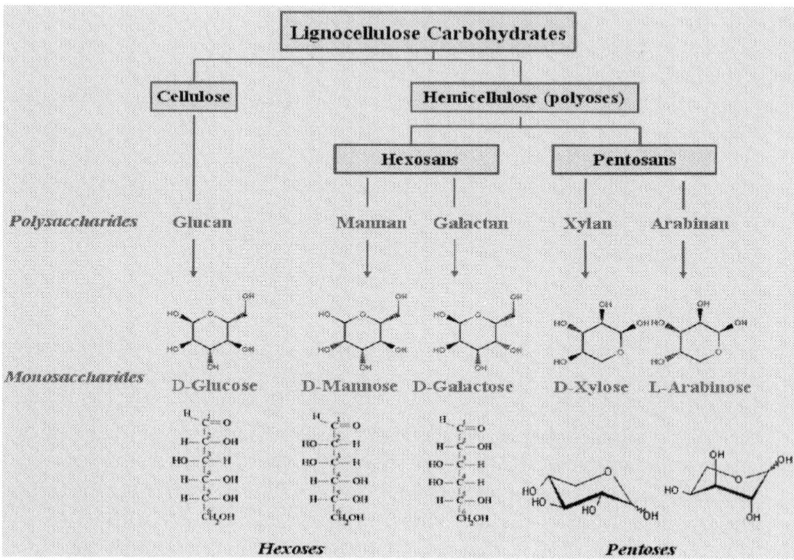

Fig. 2. Chemical structure of lignocellulose carbohydrates

2.2 Hemicellulose structure in wood

The composition and structure of hemicellulose are more complicated than that of cellulose and can vary quantitatively and qualitatively in various woody species (Fengel & Wegener, 1984). Typically, softwoods have more mannose and galactose and less xylose and acetyl groups than hardwoods (Table 2). The side-chain groups in xylan differ depending on the plant origin. The hardwood xylans as complex heteropolysaccharides comprising β-1,4-linked D-xylopyranose units are highly substituted (Sjöstrom, 1993). The xylopyranose unit of the xylan main chain can be substituted at the C2 and/or C3 positions with acetic acid (at both C2 and C3 position in hardwoods), 4-O-methylglucuronic acid (at C2 position in both hardwoods and softwoods), and arabinose (at C3 position in softwoods). The latter may be further esterified by phenolic acids which crosslink xylan and lignin in LCCs in the cell wall matrix.

In hardwoods, *O*-acetyl-4-*O*-methylglucuronoxylan is the main hemicellulose component constituting 20-30% of the wood material. The uronic acid groups in hardwood xylans are not evenly distributed, with 2-3 substituents per one xylose unit. In softwoods, every eight xylose residues are substituted with arabinose by α-1,3-glycosidic linkages whereas the ratio of xylose to glucuronic acid is 4:1.

The galactoglucomannans can be classified into two fractions with different galactose contents – galactose-poor fraction and galactose-rich fraction with a corresponding galactose/glucose/mannose ratio of 0.1/1/3, and 1/1/3, respectively, and acetyl content of 6% in both fractions. The softwood xylan is a linear polymer of D-xylopyranose units slightly branched with 1-2 side chains of arabinofuranose and glucoronic acid per molecule. The degree of substitution of hardwood xylan with acetyl groups can vary from 8 to 17%

corresponding to 3.5-7 acetyl groups per 10 xylose units, and on average every second xylose unit is acetylated.

Hemicellulose	Amount (%)	Units	Bond	Molar ratios	Solubility	DP
Hardwood						
Acetylglucuronoxylan	15-30	β-D-Xylp	1-4	10	Alkali	200
		4-O-Me-α-D-GlcpA	1-2	1	DMSO*	
		Acetyl		7		
Glucomannan	2-5	β-D-Manp	1-4	1-2	Alkaline	200
		β-D-Glcp	1-4	1	borate	
Softwood						
Acetylgalactoglucomannan	15-23	β-D-Manp	1-4	3	Alkali	100
		β-D-Glcp	1-4	1	Water*	
		α-D-Galp	1-6	1		
		Acetyl		1		
Arabinogalactan	2-5	β-D-Galp	1-3	6	Water	200
			1-6			
		α-L-Araf	2/3	1-6		
		β-L-Arap	1/3	1-3		
		β-D-GlcpA	Little	1-6		
Arabinoglucuronoxylan	5-10	β-D-Xylp	1-4	10	Alkali	100
		4-O-Me-α-D-GlcpA	1-2	2	DMSO*	
		α-L-Araf	1-3	1.3	Water*	

*Partially soluble

Table 2. Hemicellulose structure in wood

2.3 Hemicellulose behavior during pulping

During kraft pulping of wood chips, xylan and mannan are partially depolymerized, debranched and solubilized in the cooking liquor (Casebier & Hamilton, 1965). Subsequent losses of hemicellulose occur during the heating period of the kraft cook thereby about 40% of xylan is lost and glucuronic acid is converted to hexenuronic acid by ß-elimination reactions. By the end of the cook, 60-70% of the glucuronosyl and 10% of the arabinosyl substituents in softwood xylan are removed. Due to a pH drop in the pulping liquor caused by debranched acetyl residues towards the end of pulping, part of dissolved xylan, lignin and lignin-xylan complexes are reprecipitated back onto the fiber surface. The extent of this readsorption depends on the alkaline cooking conditions and wood species, however, the reprecipitated xylan has a low molecular weight without side-chain groups and a high degree of crystallinity (Gustavsson & Al-Dajani, 2000). For instance, half the xylan content of pine kraft pulp is estimated as relocated xylan whereas up to 14% of birchwood xylan can be reprecipitated during kraft pulping.

During acid sulfite pulping, redeposition of xylans onto the fiber surface has not been observed. The possible reasons for this would be that the harsh cooking conditions and presence of acid-resistant residual acetyl and 4-O-methylglucuronic acid groups act as barriers against the adsorption and intercrystallization of xylan onto the cellulose micromolecules. Significant amounts of xylans are hydrolyzed and solubilized in the sulfite pulping process. For instance, in sulfite cooking of birch only 45% of the original xylan remains in pulp after 20 min and its original DP of 200 is reduced to less than 100.

The bonds between the pentose units (arabinose and xylose) are hydrolyzed much more rapidly than the glycopyranosidic bonds. However, the glucuronic acid-xylose and xylose-acetic acid linkages are relatively more resistant to the acid hydrolysis conditions and little cellulose is lost in the sulfite cook (Kerr & Goring, 1975). The degradation products of the hemicellulose acid hydrolysis appear in the cooking liquor in the following approximate order: arabinose> galactose> xylose> mannose> glucose> acetic acid> glucuronic acid. In the above order, glucose is derived mostly from the glucomannan rather than cellulose polymer. Thus, the residual xylan in sulfite pulps is less accessible since it is localized mainly in the secondary cell walls, although the xylan distribution across the cell wall is more uniform than in kraft pulps.

3. Hemicellulose-degrading enzymes

Due to its complex structure, the complete breakdown of naturally occurring branched hemicelluloses requires the action of several enzymes with different functions. These are classified in two groups, hydrolases and esterases, based on the nature of linkages that they can cleave. The glycosyl hydrolases are involved in the enzymatic hydrolysis of the glycosidic bonds of hemicellulose. Due to the complex nature of hemicelluloses and their enzymatic hydrolysis, only xylan-degrading enzymes will be presented and discussed in this paper.

3.1 Enzymatic hydrolysis of xylan

Of major importance are the endo-β-1,4-xylanases or 1,4-β-D-xylan xylanohydrolases (3.2.1.8) that can randomly hydrolyze internal xylosidic linkages on the backbone of xylan polysaccharide. The main products formed from xylan hydrolysis by xylanase are xylobiose, xylotriose and substituted xylooligosaccharides depending on the mode of action of the particular enzyme (Table 3). The xylooligomers liberated by xylanase are converted to xylose by 1,4-ß-D-xylosidase (EC 3.2.1.37). The so-called accessory enzymes such as acetyl xylan esterases, phenolic acid esterases, arabinofuranosidases and glucuronidases cleave side groups from the xylan backbone. All xylanolytic enzymes act synergistically in xylan hydrolysis (Fig. 3).

Xylanases can be classified structurally into two major groups, family 10 and family 11. Family 10 enzymes have a relatively high molecular weight whereas family 11 xylanases are relatively low molecular weight with low or high pI values. The release of reducing sugars from xylan however has not been shown to correlate to the family belonging of enzyme. The enzyme-substrate interaction is dependent on substrate specificity and kinetic properties of enzyme and can be influenced by pH, presence of xylan binding domain and ionic strength of protein and xylan molecule. Since xylan is negatively charged due to the presence of glucuronic acid side-chain groups, the efficiency of binding of enzyme to xylan is affected by the pH of reaction and pI of protein. For instance, if pH is below the pI value, the enzyme can be completely bound to the polysaccharide. The xylanolytic enzyme system of a variety of microorganisms have been extensively investigated and several exhaustive reviews have appeared (Coughlan & Hazelwood, 1993; Viikari et al., 1993; Wong et al., 1988).

Enzyme	Mode of action
Endo-xylanase	Hydrolyses interior β-1,4-xylose bonds of xylan backbone
Exo-xylanase	Releases xylobiose from xylan backbone
β-Xylosidase	Releases xylose from xylobiose
α-Arabinofuranosidase	Hydrolyses α-arabino-furanose from xylan
α-Glucurosidase	Releases glucuronic acid from glucuronoxylans
Acetyl xylan esterase	Hydrolyses acetyl ester bonds in acetyl xylans
Ferulic acid esterase	Hydrolyses feruloyl ester bonds in xylans
ρ-Coumaric acid esterase	Hydrolyses ρ-coumaryl ester bonds in xylans

Table 3. Xylan-degrading enzymes

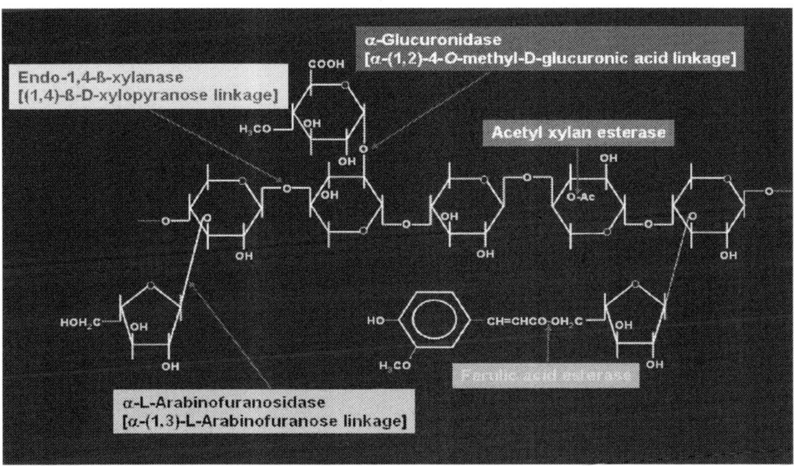

Fig. 3. Enzymatic hydrolysis of xylan

3.2 Production of xylan-degrading enzymes

The optimization of fermentation techniques and isolation of more efficient microbial strains has led to a significant increase in the production rates of xylanase. Fungal systems are excellent xylanase producers, but often co-secrete cellulases which can adversely affect pulp quality. One way of overcoming this is by using suitable separation methods to purify xylanases from contaminating cellulase activity. This approach however is expensive and impractical. By applying appropriate screening methods and selection of growth conditions, it is possible to isolate naturally occurring microorganisms which produce totally cellulase-free xylanases or contain negligible cellulase activity. Alternatively, genetically engineered organisms could be used to produce exclusively xylanase. Most xylanases studied are active in slightly acidic conditions (pH 4-6) and temperatures between 40 and 60°C. The current trend is however to produce enzymes with improved thermostability and activity in alkaline conditions to fit operations at harsh industrial conditions. ß-Xylanases are produced by many microorganisms on xylan-rich substrates (Wong et al., 1988).

3.2.1 Solid state fermentation

Currently, most commercial enzymes are mainly produced in a conventional submerged fermentation (SF) process, which is an inherently expensive operation best suited for high value antibiotics and other pharmaceutical products. Solid substrate fermentation (SSF) is an economically viable alternative for enzyme production which offers numerous advantages over the SF systems as many enzymes and other biochemicals can be produced by SSF at a fraction of the cost for SF production (Szakacs et al., 2001).

The SSF allows the direct use of *in-situ* enzymes (i.e. xylanase for pulp pretreatment and bleaching) without their prior downstream processing. The substrate (i.e. paper pulp which contains xylan), which is initially used as a carbon source for enzyme production, subsequently becomes the target substrate of enzyme (xylanase) action. This approach could certainly improve the economics and enhance the efficiency of the biobleaching technology due to the operational simplicity of SSF, high volumetric productivity and concentration of enzyme and production of substrate-specific enzymes in a water-restricted environment (Szendefy et al., 2006). Advantages include high concentration of the product and simple fermentation equipment as well as low effluent generation and low requirements for aeration and agitation during enzyme production (Pandey et al., 1999). Due to the considerably lower production costs, the SSF xylanase has been shown to be more cost-efficient when compared to commercial liquid products.

3.2.2 Spent sulfite liquors

Spent sulfite liquor (SSL) results from the delignification of wood chips in an aqueous solution of acid bisulphites with an excess of SO_2, resulting in the solubilisation of lignin and leaving the wood cellulose largely undegraded (Mueller & Walden, 1970). The resultant black liquor referred to as SSL contains 50 to 65 % lignosulphonates, 15 to 22 % total sugars and 2 to 5 % volatile acids such as acetic acid. The sugars found in the SSL include xylose, mannose, galactose, arabinose and glucose. The SSL is therefore a concentrated waste with high BOD and COD levels and needs treatment prior to disposal. The utilization and recovery of the valuable organics in this effluent would, therefore, be more desirable than its simple discharge. The microbial utilisation of SSL has been studied for production of various metabolites such as lactic acid, single-cell yeast protein (Mueller & Walden, 1970) and ethanol (Kosaric et al., 1981). Recently, the use of this inexpensive carbon source as inducer of xylanase activity has also been demonstrated (Chipeta et al., 1981). Potential advantages include reduced xylanase production costs and development of effluent-free technology that impact positively on the environment.

3.3 Application of xylan-degrading enzymes

Xylan-degrading enzymes, and in particular xylanases, have a great potential in industrial processes such as saccharification of lignocellulosic biomass to fermentable sugars for production of biofuels and biochemicals, bread making, clarification of beer and juices, enzymatic retting of flax, surface softening and smoothing of jute-cotton blended fabrics

(Royer & Nakas, 1989). Nethertheless, the most important application of these enzymes to date is their use in the pulp and paper industry. Xylanases have been reported to enhance inter-fiber bonding through fibrillation without reducing pulp viscosity. Xylanase-treated pulps have shown improved beatability and brightness stability. When applied together with cellulases, xylanases can improve the drainage rates of recycled fibers and can facilitate the release of toners from office waste and the following flotation and washing steps. The xylanase production on large scale constitutes approximately 50% of the total enzyme market and the demand for xylanases grows about 25% per year, with a major application in bleaching of paper pulps.

The use of xylanases at pulp and paper mills to facilitate bleaching (biobleaching) and improve fiber properties is one the most important large-scale biotechnological applications of recent years (Onysko, 1993; Viikari et al., 1994). The enzymatic improvement in pulp bleachability depends on a number of factors such as the wood source, pulping and bleaching processes as well as properties and substrate specificity of the enzyme. Factors such as inhibitory effect of residual pulping and bleaching chemicals in pulp as well as degradation end products on xylanase efficiency, presence of xylan-lignin and xylan-cellulose bonds may as well impact on extent of xylan hydrolysis and pulp bleachability. Restrictions in the enzymatic removal of xylan from pulp have been assigned to retarded accessibility and chemical modification of residual hemicellulose. Accessibility problems arise from the fact that chemical pulping and bleaching apparently remove the more accessible portion of xylan from the cell walls, leaving the remaining part in locations, that are less accessible to xylanase. Xylanases should contain no or very low cellulase activities as cellulases prove detrimental to yield and strength properties of pulp. The bleaching efficiency of xylanase is measured either as the reduction in the amount of chemicals used for bleaching of pulp or the brightness gain induced by the enzyme. As the biobleaching effect is dependent on the amount of enzyme used, the enzyme production costs should be kept as low as possible to ensure a cost-effective bleaching process. The major benefits from the enzyme bleaching are: 1) Reduced bleaching costs; 2) Reduced chemical consumption; 3) Increased pulp throughput; and 4) Reduced pollution (Christopher, 2004). A few hypotheses exist to explain the phenomenon of xylanase-aided bleaching of pulp, although the exact mechanism is not completely understood. It should be noted that the proposed mechanisms for biobleaching are not mutually exclusive and more than one model can be involved depending on pulp type, on one side, and substrate specificity of xylanase to a specific xylan type in pulp, on another (Wong et al., 1997).

- The initial model proposed suggested that xylanases attack and hydrolyze mainly xylan redeposited on the fiber surface thereby enabling the bleaching chemicals a better and smoother access to residual lignin (Kantelinen et al., 1993). During kraft pulping, pulp xylan is first solubilized and later on part of it is redeposited back onto the pulp fibres. Xylanase acts on these reprecipitated xylans by partially hydrolysing them to facilitate extraction of lignin during pulp bleaching (Fig. 4).
- The second hypothesis suggests that xylanases can partly hydrolyze xylan that is involved in lignin-xylan complexes thereby reducing the size of these complexes and improving their mobility and extractability from the cell walls. Indirect evidence does exist that lignin-carbohydrate bonds are formed during biosynthesis and aging of wood as well as during kraft pulping and that xylose is released as the main sugar component

of isolated lignin-carbohydrate complexes. The biobleaching effect appeared to be accompanied by a decrease in the degree of polymerization of xylan and a slight reduction in xylan.

- It has also been reported that xylan-chromophore associations can be generated during alkaline pulping which contribute to pulp color and brightness reversion of pulps. A direct brightening effect has been observed following xylanase pretreatment of pulp. This could be due to a direct removal of lignin fragments involved in lignin-xylan complexes and/or removal of xylan derived chromophore structures. This hypothesis is supported by the recent findings that during the kraft cook the methylglucuronic acid of xylan can be modified to hexenuronic acid giving rise to double bond chromophore-type formations.

- Xylanases may also be able to disrupt to an extent the physical interlinking between xylan and cellulose within the fiber matrix thereby improving the fiber swelling and generating macropores to facilitate lignin removal. The biobleaching effect observed with some hardwood sulfite pulps may also be caused by an improved pulp porosity. This suggestion is based on the fact that in acid sulfite pulps, in contrast to kraft pulps, xylan is not reprecipitated on the fiber surface but is largely entrapped across the fiber the cell walls.

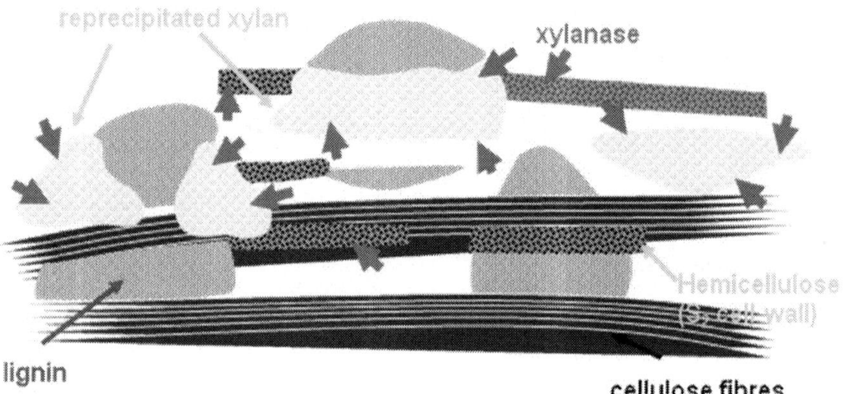

Fig. 4. Mechanism of xylanase-aided bleaching of paper pulps

3.4 Bioproducts from xylan

Hemicelluloses including xylan need to have a certain degree of purity before they can be utilized in various industrial processes. The xylan source and recovery process (extraction) directly impact the physical and chemical properties of the recovered polysaccharide and determine its applicability. Xylans can be extracted from lignocellulosic materials or partially delignified pulps. Fractionation from lignified materials yields polysaccharides with major proportions of lignin, whereas higher purity xylans are obtained when isolated from pulps, especially bleached pulps (Puls & Saake, 2004). Table 4 summarizes the major extraction, fractionation, precipitation and quantification methods and their impact xylan deacetylation and salvation.

Polymer	Extraction	Deacetylation	Solvation	Precipitation	Fractionation	Quantification
Xylan	DMSO KOH NaOH Alkaline peroxide Water (steam)	NaOH KOH Water (partial)	DMSO	Acidification (CH_3COOH) Solvents (C_2H_5OH)	Increasing gradient of alkali	Acid/enzyme hydrolysis Sugar/lignin analysis

Table 4. Xylan extraction methods

However, the properties of xylans have not been fully characterized, defined and exploited. Although annual plants have been proven a rich source of xylan, because of the difficulties in extraction and purification of xylans and hemicellulose in general, an efficient isolation process has never been realized (Ebringerova & Heize, 2000). Table 5 summarizes some of the most important current and potential applications of xylan in the pulp and paper, pharmaceutical, chemical, food and fermentation industries.

3.5 Bioproducts from xylose

3.5.1 Ethanol

For the economic production of ethanol from lignocellulosics, the fermentation of both glucose and xylose is an economic necessity. Unfortunately most yeasts including *Saccharomyces cerevisiae* do not ferment xylose. On the other hand, bacterial fermentations are associated with low ethanol yields, slow fermentation rates, byproduct formation (acids) which requires additional product separation, contamination problems due to the neutral pH requirements for bacterial growth, bacterial sensitivity to inhibitors and intolerance to high ethanol concentrations. To overcome these problems, two major strategies have been developed (Fig. 5).

The first strategy is to introduce pentose-utilizing capability into efficient ethanol producers such as *Saccaromyces* and *Zymomonas* (Matsushika et al., 2009). Genes encoding for xylose reductase, xylose isomerase, xylose isomerise, xylulokinase, transaldolase and transketolase (Fig. 5 - blue print) are inserted to enable the pentose-phosphate pathway which enables xylose to enter the glycolysis pathway of glucose fermentation to ethanol.

Pulp and Paper	Pharmaceutical	Chemical	Food	Fermentation
• Beater additive improved swelling, porosity, drainage, strength • Fiber coating • Wood resin stabilizer	• Anticoagulant • Anti-cancer agent • Cholesterol-reducing agent • Wound treatment agent • HIV inhibitor • Tabletting material • Dietary fiber	• Thermoplastic material • Polypropylene filler • Paint formulations • Gel-forming material • Chiral polymer building blocks	• Xylose • Xylitol • Biodegradable polymers – plastics, films, coatings with increased hydrophobility and water resistance (acety xylans)	• Enzymes–xylanase, xylose isomerise • Biopolymers–polyhydroxy-alkanoates

Table 5. Major large-scale applications of xylan

Using this strategy for *Z. mobilis*, Zhang et al. (1995) achieved 85% of the theoretical ethanol yield on xylose. The second strategy targets to divert the carbon flow in *E. coli* from native fermentation products to ethanol by introducing pyruvate decarboxylase and alcohol dehydrogenase (Fig. 5 – red print). On mixed sugars (xylose and glucose), a recombinant *E. coli* produced 103-106% of the theoretical yield of ethanol (Tao et al., 2001). However, the most common problems with recombinant microorganisms that still need to be resolved are their instability, slower production rates and reduced robustness compared to the wild strains (Eliasson et al., 2000).

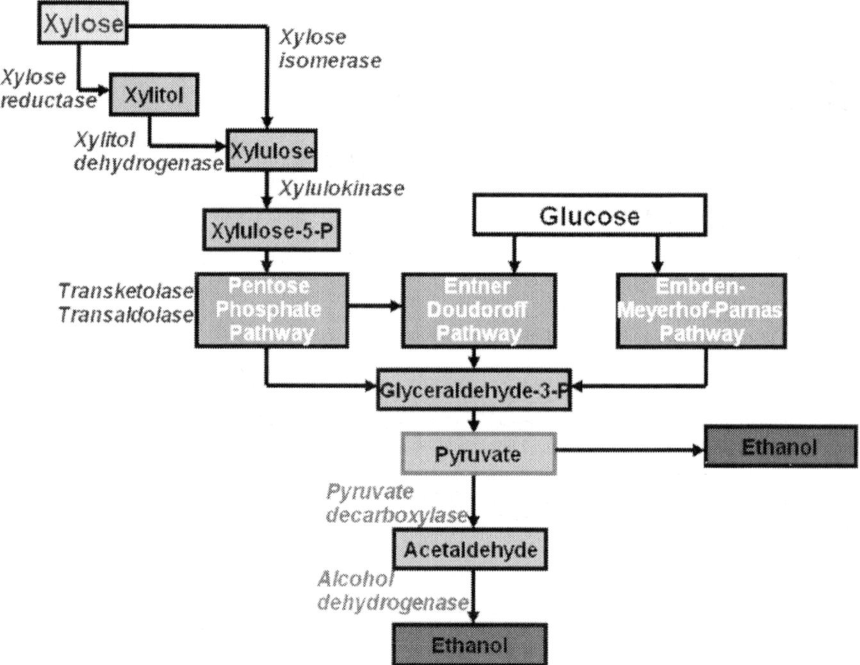

Fig. 5. Fermentation pathways of xylose to ethanol

3.5.2 Organic acids

In the last decade, microbially-produced organic acids (Mattey, 1992) find increased use in the food industry and as raw materials for manufacture of biodegradable polymers (Magnusson & Lasure, 2004). For instance, the production of D-lactic acid as well as L-lactic acid is of significant importance for the practical application of polylactic acid, which is an important raw material for bioplastics that can be produced from biomas (Okano et al., 2009) Organic acids are used in food preservation because of their effects on bacteria (Dibner & Butin, 2002). The non-dissociated (non-ionized) organic acids can penetrate the bacteria cell wall and disrupt the normal physiology of certain types of bacteria such as *E. coli*, *Salmonella* and *Campylobacter* species that are pH-sensitive and cannot tolerate a wide internal and external pH gradient. Upon passive diffusion of organic acids into the bacteria, where the pH is near or above neutrality, the acids dissociate and the cations lower the bacteria internal pH, leading to situations that impair or stop the growth of bacteria. Furthermore, the anions of the dissociated organic acids accumulate within the bacteria and disrupt their metabolic functions leading to osmotic pressure increase that is incompatible with the bacterial survival. For example, lactic acid and its sodium and potassium salts are widely used as antimicrobials in food products, in particular, meat and poultry such as ham and sausages. Tables 6 and 7 summarize the major production organisms, substrates and uses of organic acids.

3.5.3 Xylitol

Xylitol is produced chemically by hydrogenation of xylose, which converts the sugar aldehyde into a primary alcohol (Karimkulova et al., 1989). Hydrogenation is carried out at high pressures (up to 50 atm), high temperature (80-140°C) using expensive catalysts (Nickel Raney) and expensive purification processes (Mikkola et al., 2000). The xylitol yields are low – on average 50-60% from xylan. The drawbacks of the chemical process can be overcome by using a biological route of xylitol production that is carried out by microorganisms at low temperature (30-35°C). The microbial conversion employs naturally fermenting yeasts (*Candida*) such as *C. tropicalis and C. guillermondi* that yield of 65-90% from xylan.

Organic acid	Production organism	Substrate	State
Citric	*A. niger*	Sugar cane molasses, corn syrups, lignocellulose, agri- and food waste	Commercial
Gluconic	*A. niger*	Glucose, glucose corn syrups	Commercial
Lactic	*L. delbrueckii* *A. oryzae*	Xylose, glucose, starch, cellulose, newspaper, MSW (xylose, mannose)	Commercial
Itaconic	*A. terreus*	Sugar cane molasses, corn syrups, xylose	Commercial
Fumaric	*Rhyzopus* spp.	glucose, sucrose, sugar cane molasses, corn syrups, starch, xylose	Experimental
Malic	*Brevibacterium*	Fumaric acid	Commercial
Aspartic	*E. coli*	Fumaric acid + NH_3	Commercial
Succinic	*A. succiniciproducens* *A. succinogenes* *E. coli* (recombinant)	Glucose, sugar cane molasses Glucose, xylose	Experimental

Table 6. Production of organic acids from xylose

Alternatively, recombinant strains containing a xylose reductase gene (i.e. recombinant *S. cerevisiae*) can be used with a very high production yield of 95% from the theoretical maximum. The microbially produced xylitol requires less purification than the chemical process (Prakasham et al., 2009). Due to its anti-cariogenic and anti-plaque action (Trahan, 1995), xylitol is used around the world as a sweetener in chewing gums, pastilles, and oral hygiene products such as toothpaste, fluoride tablets and mouthwashes. More than 10% of its use in sugar-free chewing gums which have a world market of more than $12 million per annum. Due to its structure, xylitol is a non-fermentable sugar alcohol with dental health benefits in caries prevention, showing superior performance to other polyols (polyalcohols). Its plaque-reducing effect is manifested by attracting and starving harmful micro-organisms because cariogenic bacteria prefer fermentable six-carbon sugars as opposed to the nonfermentable xylitol (Milgrom et al., 2006).

Possessing approximately 40% less food energy, xylitol is a low-calorie alternative to table sugar. Absorbed more slowly than sugar, it does not contribute to high blood sugar levels or the resulting hyperglycemia caused by insufficient insulin response. Its glycemic index is approximately 10-fold lower than that of sucrose (Fig. 6). This characteristic has also proven beneficial for people suffering from metabolic syndrome, a common disorder that includes insulin resistance, hypertension, hypercholesterolemia, and an increased risk for blood clots.

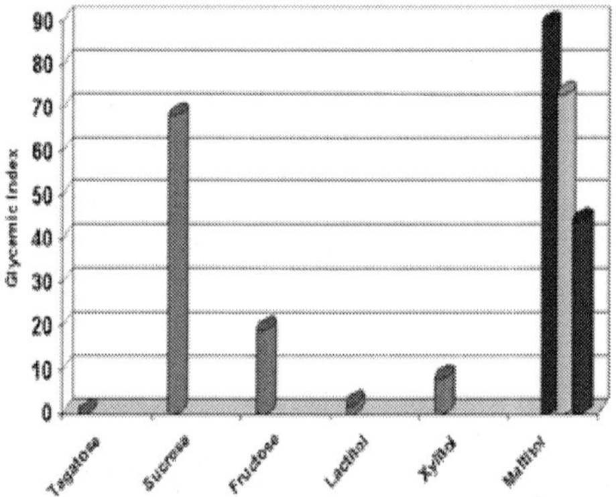

Fig. 6. Glycemic index of xylitol in comparison to other sweeteners

Xylitol also has potential as a treatment for osteoporosis - it prevents weakening of bones and improves bone density (Mattilla et al., 2002). Studies have shown xylitol chewing gum can help prevent ear infections (Uhari et al., 1998). When bacteria enter the body, they adhere to the tissues using a variety of sugar complexes. The open nature of xylitol and its ability to form many different sugar-like structures appears to interfere with the ability of many bacteria to adhere. Xylitol is also one of the building block chemicals that can be used in production of ethylene glycol, propylene glycol, lactic acid, xylaric acid, and for synthesis of unsaturated polyester resins, antifreeze, etc.

3.5.4 Enzymes

As discussed in 3.2, xylan-containing substrates, and in some instances xylose, can serve as inducers for production of xylan-degrading enzymes including xylanase. Another enzyme of importance that can be produced on these substrates is xylose isomerise (EC 5.3.1.5). This enzyme is used industrially to convert glucose to fructose in the manufacture of high-fructose corn syrups, HFCS (Bhosale et al., 1996). HFCS is produced by milling corn to produce corn starch which is first treated with alpha-amylase to produce shorter chains oligosaccharides and then with glucoamylase to produce glucose. Finally, xylose isomerase (also known as glucose isomerase) converts glucose to a mixture of about 42% fructose and

Organic acid	Application
Citric (Kubicek et al., 1980)	70% in food, confectionary and beverage products, 30% pharmaceuticals (anticoagulant blood preservative, antioxidant) & metal cleaning. Selling price decreased with market shift from pharmaceuticals to food applications (879,000 metric tons produced in 2002)
Lactic (Yang et al., 1995)	Acidulant, flavor enhancer, food preservative, feedstock for calcium stearoyl-2-lactylates (baking), ethyl lactate (biodegradable solvent) and polylactic acid plastics (100% biodegradable) for packaging, consumer goods, biopolymers (approved by FDA). Estimated US consumption 30 million lb with 6% growth pa. Potential demand 5.5 billion lb as very large volume-commodity chemical
Itaconic (Kautola et al., 1985)	Feedstock for syntheses of polymers for use in carpet backing and paper coating N-substituted pyrrolidinones for use in detergents and shampoos. Cements comprising copolymers of acrylic and itaconic acid
Aspartic (Dunn & Smart, 1950)	For synthesis of aspartame, monomer for manufacture of polyesters, polyamides, polyaspartic acid as a substitute for EDTA with potential market of $450 million per year
Fumaric (Overman & Romano, 1969)	For manufacture of synthetic resins, biodegradable polymers, intermediate in chemical and biological synthesis
Malic (Peleg et al., 1989)	Acidulant in food products, citric acid replacement, raw material for manufacture of biodegradable polymers, for treatment of hyperammoemia, liver dysfunction, component for aminoacid infusions
Succinic (Zeikus et al., 1999)	Acidulant, pH modifier, flavoring and antimicrobial agent, ion chelator in electroplating to prevent metal corrosion, surfactant, detergent, foaming agent, for production of antibiotics, amino acids, pharmaceuticals. Market potential of 270,000 t in 2004, US domestic market estimated at $1.3 billion per year with 6-10% annual growth

Table 7. Applications of organic acids

50–52% glucose (HFCS-42) with some other sugars mixed in. This 42–43% fructose-glucose mixture is then subjected to a liquid chromatography step, where the fructose is enriched to about 90% and then back-blended with 42% fructose to achieve a 55% fructose final product (HFCS-55). While the relatively inexpensive alpha-amylase and glucoamylase enzymes are added directly to the slurry and used only once, the more costly xylose isomerase is packed into columns and used repeatedly until it loses its activity. Thus, production of HFCS using xylose isomerase is the major application of immobilized enzyme technology (Parker et al., 2010).

The most widely used varieties of high-fructose corn syrup are HFCS-55 (mostly used in soft drinks) and HFCS-42 (used in many foods and baked goods). In the US, HFCS is among the sweeteners that have primarily replaced sucrose. Factors for this include governmental production quotas of domestic sugar, subsidies of US corn, and an import tariff on foreign sugar, all of which combine to raise the price of sucrose to levels above those of the rest of the world, making HFCS less costly for many sweetener applications. Pure fructose is the sweetest of all naturally occurring carbohydrates and 1.73 times as sweet as sucrose (Hyvonen & Koivistoinen, 1982). Fructose has the lowest glycemic index (GI = 19) of all the natural sugars and may be used in moderation by diabetics. In comparison, ordinary table sugar (sucrose) has a GI of 65 and honey has a GI of 55. Per relative sweetness, HFCS-55 is comparable to sucrose. Currently, HFCS dominate industrial sugar market in the US. The average American consumed approximately 17.1 kg of HFCS in 2008 versus 21.2 kg of sucrose. In Japan, HFCS consumption accounts for one quarter of total sweetener consumption. The world market for HFCS was 5 million tons in 2004.

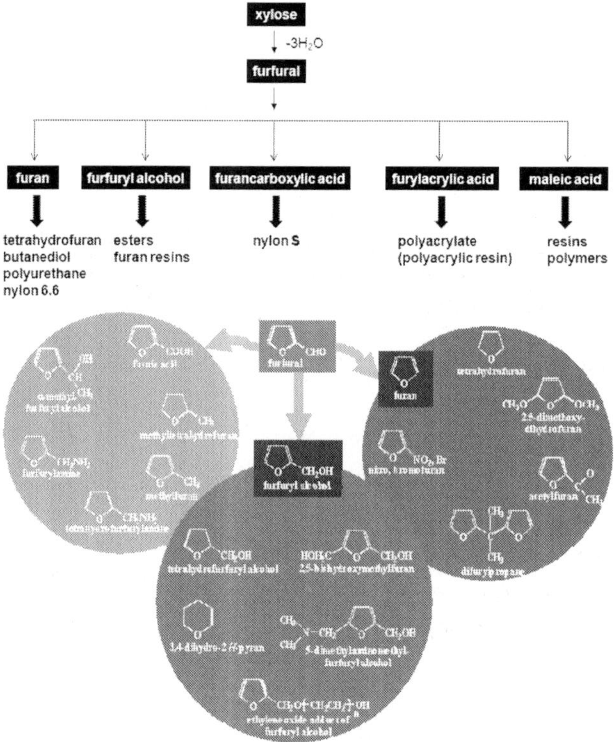

Fig. 7. Furfural-based applications

3.5.5 Furfural

When heated with sulfuric acid, hemicellulose (xylan) undergoes hydrolysis to yield monosugars (xylose). Under the same conditions of heat and acid, xylose and other five carbon sugars undergo dehydration, losing three water molecules to become furfural (Fig 7). Furfural and water evaporate together from the reaction mixture, and separate upon condensation. Biomass agri-wastes like cornstalks, corncobs and the husks of peanuts and oats are rich in xylan and about 10% of the mass these plant residues can be recovered as furfural (Zeitsch, 2000). Furfural represents a renewable building block chemical which is currently regaining attention as a biobased alternative for the production of industrial and household chemicals (Mamman et al., 2008) - from antacids and fertilizers to plastics and paints (Fig. 6). The global production capacity is about 450,000 tons as of 2004. China is the biggest supplier of furfural, and accounts for around half of the global capacity. The world production of furfural in 2005 was about 250,000 t/a, at a stable price of $1,000/t and it is being projected to 225 thousand metric tons per annum (Win, 2005).

4. Hemicellulose-based biorefinery

A diagram of a biorefinery based on generation of primary and secondary bioproducts from hemicelluloses is presented in Fig. 8. It illustrates the enormous potential that hemicelluloses have to produce high-value products that can enhance the economics of the integrated production of biofuels, biochemicals and biopolymers (Zhang, 2008). There are however technological and socio-economic challenges that still need to be overcome to make these processes viable (Carvalheiro et al., 2008). The technological challenges are related to optimization of process conditions to maximize biorefinery-derived value such as 1) improvements in extraction efficiency of hemicellulose for minimal sugar degradation while preserving the pulp and paper properties; 2) improvements in pentose fermentation and tolerance of microbial producers to inhibitors and ethanol. A further process integration would reduce the number of processing steps, decrease energy needs and reuse process streams. The socio-economic challenges need to address the 1) complex systems of policies and regulations in different countries and make them more compatible; 2) environmental impact of biomass removal; 3) pressure from environmental groups on policy makers; and 4) unstable commodity prices.

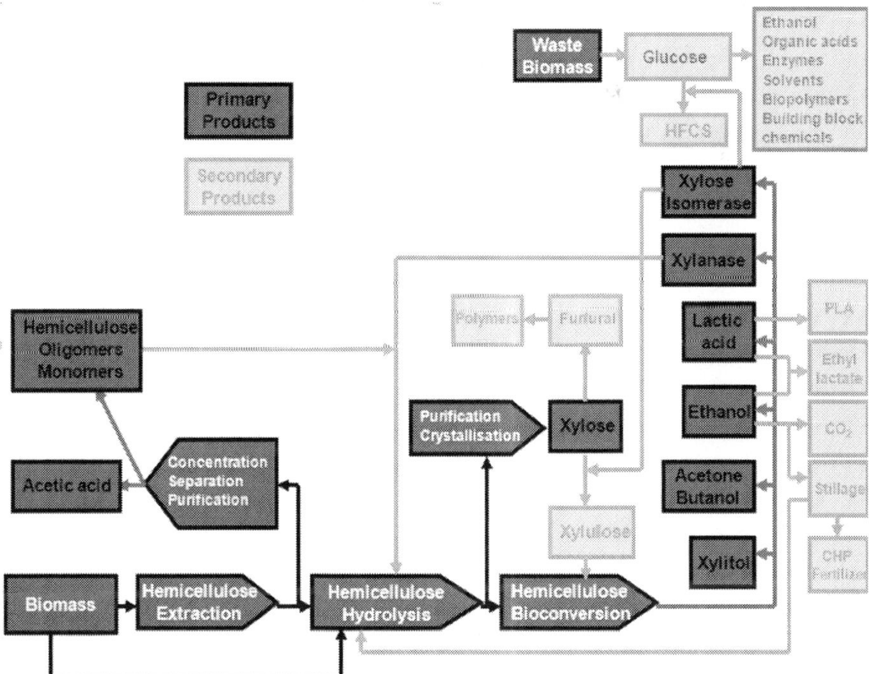

Fig. 8. Schematic of a hemicellulose-based biorefinery

5. Conclusion

Hemicellulose is the second most abundant polysaccharide after cellulose and comprises about 25-30% of the lignocellulosic biomass, which has an estimated annual production on earth of about 60 billion tons. The US Department of Energy has recently projected that annually more than 1.3 billion tons of biomass could be produced on a sustainable basis which could provide about 30% of the country's demand for transportation fuels. About a third of the biomass resources in US are wood-based. Due to the increasing off-shore competition, global movement and incentives for green fuels and chemicals, the North American pulp and paper and other fiber processing industries need to create additional revenues and diversify their products and markets to remain competitive. To achieve this, these forest-based industries need to evolve into integrated forest biorefineries. Based on the billion ton vision, nearly 400 million tons of hemicellulose are available in US per annum for bioprocessing to fuels and chemicals. In addition, every year approximately 15 million tons of hemicellulose are produced by the pulp and paper industry alone, and according to preliminary results, this can yield in excess of 2 billion gallons of ethanol and 600 million gallons of acetic acid, with a net cash flow of $3.3 billion. Unfortunately, during the current fiber processing of woody feedstocks, hemicellulose is not efficiently utilized and often discarded as agri-waste or industrial effluents with limited usage. However, a number of bioproducts including biofuels, biochemicals and biomaterials can be produced from biomass hemicellulose to enhance the value extracted from wood fiber and improve the process economics in a forest-based biorefinery.

6. Acknowledgement

Financial support by the Center for Bioprocessing Research and Development (CBRD) at the South Dakota School of Mines & Technology (SDSM&T), the South Dakota Board of Reagents (SD BOR), the South Dakota Governor's Office for Economic Development (SD GOED), and the U.S. Air Force Research Laboratory (AFRL) is gratefully acknowledged.

7. References

Bhosale, S.H, Mala, B.R. & Deshpande, V.V. (1996). Molecular and industrial aspects of glucose isomerase. *Microbiol Rev*, Vol. 60, pp. 280-300

Biobased industrial products: Research and commercialization priorities. (2000). *National Academy Press*, Washington, DC, <Available at:
<http://www.nap.edu/books/0309053927/html>

Casebier R.L. & Hamilton J.K. (1965). Alkaline degradation of glucomannans and galactoglucomannans. *Tappi J*, Vol. 48, pp. 664-669

Carvalheiro, F., Luís C. Duarte, L.C. & Gírio, F.M. (2008). Hemicellulose biorefineries: a review on biomass pretreatments. *J Sci & Ind Res*, Vol. 67, pp.849-864

Chipeta, Z.A., du Preez, J.C., Szakacs, G. & Christopher, L. (2005). Xylanase production by fungal strains on spent sulphite liquor. *Appl Microbiol Biotechnol* , Vol 69, pp. 71-78

Christopher, L. (2004). Xylanases: properties and applications. In: *Concise encyclopedia of bioresource technology*. Pandey, A. Ed., pp. 601-609, The Haworth Press, Binghamton, New York, USA

Coughlan M.P. & Hazlewood G.P. (1993). ß-1,4-D-Xylan-degrading enzyme systems: biochemistry, molecular biology and applications. *Biotechnol Appl Biochem*, Vol 17, pp. 259-289

Dibner, J.J. & Butin, P. (2002). Use of organic acids as a model to study the impact of gut microflora on nutrition and metabolism. *J Appl Poultry Res*, Vol. 11, pp. 453-463

Dunn, M.S. & Smart, B.W. (1950). DL-Aspartic acid. *Org Synth*, Vol. 30, p. 7

Ebringerova, A. & Heize, T. (2000). Xylan and xylan derivatives - biopolymers with valuable properties 1. Naturally occuring xylans, structures, isolation procedures and properties. *Macromol Rap Commun*, Vol. 21, pp. 542-456

Eliasson, A., Christensson, C., Wahlbom, C.F. & Hahn-Hägerdal, B. (2000). Anaerobic xylose fermentation by recombinant *Saccharomyces cerevisiae* carrying *XYL1*, *XYL2*, and *XKS1* in mineral medium chemostat cultures. *Appl Environ Microbiol*, Vol. 66, pp. 3381-3386

Fengel, D. & Wegener, G. (1984). *Wood-chemistry, ultrastructure, reactions*. Walter de Gruyter, Berlin, Germany

Gustavsson, C.A.S. & Al-Dajani, W.W. (2000). The influence of cooking conditions on the degradation of hexenuronic acid, xylan, glucomannan, and cellulose during kraft pulping of softwood. *Nord Pulp Pap Res*, Vol. 15, pp. 160-167

Hyvonen, L. & Koivistoinen, P. (1982). Fructose in food systems. In: *Nutritive sweeteners*. Birch, G.G. & Parker K.J., Eds., pp. 133-144, Applied Science Publishers, London, UK & New Jersey, USA

Kantelinen, A., Hortling, B., Sundquist, J., Linko, M. & Viikari, L.(1993). Proposed mechanism of the enzymatic bleaching of kraft pulp with xylanases. *Holzforschung*, Vol. 47, pp. 318-324

Kato, S. (1981). Ultrastructure of the plant cell wall: biochemical view point. *Encyclop plant physiol*, Vol. 30, pp. 409-425

Kautola, H., Vahvaselka, M., Linko, Y.Y. & Linko, P.(1985). Itaconic acid production by immobilized *Aspergillus terreus* from xylose and glucose. *Biotechnol Lett*, Vol. 7, pp. 167-172

Karimkulova, M.P., Khakimov, Y.S. & Abidova, M.F. (1989). Activity of modified catalysts in hydrogenation of xylose to xylitol. *Chem Nat Comp*, Vol. 25, pp. 370-371

Kerr, A.J. & Goring, D.A.I. (1975). The role of hemicellulose in the delignification of wood. *Can J Chem*, Vol. 53, pp. 952-959

Kosaric, N., Ho, K.K. & Duvnjak, Z. (1980). Effect of spent sulphite liquor on growth and ethanol fermentation efficiency of *Saccharomyces ellipsoideus*. *Water Poll Res J Can*, Vol. 16, pp. 91-98

Kubicek, C.P., Zehentgruber, O., El-Kalak, H. & Röhr, M. (1980). Regulation of citric acid production by oxygen: Effect of dissolved oxygen tension on adenylate levels and respiration in *Aspergillus niger*. *Eur J Appl Microbiol Biotechnol*, Vol. 9, pp. 101-115

Magnuson, J.K. & Lasure, L.L. (2004). Organic acid production by filamentous fungi. In: *Advances in fungal biotechnology for industry, agriculture, and medicine*, Tkacs, J.S. &

Lange, L., Eds., pp. 307-340, Kluwer Academic/Plenum Publishers, New York, USA

Mamman, A.S., Jong-Min Lee, J.-M., Yeong-Cheol Kim, Y.-C., In Taek Hwang, I.T., No-Joong Park, N.-J., Young Kyu Hwang, Y.K., Jong-San Chang, J.-S. & Jin-Soo Hwang, J.-S. (2008). Furfural: Hemicellulose/xylosederived biochemical. *Biofpr*, Vol. 5, pp. 438-454

Matsushika, A., Inoue, H., Kodaki, T. & Sawayama, S. (2009). Ethanol production from xylose in engineered *Saccharomyces cerevisiae* strains: current state and perspectives. *Appl Microbiol Biotechnol*, Vol. 84, pp. 37-53

Mattey, M. (1992). The production of organic acids. *Crit Rev Biotechnol*, Vol. 12, pp. 87-132

Mattila, P.T., Svanberg, M.J., Jämsä, T. & Knuuttila, M.L. (2002). Improved bone biomechanical properties in xylitol-fed aged rats. *Metabolism*, Vol. 51, pp. 92-96

Mikkola, J.-P., Salmi, T., Sjöholm, R., Mäki-Arvela, P. & Vainio, H. (2000). Hydrogenation of xylose to xylitol: three-phase catalysis by promoted raney nickel, catalyst deactivation and in-situ sonochemical catalyst rejuvenation. *Stud Surf Sci Catal*, Vol. 20, pp. 27-32

Milgrom, P., Ly, K.A., Roberts, M.C., Rothen, M., Mueller, G. & Yamaguchi, D.K. (2006). Mutans streptococci dose response to xylitol chewing gum. *J Dent Res*, Vol. 85, pp. 177-181.

Mueller, J.C. & Walden, C.C. (1970). Microbiological utilisation of sulphite liquor. *Process Biochem*, Vol. 6, pp. 35-42

Okano, K., Yoshida, S., Yamada, R., Tanaka, T., Ogino, C., Fukuda, H. & Kondo, A. (2009). Improved production of homo-D-lactic acid via xylose fermentation by introduction of xylose assimilation genes and redirection of the phosphoketolase pathway to the pentose phosphate pathway in L-lactate dehydrogenase gene-deficient *Lactobacillus plantarum*. *Appl Environ Microbiol*, Vol. 75, pp. 7858-7861

Onysko, K.A. (1993). Biological bleaching of chemical pulps: A review. *Biotech Adv*, Vol. 11, pp. 179-198

Overman, S.A. & Romano, A.H. (1969). Role of pyruvate carboxylase in fumaric acid accumulation by *Rhizopus nigricans*. *Bacteriol Proc*, Vol. 69, p. 128

Pandey, A., Selvakumar, P., Soccol, C.R & Nigam, P. (1999). Solid state fermentation for the production of industrial enzymes. *Curr Sci*, Vol. 77, pp. 149-162

Parker, K., Salas, M. & Nwosu, V.C. (2010). High fructose corn syrup: Production, uses and public health concerns. *Biotechnol Mol Biol Rev*, Vol. 5, pp. 71-78

Peleg,Y., Barak, A., Scrutton, M.C. & Goldberg, I. (1989). Malic acid accumulation by *Aspergillus flavus*. 3. 13C-NMR and isoenzyme analyses. *Appl Microbiol Biotechnol*, Vol. 30, pp. 176-183

Perlack, R.D., Wright, L.L., Turhollow, A.F., Graham, R.L., Stokes, B.J. & Erbach, D.C. (2005). Biomass as feedstock for a bioenergy and bioproducts industry: The technical feasibility of a billion-ton annual supply. U.S. Department of Energy, Available at: < http://www.osti.gov/bridge>

Prakasham, R.S., Sreenivas, R.R. & Hobbs, P.J. (2009). Current trends in biotechnological production of xylitol and future prospects. *Curr Trends Biotechnol Pharm*, Vol. 3, pp. 8-36

Puls, J. & Saake, B. (2004). Industrially isolated hemicelluloses. In: *Hemicelluloses: science and technology*. Gatenholm, P. & Tenkanen, M., Eds., pp. 24-37, American Chemical Society, Washington DC, USA

Royer, J.C. & Nakas, J.P. (1989). Xylanase production by *Trichoderma longibrachiatum*. Enzyme Microb Technol, Vol. 11, pp. 405-410

Sjostrom, E. (1993). Wood chemistry. Fundamentals and applications. Academic Press, San Diego, USA

Szakacs, G., Urbanszki, K. & Tengerdy, R.P. (2001). In: *Glycosyl hydrolases for biomass conversion*. Himmel, M.E.,Baker, J.O. & Saddler, J.N., Eds., pp. 190-203, American Chemical Society, ACS Symposium Series 769, Washington DC, USA

Szendefy, J., Szakacs, G. & Christopher, L. (2006). Potential of solid-state fermentation enzymes of *Aspergillus oryzae* in biobleaching of paper pulp. *Enzyme Microb Technol*, Vol. 24, 1149-1156

Tao, H., Gonzalez, R., Martinez, A., Rodriguez, M., Ingram, L.O., Preston, J.F. & Shanmugam, K.T. (2001). Engineering a homo-ethanol pathway in *Escherichia coli*: Increased glycolytic flux and levels of expression of glycolytic genes during xylose fermentation. *J Bacteriol*, Vol. 183, pp. 2979-2982

Trahan, L. (1995). Xylitol: a review of its action on mutans streptococci and dental plaque-its clinical significance. *Int Dent J*, Vol. 45, pp. 77-92

Uhari, M., Kontiokari, T. & Niemelä, M. (1998). A novel use of xylitol sugar in preventing acute otitis media. Pediatrics, Vol. 102, pp. 879–974

Viikari, L., Tenkanen, M., Buchert, J., Rättö, M., Bailey, M., Siika-aho, M. & Linko M. (1993). Bioconversion of forest and agricultural plant residues. In: *Hemicellulases for industrial applications*. Saddler, J.N, Ed., pp. 131-182, CAB International, Wallingford, UK

Viikari, L., Kantelinen, A., Sundquist, J. & Linko, M. (1994). Xylanases in bleaching: From an idea to the industry. *FEMS Microbiol Rev*, Vol. 13, pp. 335-350

Win, D.T. Furfural – gold from garbage. (2005). *AU J Technol*, Vol. 58, pp. 185-190

Wong, K.Y.W., Tan, L.U.L. & Saddler, J.N. (1988). Multiplicity of β-1, 4-xylanase in microorganisms: functions and applications. *Microbiol Rev*, Vol. 52, pp. 305-317

Wong, K.Y.W., De Jong, E., Saddler, J.N. & Allison, R.W. (1997). Mechanism of xylanase aided bleaching of kraft pulp. *Appita J*, Vol. 50, pp. 509-518

Yang, C.W., Lu, Z.J. & Tsao, G.T. (1995). Lactic acid production by pellet-form *Rhizopus oryzae* in a submerged system. *Appl Biochem Biotechnol*, Vol. 51, pp. 57-71

Zeikus, J.G., Jain, M.K. & Elankovan, P. (1999). Biotechnology of succinic acid production and markets for derived industrial products. *Appl Microbiol Biotechnol*, Vol. 51, 545-552

Zeitsch, K.J. (2000). *The chemistry and technology of furfural and its many by-products*. Elsevier, Köln, Germany

Zhang, M., Eddy, C., Deanda, K., Finkelstein, M. & Picataggio, S. (1995). Metabolic engineering of a pentose metabolism pathway in ethanologenic *Zymomonas mobilis*. *Science*, Vol. 267, pp. 240-243

Zhang, Y.-H.P. (2008). Reviving the carbohydrate economy via multi-product lignocellulose biorefineries. *J Ind Microbiol Biotechnol*, Vol. 35, pp. 367-375

Zheng, Y., Pan, Z. & Zhang, R. (2009). Overview of biomass pretreatment for cellulosic ethanol production. *Int J Agric & Biol Eng*, Vol. 2, pp. 51-58

15

Incorporating Carbon Credits into Breeding Objectives for Plantation Species in Australia

Miloš Ivković[1], Keryn Paul[1] and Tony McRae[2]
*[1]Commonwealth Scientific and
Industrial Research Organisation, Canberra, ACT
[2]Southern Tree Breeding Association Inc., Mount Gambier, SA
Australia*

1. Introduction

Forest tree plantations can capture more carbon than most other land uses. For example, a radiata pine (*Pinus radiata* D. Don) or blue gum (*Eucalyptus globulus* Labill.) plantations in the Green Triangle region in south-eastern Australia can on average produce more than 20 $m^3ha^{-1}y^{-1}$ of wood, with a 19% and 25% rate of carbon sequestration, respectively (Polglase et al. 2008). As carbon trade is becoming a reality, there is a need to incorporate carbon credits into the economic models of plantation forestry for those major and some emerging species. Carbon sequestration needs to be considered as a biological trait and bio-economic models need to include its value. In this chapter we will show how carbon sequestration can be incorporated into the breeding objectives of Australian tree improvement programs for commercial plantation species.

The procedure for the development of a bio-economic model and setting up breeding objectives for plantation trees includes the following major steps (Ivković et al., 2010):

- specification of the production system, which includes, silvicultural management regime, and harvesting, transportation and processing systems;
- identification of the wood-flows, and sources of income and cost in a specified production system (i.e. plantation growing, processing or integrated production);
- determination of how different biological traits influence wood-flows, and incomes and costs in the production system (i.e. identification of the critical traits affecting production system to consider as breeding-objective traits);
- derivation of the economic value, or weight, defined as the value of unit trait change for each breeding objective trait for existing and future plantation production systems.

The economic weights for breeding objective traits are used to develop an optimal selection index that maximises the expected genetic gain and profitability of production (e.g. Ivković et al., 2006a). To incorporate carbon credits into breeding objectives and to derive economic weights for plantation species in Australia it is necessary to:

- consider current breeding objectives for major plantation species including *P. radiata, E. globulus* and other minor species;

- consider currently proposed carbon pooling and accounting systems for reforestation; and

- incorporate carbon credits for sequestration into existing breeding objectives for *P. radiata*, *E. globulus* and other minor plantation species

We used *P. radiata* as a model species for incorporating carbon credits into breeding objectives. The model should be applicable to other plantation species. Sensitivity analyses involving different carbon prices, discount rates, timber prices, and carbon accounting options were performed.

2. Current breeding objectives

There has been a substantial amount of work on developing bio-economic models and breeding objectives for *P. radiata* structural timber, and to a lesser extent for pulp and paper production (e.g. Apiolaza and Garrick, 2000; Greaves, 1999; Ivković et al., 2006a, 2006b). Only one study addressed the issue of genetic gains for carbon sequestration, a preliminary study of radiata pine plantations in New Zealand (Jayawickrama, 2001). In *E. globulus*, bio-economic models have been developed for pulp production (e.g. Borralho et al., 1993, Greaves et al., 1997). There has been only one study on integrating carbon into the breeding objectives for *E. globulus* (Whittock et al., 2007). There has also been recent interest in using *E. globulus* and *E. nitens* for solid timber production and the development of bio-economic models for such production systems (e.g. Nolan et al. 2005). Such bio-economic models are being developed for other (minor) tree species, as well (e.g. *E. cloeziana*, *E. dunnii*, *E. saligna*, *E. obliqua*, *E. pilularis*, *Corymbia maculata*, *C. citriodora*, and *E. diversicolor*). However, there are no formally defined breeding objectives for the alternative production systems and/or minor plantation species.

2.1 *Pinus radiata*

Preliminary estimates of gains in carbon sequestration, obtainable through genetic improvement, for radiata pine plantations in New Zealand were calculated by Jayawickrama (2001). Data on height, diameter and wood density from New Zealand progeny trials were used to estimate heritability and the phenotypic variance for stem dry-weight production. The amount of carbon sequestered in the stem, branches and roots of typical stands of unimproved radiata pine was simulated using the C_Change simulator (Beets, 2006; Beets et al., 1999). The amounts of carbon sequestered at the end of one rotation, and the long-term average over successive rotations were estimated. Carbon sequestered under a standard, "direct sawlog" management regime differed among regions, from 211 tonnes (t) at the end of a 28-year rotation in Canterbury to 322 t on the East Coast. The highest simulated genetic gain in long-term carbon sequestration for the standard regime was 29 t ha^{-1}y^{-1}, for a stand grown in the East Coast region, using seed from the best 10 of 1000 ranked parents.

Based on the highest estimated rate of carbon sequestration, and a value of NZ$20 per t of carbon, the extra carbon credits attributable to genetic improvement, under the direct sawlog regime, would be worth NZ$307 per ha. However, a low-cost "plant-and-leave" regime sequestered even more carbon than the standard sawlog regime. Therefore, owners of large post-1990 radiata pine plantings on converted farmland or pasture could gain large financial benefits from genetic improvement and/or management targeting increased carbon sequestration. Plantation growers in New Zealand are likely to aim for multiple objectives rather than for carbon sequestration alone and the trees would be selected on an optimised combination of objective traits.

In Australia the Southern Tree Breeding Association (STBA), uses a 'bio-economic' model based on industry data and various sawmill performance studies to assess the relative importance of breeding objective traits (Ivković et al., 2006a, 2006b). A generic bio-economic model similar to that used by the STBA was created for the purpose of this study (**Table 1.**).

		Effect of 10% Trait Increase			
	Base	**MAI**	**SWE**	**BRS**	**MOE**
Wood flows (m³/ha)					
pulplog	166	173	181	178	166
sawlog	555	620	540	543	555
green sawn timber	269	296	260	263	269
dry structural timber	182	200	176	178	182
sawmill residue (chip)	255	280	250	250	254
Total harvested volume	**721**	**793**	**721**	**721**	**721**
Costs NPV ($/ha)					
establishment	-2,047	-2,047	-2,047	-2,047	-2,047
ann. maintenance	-1,241	-1,241	-1,241	-1,241	-1,241
harvest	-1,628	-1,881	-1,628	-1,654	-1,628
transport	-973	-1115	-973	-973	-973
green mill	-2,149	-2,388	-2,093	-2,101	-2,149
dry mill	-1,673	-1,842	-1,619	-1,637	-1,673
Total costs	**-9,712**	**-10,514**	**-9,602**	**-9,654**	**-9,712**
Income NPV ($/ha)					
pulplog	1,952	2,030	2,057	2,013	1,952
sawlog	6,151	7,184	5,976	6,020	6,151
Total stumpage	**8,103**	**9,214**	**8,033**	**8,034**	**8,103**
sawn timber	10,724	11,815	10,378	10,402	11,547
other sawn products	1,046	1,152	1,012	1,023	1,046
sawmill residue chip	1,256	1,406	1,230	1,228	1,256
Total income	**14,979**	**16,403**	**14,677**	**14,667**	**15,801**
Income NPV – Costs NPV($/ha)					
Plantation growing	**2,214**	**2,930**	**2,144**	**2,119**	**2,214**
Percent NPV change		**32%**	**-3%**	**-4%**	**0%**
Integrated production	**5,267**	**5,889**	**5,075**	**5,013**	**6,089**
Percent NPV change		**12%**	**-4%**	**-5%**	**16%**

Table 1. Summary of the wood flows, costs and incomes per hectare for a generic integrated radiata pine production system at the base level and after increase in mean annual increment (MAI), sweep (SWE), branch size (BRS), and modulus of elasticity (MOE). Net present value (NPV) was calculated at 7% discount rate over one 30-year rotation.

The breeding objective traits, namely, mean annual increment (MAI), stem straightness or sweep (SWE), branch size (BRS), and wood stiffness measured by modulus of elasticity (MOE), were chosen based on their significant economic importance. For plantation growers, MAI (i.e. harvest volume) was the main driver of profit. For integrated production systems, MOE and MAI were more balanced in their relative importance. In section 4 of this chapter, we will incorporate carbon credits into this bio-economic model, which is expected to influence the breeding objective and decisions regarding selection of trees for establishing new plantations (e.g. Maclaren et al., 2008). Carbon sequestration in plantations is directly related to the amount of dry-weight of biomass accumulated and therefore the relative economic weights of traits related to growth (i.e. MAI and BRS) and wood density (i.e. MOE) and therefore dry weight yield of biomass.

2.2 *Eucalyptus globulus*

Economic breeding objectives for the production of kraft pulp from plantation grown *E. globulus* have been defined previously by several authors (e.g. Borralho et al., 1993; Greaves et al., 1997). The authors identified the same three biological traits (clearfall volume, wood basic density and kraft pulp yield) as having the greatest economic value. Those models considered only costs and/or incomes within a single rotation. More recently, Whittock *et al.* (2007) developed a model on a plantation estate scale, consisting of multiple stands and age classes, accounting for carbon sequestration in biomass and carbon revenues. The authors considered an *E. globulus* plantation estate that has been established on ex-pasture sites with the major expansion of the estate occurring after 1990. They examined the impact of carbon revenues associated with clearfall volume and basic density in an export chip production system. Income was calculated based on sale of wood chips, and carbon revenues were directly proportional to the biomass accumulation in the plantation estate. Whole tree growth was proportioned to merchantable volume increment and biomass allocated to roots, stem, branches, leaves and bark following methods by Madiera *et al.* (2002). Thinned material, stumps and roots were assumed to decay linearly over a 7-year-period.

In the study the tradable unit of CO_2 was the biomass equivalent of one metric ton of CO_2 (tCO_2e). Carbon was assumed to be 46% of oven-dry biomass, and a ton of carbon equivalent to 3.67 tCO_2. Carbon in wood products was considered lost immediately after harvest. Total carbon dioxide equivalent (CO_2e) accumulation was in the order of ~146 t CO_2e ha^{-1}, of which 62 t CO_2e ha^{-1} is tradable in 2012 (the 1st Kyoto Protocol commitment period) and a further 30 t CO_2e ha^{-1} is tradable in 2016 (a hypothetical second Kyoto protocol commitment period). The correlated response among breeding objectives with and without carbon revenues never fell below 0.86 in sensitivity analysis, and the mean was 0.93. Where economic breeding objectives for the genetic improvement of *E. globulus* for pulpwood plantations are based on maximizing net present value by increasing biomass production, the consideration of carbon revenues in economic breeding objectives will have a minimal effect on the relative economic weights of the key economic traits, wood basic density and standing volume at harvest.

2.3 Other production systems and plantation species

The hardwood industry in Australia has been affected by a reduction in the harvest of native forests and competition from lower priced plantation softwood products. However,

Eucalyptus species have desirable characteristics that most softwoods cannot match and the hardwood industry is moving to higher price, appearance and niche structural products (Nolan et al., 2005). Eucalypt plantations for solid wood, or for dual pulp and solid wood production are emerging as a new industry and more than 100,000 ha of eucalypt plantations are currently managed for sawlog production (DAFF 2007). The first step in the process of tree improvement is to clearly set the breeding and silvicultural objectives and formally determine the economic importance of different biological tree traits.

In eucalypt plantations for solid wood, different production systems are used, such as, unthinned and unpruned; thinned and unpruned, and thinned and pruned (i.e. sawlog regime) for different species. The choice of species and silviculture is mostly determined by the requirement to meet commitments to customers and minimise business risk. Further development of breeding objectives that will maximise returns on investment for different *Eucalyptus* species and production systems is still a priority of the emerging industry. Formal methodology of breeding-objective development, incorporating carbon credits, needs to be applied and the developed breeding objectives adopted by the industry. Yet, there has been only limited work reported on development of bio-economic models involving the plantation grown eucalypt resource (Hamilton 2007).

3. Carbon crediting mechanisms

Under Article 3.3 of the Kyoto protocol (UNFCCC 1997), to receive credits a forest must be planted after 1 January 1990 on land that was not previously forested. Article 3.4 of the Kyoto Protocol specifies that carbon sequestration due to "additional forest management activities" in existing managed forests could also be used to meet emission reduction targets. The Australian, New Zealand and Canadian Governments elected not to account for additional forest management activities, because managed forests may be a source of emissions, and the cost of sequestration measurement may exceed the benefit. However, the Australian Government may decide to cover in a future ETS all managed forests or only plantations.

For the ETS the Australian National Carbon Accounting System (NCAS) would generate carbon emission and removal estimates based on satellite images, climate and soil data, and extensive databases on forest type and management. The National Carbon Accounting Toolbox (NCAT) would use modelling capability of the NCAS to generate emission and removal estimates for reforestation at the project level.

To receive permits, forest growers would need to satisfy reporting and other obligations designed to ensure that the correct number of permits are issued and surrendered. They would also need to indicate the date of forest establishment and provide information about the location of the forest. They would need to provide maps based on NCAS data to assist stakeholders to determine the eligibility of their forests. An approach to reporting similar to the existing New South Wales Greenhouse Gas Reduction Scheme (http://greenhousegas.nsw.gov.au/) would be then used, which includes annual reporting with full verification at periodic intervals (i.e. every five years or following each international commitment period). Forest growers would be required to report any major changes to the emissions estimation plan as a result of changes to forest management or natural disturbances. The regulatory organisation would publish information about all forest registrations.

The Australian Government, Department of Climate Change and Energy Efficiency proposed two approaches to crediting reforestation activities: "full crediting" and "average crediting". The full crediting approach reflects real annual changes in greenhouse gas emissions and removals as reported in Australia's national forest inventory. However, this approach can create risks that, in any given year, severe droughts or fires could significantly and unexpectedly reduce the total number of permits available to forest growers and subsequently to the market. This approach also involves high compliance costs, as permits must be issued on an ongoing basis. On the other hand, this approach would expose forest entities to the full marginal carbon price at all times. The forests included in the ETS would, therefore, be managed for optimal carbon storage and wood production.

The Australian Government would more likely apply the more conservative average crediting approach. For a plantation the permit limit would be based on the average cumulative net greenhouse gas removals calculated at the end of rotation (i.e. prior to harvest) over, for example two long rotations (about 70 years). The net greenhouse gas removals would be based on the forest grower's initial estimation plan and updated as necessary. Although forest growers would not be exposed to the full marginal carbon costs at all times, this approach would reduce the risks of non-compliance. The approach also involves lower compliance costs as permits would generally not need to be surrendered on harvest or following fire and then re-issued as the forest is re-established.

To account for natural disturbances such as fire, disease outbreak, insect attack, storms or severe drought, the permit limit could be reduced by an amount proportional to the risk, i.e., the "risk of reversal buffer". A delay in applying the risk of reversal buffer would mean that the forest entity would receive the full allocation of permits during forest establishment when costs are greatest. In addition, during the early years of forest growth the amount of total carbon storage and therefore potential carbon that could be lost due to natural disturbance is relatively low. This approach would generally remove the need to require the surrender of permits in the event of natural disturbances. Permits would be issued for each tonne of net greenhouse gas removals and would only be issued after trees have grown. The permits would be issued up to a limit, incorporating a risk of reversal buffer. An example of the average crediting approach is given in **Fig. 1**.

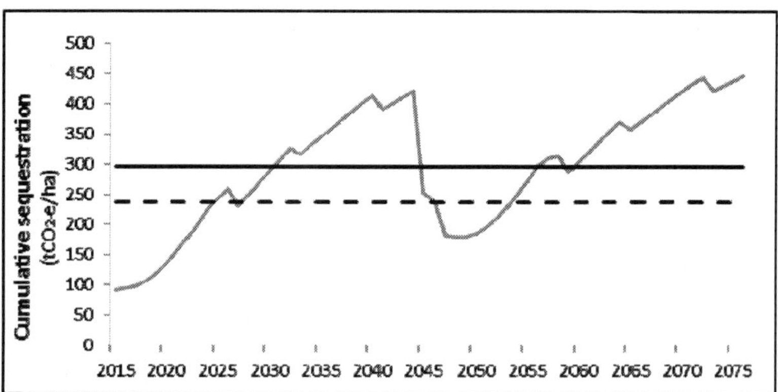

Fig. 1. Diagram of the cumulative carbon sequestration over two rotations and the average crediting approach for a hypothetical *P. radiata* plantation, generated using the FullCam package (Richards et al., 2005). The solid horizontal line represents the permit limit and the dashed line marks the risk of reversal buffer.

4. Incorporating carbon credits into plantation production systems – Example of radiata pine in Australia

Besides volume production and wood quality, breeding and more generally silviculture need also to optimise carbon sequestration. To incorporate carbon credits into breeding objectives we developed a bio-economic model that described the role of carbon crediting in profitability of a radiata pine production system. The model considered the current and future plantation resource, actual and intended silvicultural management regimes, costs and incomes, different discount rates and carbon prices, and different carbon accounting options. Using the methodology previously developed for radiata pine in Australia (Ivković *et al.* 2006a), the objectives were to:

1. Determine future areas of new (i.e. converted from other land uses) and replanted radiata pine plantations that will be established, and on which improved trees will be deployed;
2. Develop a bio-economic model incorporating credits for carbon sequestration in radiata pine plantations;
3. Estimate economic weights for breeding (and silviculture) objective traits based on the production system model incorporating carbon credits;
4. Perform sensitivity analyses by altering the main production system and carbon accounting model assumptions.

4.1 Future area of new plantations

The coverage of additional forest management activities in the future ETS would determine if the genetic improvement will contribute to carbon sequestration in newly established areas only, or also in re-established existing plantations. Currently, in Australia, there is a steady trend of relatively low establishment of new areas of softwood plantations, including radiata pine plantations. Long-term predictions are generally difficult, and the projections of plantation area were generally overestimated (e.g. Ferguson et al. 2002), when compared with the actual plantation area expansion reported by the NPI (2010). The development of the softwood plantation estate was strongly supported by the government during 1970s and 1980s. However, ownership of state owned plantations has been changing to private. In 2009, the area of new plantations established on land not previously used for plantation forestry was 43,231 hectares of hardwoods and 6,427 hectares of softwoods.

As the existing short rotation pulpwood plantations in Australia are harvested, funds will be increasingly redirected towards replanting rather than hardwood estate expansion. Meanwhile, the inability to attract investments for the longer rotation plantations will limit the expansion of the softwood sector. There may be some scope for limited expansion of the softwood plantations to areas that do not prove suitable for second rotation hardwood replanting. The trend of a steady but low rate of establishment of softwood plantation estate is expected to continue. The trend for the softwoods has followed approximately the low estimate of plantation expansion provided by the "Plantations of Australia: Wood Availability 2001-2044" report (Ferguson et al., 2002).

For the purpose of modelling in this study, we assumed an average annual softwood plantation expansion of 6,500ha (min=5,000 and max=10,000), of which radiata pine would comprise 75%. We also assumed a normal, regulated, radiata pine plantation estate of

750,000 ha that will expand over the next two rotations (i.e. 60 years). Such a regulated, so called "normal", forest had equal areas in each age-class up to the rotation length of 30-years. The regulated plantation estate initially had an annual area of clearfelling and planting of 25,000 ha. Forest productivity across the carbon pool was assumed constant and sustainable (i.e. non-declining yield or volume harvested is equal in annual volume increment from growth).

4.2 Bio-economic model

We assumed an integrated forestry production system including plantation growing and sawmill processing components, as described in Ivković et al. (2006a). The bio-economic model had a structure which is summarised in **Table 1.**, but it is a detailed model with approximately 250 production system components (i.e. input parameters). For example, for plantation growing wood flows, details included harvested volume by operation (i.e. 3 thinnings and clear fall), by roundwood category and by 5 cm sawlog diameter classes. For sawmill production systems the wood flow details included sawlog volumes by 5 cm diameter classes and bark percentages, green mill productivity and recovery by diameter and by heart- and sap-wood classes, green mill product output (volumes of green board, chip, fines and shavings), kiln productivity and recovery (i.e. shrinkage volume loss), dry mill productivity and recovery (i.e. planing and docking volume loss), finished product volumes etc.

Variable costs for a forest grower included land rental costs, establishment and maintenance costs (i.e. site preparation, plants and planting, fertilisation, weed and wildling control, annual land and maintenance costs), harvesting, chipping and haulage costs by product type (i.e. pulplog, whole tree chip, preservation sawlog, plylog and recovery log). Variable costs for a sawmill included sawlog procurement (mill gate), mill yard costs, debarking, green and dry mill, kiln cost by board class, and chipping costs.

Incomes for a forest grower included royalties for whole tree chip, pulpwood, preservation, and sawlog by 5 cm diameter classes. Incomes for a sawmill are obtained from sale of finished product (i.e. structural machine graded and visual grades), boards standard and wide, flooring, linings, mouldings, scantlings, treated material etc. and sawmill residues, less marketing costs.

On the base model structured as described above the effects of four breeding objective traits were superimposed. Those four traits had different effects on different production system components and those effects are described below. A discounted cash-flow analysis was then applied to calculate net present value (NPV) of the system before and after a 10% increase of breeding objective traits.

Mean annual increment (MAI), was defined as the average annual increment in volume in cubic metres per hectare per year (m^3 ha^{-1} y^{-1}) evaluated at the end of rotation. To estimate the effects of increasing MAI on the total merchantable volume and sawlog distribution obtained from thinnings and final felling, industry data and the South Australian yield tables (Lewis *et al.* 1976) were used for interpolation. An increase in MAI affected green timber recovery due to changes of diameter distribution. It also impacted carbon revenues, and in turn the economic weight for volume production relative to basic density, within the modelled production system. Whole tree growth was proportional to merchantable volume increment and allocation of biomass to roots, stem, branches, leaves and bark (see next section).

Stem straightness or sweep (SWE) was defined as the maximum deviation of log axis from straight line over a length of the log in units of millimetres per metre (mm/m). Sweep was not considered to have an impact on carbon revenues because the trait does not affect the tree biomass.

The average branch size (BRS) had a major effect on log and timber grade estimated. The effect of branch size on carbon revenues was through its effect on proportion of debris after harvest.

Stiffness of clear wood (measured as Modulus of Elasticity, MOE), had a major effect on timber grade. The effect of MOE on carbon revenue was through its relationship with wood density. It was assumed that there was a slope of $b=0.8$ for relationship between MOE and wood density (Ivković et al. 2009). Green volume multiplied by basic wood density (defined as ratio of dry weight over green volume) determines the dry weight biomass production. Wood density was included in log specifications and therefore affected carbon accounting.

4.2.1 Carbon stocks and carbon revenue models

The FullCAM Carbon Accounting Model (Version 3.0, Richards et al., 2005) models carbon stocks for pasture-to-plantation conversion and over subsequent rotations. The package is a component of the National Carbon Accounting Toolbox Version 1.0 (http://www.climatechange.gov.au/en/government/initiatives/ncat/ncat-toolbox-cd.aspx). FullCAM calculates the gross annual increments in stem volume from yield tables and multiplies the increments by the wood density of the corresponding annual growth sheaths. Wood density can be either estimated from breast height outerwood density samples, or predicted from the site mean air temperature, nitrogen fertility, tree stocking and stand age (Beets et al., 2007a). The product of growth increment and wood density determines annual increments in stem wood carbon. Expansion factors are used to convert those increments in stem wood carbon to estimates of the increments of other tree components, including stem bark, branches, foliage, and roots (Beets et al., 1999; Beets et al., 2007b).

Biomass losses due to tree component mortality are then estimated. For needles, litter fall is based on the needle retention score for the plot. Branch mortality is based on tree stocking and crown occupancy. Thinning and harvesting related losses are based on the estimated reduction in live stem volume. Decay functions are applied to the dead material, thus affecting estimates of the required carbon pools. The modelling system provides point estimates of carbon stocks for each of the required pools. Given complete stand tending regime, the modelling system outputs a carbon yield table over a full rotation and predicts carbon stock changes for the stand.

In this study, the research version of FullCAM was used to predict carbon stock yield on a hypothetical one hectare plot in the Australian Green Triangle region. Estimates of carbon stocks and the effects of carbon credits were obtained over 2 rotations (1. pasture to pine plantation and 2. second rotation of pine plantation). Those carbon stocks were matched with a generic production system described above.

Costs related to carbon accounting can vary widely (and inversely) to the scale of the enterprise, but they generally include: initial set-up cost, registration and lawyers fees (assumed ~\$10 ha^{-1}), ongoing monitoring and accounting annual costs, (assumed ~\$40 ha^{-1}yr) and carbon pooling costs (assumed ~\$2.50 ha^{-1}yr^{-1}).

For forestry systems that include harvested products, but are also managed for carbon, the average stock approach is usually adopted to estimate the value of carbon (Maclaren, 2000; Baalman and O'Brien, 2006). For forests managed in a regulated way over multiple rotations the carbon stock that may be claimed for credits is approximately half the carbon stock at harvest (Fig. 1). The approach yields results equivalent to a carbon pool that is managed as a normal forest with an equal forest area in every age-class up to the rotation length. For example, a 30,000 ha forest estate with stands managed on a 30-year rotation, would have 1,000 ha stands in each age-class. The assumptions were also that the forest productivity across the carbon pool is the same and that the forest is managed to yield a constant sustainable harvest (non-declining yield), based on volume control, i.e., the annual forest volume harvested.

While most of the carbon in a plantation is below ground, carbon stock per ha and its partitioning to roots and woody biomass depends on survival, tree age, irrigation regime, and nutrient status. Nevertheless, an increase in woody biomass results in an increase in the carbon stored in the plantation. In this study, carbon revenues were considered to be directly proportional to the biomass accumulation in the plantation estate. The effect of breeding objective traits (MAI, SWE, BRS and MOE) on a radiata pine production system was determined as described above. Effects of those breeding objective traits on carbon revenues were proportional to their effects on dry bio-mass production per hectare. The tradable unit of CO_2 was the biomass equivalent of one metric ton of CO_2 (1 tCO$_2$e). In the FullCAM model carbon was assumed to make up 50% of oven dry tree biomass, and every ton of biomass carbon is equivalent to 3.67 tCO$_2$. A base price of \$25.0 t^{-1} CO$_2$e was assumed. In the base model, carbon in wood products was not considered, and all carbon in biomass sold was lost to the system immediately upon harvest.

To estimate the increase of carbon storage in wood products the TimberCAM version 1.15.5 model was used (CRC for Greenhouse Accounting, 2004). The model tracks the carbon stored in wood products through their life cycle from harvest through to manufacture, service and disposal. Harvested log removals were allocated to various wood product categories, which all had an estimated service life. All carbon was accredited to the forest grower, regardless of the fate of wood product. Volumes of each major wood product were as listed in Table 1, and described in more detail in Ivković et al. (2006a).

4.3 Economic weights for breeding objective traits

Introduction of carbon credits (scenarios NCP vs. MCP defined in the caption of Table 2) increased the base NPV per hectare from \$2,485 to \$3,818 and from \$5,913 to \$7,246 for plantation growing and integrated production systems, respectively. The relative importance of genetic improvement in different traits also changed. For the plantation growing component of the production system the importance of MAI and MOE (via wood density) increased relative to that of SWE and BRS. For an integrated production system, including both plantation growing and sawmill processing, there was also an increase in the importance of MAI and MOE relative to the importance of SWE and BRS. However, the difference was less than for the plantation growing, because of the high influence of value adding in the sawmill component of the production system (Table 2).

Production system	Scenario	Base	MAI	SWE	BRS	MOE
Plantation Growing	NCC	$2,485	$3,290 32%	$2,424 -2.5%	$2,378 -4.3%	$2,485 0%
	MCP	$3,818	$4,830 27%	$3,739 -2.1%	$3,721 -2.5%	$3,984 4.4%
Integrated Production	NCC	$5,913	$7,111 20%	$5,732 -3.1%	$5,628 -4.8%	$7,252 23%
	MCP	$7,246	$8,652 19%	$7,065 -2.5%	$6,971 -3.8%	$8,751 21%

Table 2. The average NPV per hectare of a generic radiata pine production system over two 30-year rotation before (Base) and after a 10 % increase in the four breeding objective traits at a discount rate of 7%. The values are for production system without carbon credits (NCC) and with a medium carbon price of $25 × tCO2e^{-1} incorporated (MCP).

The change in NPV ha^{-1} y^{-1} per unit trait change, i.e., economic weight (EW) for MOE increased greatly (i.e., from $0 to $145), after carbon credits were accounted for in the model (scenarios for plantation growing NCC vs. MCP defined in the caption of Table 3). However, the EW for MAI for integrated production system showed higher rate of increase (from $499 to $585) than that of MOE (from $1,164 to $1,309), indicating that carbon revenue can change the relativity of EWs for MAI and MOE. At the same time the EW of SWE remained unchanged while that of BRS became less negative both for plantation growing and integrated production (Table 3). These changes in EWs were examined in more detail in the next section.

4.3.1 What-if scenarios

When carbon price was increased from $10 per tCO$_2$e (scenario LCP, Table 3) to $40 per tCO$_2$e (scenario HCP, Table 3), it increased EW of MOE for plantation growing significantly (i.e., from $58 to $231). This was because there was no premium for wood quality of logs and the EW for MOE of a plantation grower was based exclusively on carbon price. However, for an integrated production system the EW of MAI increased at a higher rate (i.e., from $533 to $637) than that of MOE (i.e., from $1,222 to $1,367), which affected the relativity of two EWs. This was because of the very high absolute value of EW for MOE in an integrated production system. The increase in carbon price did not affect the EW for SWE, but it had made the EW for BRS slightly less negative (Table 3).

Discount rate had less influence on cash flows that occurred earlier in the rotation(s), such as carbon credits, than on revenue from the final harvest. For plantation growing the EW for MOE was based exclusively on carbon credits, but when discount rate increased from 4% (scenario LDR, Table 3) to 10% (scenario HDR, Table 3) it influenced the EW of MOE (i.e. a decrease from $203 to $108) less than that of MAI (i.e. a decrease from $785 to $247). Therefore the relative weight of MOE to MAI increased from 26% to 44%. For integrated

Production system	*Scenario*	**MAI**	**SWE**	**BRS**	**MOE**
Plantation Growing	NCC	$335	$-60	$-215	$0
	LCP	$369	$-60	$-207	$58
	MCP	$421	$-60	$-195	$145
	HCP	$473	$-60	$-182	$231
	LDR	$785	-$141	-$514	$203
	HDR	$247	-$29	-$82	$108
	LTP	$228	-$3	-$77	$145
	HTP	$614	-$124	-$312	$145
	WPI	$425	-$75	-$194	$153
Integrated Production	NCC	$499	$-177	$-572	$1,164
	LCP	$533	$-177	$-564	$1,222
	MCP	$585	$-177	$-551	$1,309
	HCP	$637	$-177	$-539	$1,396
	LDR	$1,574	-$530	-$1,504	$3,414
	HDR	$247	-$63	-$223	$593
	LTP	$253	-$36	-$241	$858
	HTP	$916	-$318	-$862	$1,806
	WPI	$589	-$206	-$550	$1,338

Table 3. Economic weights - the average NPV ha[-1] per unit trait change for: mean annual increment (MAI in $m^3y^{-1}ha^{-1}$), sweep (SWE in mm×m[-1]), branch size (BRS in cm), and modulus of elasticity (MOE in GPa) at a discount rate of 7%. The values are for a production system with no carbon credits (scenario NCC) and with low carbon price of $10×tCO_2e^{-1}$ (LCP), medium carbon price of $25×tCO_2e^{-1}$ (MCP), high carbon price of $40×tCO_2e^{-1}$ (HCP), low discount rate of 4% (LDR), high discount rate of 10% (HDR), low timber price (60% of prices given in Table 1), high timber price (140% relatve to prices given in Table 1) and with carbon storage in wood products included (WPI).

production the relative increase was much less (from 217% to 240%). For SWE and BRS, the traits less affected by carbon credits, EWs increased relative to those of MAI and MOE with the increase in discount rate. On the other hand, for a single-rotation production system that did not include carbon credits the EWs, all traits decreased at relatively the same rate with an increase in discount rate (Ivković et al., 2006b).

The increase in round-wood and timber price from 60% to 140% of the base price for plantation growing given in Table 1. did not affect the EWs for MOE ($145), however, the EW for MAI increased from $228 to $614 (Table 3). Therefore the importance of MOE relative to MAI decreased from 64% to 24%. For the integrated production, the increase in EW was greater for MOE than for MAI, i.e., from $858 to $1,806 and from $253 to $916, respectively. The EWs for form and branching (i.e. SWE and BRS) became more negative with the increase in timber price (Table 3.). Similar results were also obtained for a single-rotation production system that did not include carbon credits (Ivković et al., 2006b).

If carbon sequestration in wood products is included in the ETS the NPV of plantation growing and integrated production would significantly increase from $2,485 to $3,908 and

$5,913 to $7,336, respectively. However, this did not seem to significantly change the relative weighting for the four breeding objective traits (scenario WPI, Table 3).

5. Conclusion

Genetically improved germplasm may be deployed both in areas of new plantation and in re-established areas previously under pine plantations. However, the latter may not be eligible to receive carbon credits under the proposed future ETS. The economic weights for breeding objective traits should be calculated as an average for the plantations that are receiving and those that are not receiving carbon credits, weighted by their respective projected plantation areas.

The area of new softwood plantation establishment is currently predicted to be small (i.e., approximately 19%) relative to the area of re-established plantations in the current Australian estate (NPI 2010). Therefore the introduction of carbon credits may only slightly affect the overall breeding objective in the short term. In the long term, consequences of the introduction are more difficult to predict because they rely on various assumptions about plantation expansion.

The analyses on a per hectare basis performed in this study show clearly an increase in the relative importance of biomass production with the introduction of carbon credits. In the case that increases in carbon sequestration resulting from tree breeding and genetic improvement will be accounted for both newly established and re-established plantation areas, the importance of MAI and MOE is expected to increase relative to SWE and BRS. However, the increase in relative value of those traits would also depend on a range of factors: such as changes in rotation length, and accounting for longevity of carbon in the wood products.

Based on the modelling of radiata pine plantations in New Zealand the optimum rotation length is expected to increase with the introduction of carbon credits (e.g. Maclaren *et al.*, 2008). For a production system with rotation length extended from 30 to 35 years the economic weights for MOE are expected to decrease relative to that of other traits (Ivković *et al.* 2006b). However, as the price of carbon increases, regimes with minimum silvicultural intervention and even longer rotations may become more profitable.

Revenue from sales of carbon credits significantly increased the profitability of the radiata pine production system on a per hectare basis. The inclusion of these potential revenues increased the relative economic weights on traits MAI and MOE (i.e., wood density) that determine dry weight of biomass and in turn carbon yield per hectare. Carbon credits are cash flows that occur early in the first rotation, as opposed to the later cash flows from the final harvest and subsequent rotations, and therefore an increase discount rate had less influence on carbon credits. However, the relative economic weights in the production system model with carbon credits behaved similarly to a production system model without carbon credits when discount rate changed (Ivković 2006b).

There is no doubt that for any given production system an increase in growth rate and wood density resulting from genetic improvement will increase carbon storage in stems, products and soil (e.g., Polglase et al., 2008). Although there may have been some losses in wood density due to negative genetic correlation between the two traits (e.g., Wu et al., 2010), genetically improved radiata pine trees can certainly capture more carbon because their biomass production is higher on average.

6. Acknowledgment

This research project was supported by funding from the Australian Government Department of Agriculture, Fisheries and Forestry under its Forest Industries Climate Change Research Fund Program.

7. References

Apiolaza, L.A. & Garrick, D.J. (2000). Breeding objectives for three silvicultural regimes of radiata pine. *Canadian Journal of Forest Research*, Vol.31, pp. 654-662

Baalman, P. & O'Brien, N.D. (2006). Crediting of carbon storage in reforestation projects under timber management. *Australian Forestry*, Vol.69, 90-94

Beets, P.N.; Robertson, K.A.; Ford-Robertson, J.B.; Gordon, J. & Maclaren, J.P. (1999). Description and validation of C_Change: a model for simulating carbon content in managed *Pinus radiata* stands. *New Zealand Journal of Forestry Science*, Vol.29, No.4, pp. 409 - 427

Beets, P.N. (2006). Description of C_Change version 2.0. Client report prepared for Ministry for the Environment, Australian Government

Beets, P.N.; Kimberley, M.O.; & McKinley, R. (2007a). Predicting wood density of Pinus radiata annual growth increments. *New Zealand Journal of Forestry Science*, Vol.37, No.2, pp. 241-266

Beets, P.N.; Pearce, S.H.; Oliver, G.R.; & Clinton, P.W. (2007b). Root/shoot ratios for deriving the below ground biomass of Pinus radiata. *New Zealand Journal of Forestry Science*, Vol.37, No.2, pp. 241-266

Borralho, N.M.G.; Cotterill, P.P.; & Kanowski, P.J. (1993). Breeding objectives for pulp production of *Eucalyptus globulus* under different industrial cost structures. *Canadian Journal of Forest Research*, Vol.23, pp. 648-656

CRC for Greenhouse Accounting (2004). TimberCAM – A carbon accounting model for wood and wood products. Users guide. September 2004.

DAFF, (2007). Australia's forest industry in the year 2020. Available from: http://www.daff.gov.au/__data/assets/pdf_file/0009/643743/2020-report-final.pdf

Ferguson, Ian (2002). Plantations of Australia : wood availability, 2001-2044, National Forest Inventory Australia, Bureau of Rural Sciences, Canberra

Greaves, B.L.; Borralho, N.M.G.; & Raymond, C.A. (1997). Breeding objective for plantation eucalypts grown for production of kraft pulp. Forest Science 43: 465-475

Greaves, B.; Borralho, N.M.G.; & Raymond C. A. (2003). Early selection in eucalypt breeding in Australia optimum selection age to minimise the total cost of kraft pulp production *New Forests* 25: 201–210

Greaves, B.; Dutkowski, G.; & McRae, T. (2004). Breeding Objectives for *Eucalyptus globulus* for products other than kraft pulp, IUFRO Conference – Eucalypts in a changing world, Averio, Portugal 11-15 October 2004

Hamilton, M. G. (2007). The Genetic Improvement of *Eucalyptus Globulus* and *E. Nitens* for Solidwood Production. PhD University of Tasmania

Ivković, M.; Wu, H.X.; McRae, T.A.; & Powell, M.B. (2006a). Developing breeding objective for *Pinus radiata* pine structural wood production: bio-economic model and economic weights. *Canadian Journal of Forest Research* 36, 2921–2932

Ivković, M.; Wu, H.X.; McRae, T.A.; & Matheson, A.C. (2006b). Developing breeding objectives for *Pinus radiata* pine structural wood production: sensitivity analyses. *Canadian Journal of Forest Research* 36, 2932–2942

Ivković, M.; Gapare, W.J.; Abarquez, A.; Ilic, J.; Powell, M.B.; & Wu, H.X. (2009) Prediction of Wood Stiffness, Strength, and Shrinkage in Juvenile Wood of Radiata Pine. *Wood Science and Technology* 43: 237-257.

Ivković, M.; Wu H.X.; & Kumar S. 2010. Bio-economic modelling as a method for determining economic weights for optimal multiple-trait tree selection. *Silvae Genetica* 59(2–3): 77-90.

Jayawickrama, K. J. S. (2001). Potential genetic gains for carbon sequestration: a preliminary study on radiata pine plantations in New Zealand. *Forest Ecology and Management* 152(1/3): 313-322

Lewis, N.B.; Keeves, A.; & Leech, J.W. (1976). Yield Regulation in South Australian Pinus radiata Plantations. Woods and Forests Department, South Australia

Maclaren, J.P.; Manley, B.; & final year of forestry students (2008). Trees in the Greenhouse: The Role of Forestry in Mitigating the Enhanced Greenhouse Effect. *Forest Research Bulletin No. 219.* Scion, Rotorua, NZ

Maclaren, J.P. (2000). Trees in the Greenhouse: The Role of Forestry in Mitigating the Enhanced Greenhouse Effect. *Forest Research Bulletin No. 219.* Scion, Rotorua, NZ.

Madeira, M.V.; Fabiao, A.; Pereira, J.S.; Araujo, M.C.; & Ribeiro, C. (2002). Changes in carbon stocks in *Eucalyptus globulus* Labill. Plantations induced by different water and nutrient availability, *Forest Ecology and Management* 171 75−85

Nolan, G.; Greaves, B.; Washusen, R.; Parsons, M.; & Jennings S. (2005). *Eucalypt* Plantations for Solid Wood Products in Australia– A Review. Forest & Wood Products Research and Development Corporation

NPI (2010). National Plantation Inventory Update. Department of Agriculture, Fishery and Forestry. Bureau of Rural Sciences

NZ MAF (2007). Proposed PFSI Carbon Accounting System, Report to MAF by the PFSI Carbon Accounting Design Team

Polglase, P.; Paul, K.; Hawkins, C.; Siggins, A.; Turner, J.; Booth, T.; Crawford, D.; Jovanovic, T.; Hobbs, T.; Opie, K.; Almeida, A. & Carter, J. (2008). Regional Opportunities for Agroforestry Systems in Australia. Rural Industries Research and Development Corporation, Canberra, Australia, 98p

Richards, G., Evans, D., Reddin, A., & Leitch J. (2005). The *Fullcam* Carbon Accounting Model (Version 3.0 User Manual, National Carbon Accounting System, Department of the Environment and Heritage, March 2005

UNFCCC (1997). Kyoto Protocol to the United Nations Framework Convention on Climate Change.United Nations Framework Convention on Climate Change. Available at: http://unfccc.int/essential_background/kyoto_protocol/items/1678.php

URS (2007). Australa's forest industry in year 2020. Report for the Department of Agriculture, Fishery and Forestry. URS Australia Ltd Pty.

Washusen, R.; Reeves, K.; Hingston, R.; Davis, S.; Menz, D.; & Morrow, A. (2004). Processing Pruned and Unpruned *Eucalypt globulus* Managed for Sawlog Production. Forest & Wood Products Research and Development Corporation

Whittock, S.; Dutkowski, G.W.; Greaves, B.L., Apiolaza, L.A. (2007). Integrating revenues from carbon sequestration into economic breeding objectives for *Eucalyptus globulus* pulpwood production. Annals of Forest Science 64: 239–246

Wu, H. X. ; Eldridge, K. G. ; Matheson, A. C. ; Powell, M. B. ; McRae, T. A. ; Butcher, T. B. & Johnson, I. G. (2007). Achievements in forest tree improvement in Australia and New Zealand: 8. Successful introduction and breeding of radiata pine in Australia. Australian Forestry 70(4): 215-225

16

Irreversibility and Uncertainty in Multifunctional Forest Management Allocation

Jens Abildtrup[1], Jacques Laye[1], Maximilien Laye[2] and Anne Stenger[1]
[1]*INRA, Laboratoire d'Économie Forestière, Nancy*
[2]*Département d'Économie de l'École Polytechnique, Palaiseau*
France

1. Introduction

Forest multifunctionality refers to the valuation of both wood and non-wood forest services. Non-wood forest services are essentially environmental services such as bio-diversity, watershed protection, landscapes, carbon sequestration, or social services such as recreation. Strongly determined by natural forest attributes, forest multifunctionality is increasingly valued in consumers' preferences (Mill et al., 2007). However, most of the non-wood services are not valued in the market, although they could contribute to forest owners' benefits (Katila & Puustjarvi, 2004). Besides the private forest owner's capacity to respond to the increasing demand for non-wood services, the difficulty in quantifying the benefits coming from a multifunctional management puts in question the effective payments determining their participation (Engel et al., 2008).

In the literature, forest multifunctionality is generally modeled as a joint production process (Bowes & Krutilla, 1989; Gregory, 1955; OECD, 2001). Multifunctionality also raises the question of optimal timber rotation when considering non-wood services (Hartman, 1976). When adding a spatial dimension, forest multifunctionality may be analyzed at the larger scale of a forested landscape including more than one forest area. The question of spatial allocation of multifunctionality is crucial in the literature and at a political level (Andersson et al., 2005; Boscolo & Vincent, 2003). For instance, in a *New forestry* approach, Franklin (1989) suggests implementing multifunctionality in each forest area, as it is already experimented in tropical forest management. This conclusion turned out to be the core issue in the debate between *Old Forestry* and *New Forestry* in the United States, inducing theoretical developments on land use through multifunctionality in every forest area *versus* specialization (Helfand & Whitney, 1994; Vincent & Binkley, 1993; 1994). Empirical analyses of forest management have confirmed that there are situations where specialization may be the optimal management regime (Andersson et al., 2005; Boscolo & Vincent, 2003). Swallow et al. (1997) shows that accounting for spatial interaction due to ecological interactions also may prescribe a management which recommends specialization over space.

Added to spatial considerations, time also has an impact on forestry decisions. In particular, the irreversibility of forest management schemes condition the available strategies and future benefits. Natural resources management in a dynamic context can be analyzed by the quasi-option value approach (Arrow & Fischer, 1974; Henry, 1974). This approach takes into account the characteristics of irreversibility and uncertainty of the sequential decision process. Quasi-option value is the value of information conditional on remaining flexible enough to use this information. Indeed, it represents the difference between the value of an option when considering that information is forthcoming (*closed-loop* rationality) and the value of this option without considering that information is forthcoming (*open-loop* rationality). Quasi-option value has been used in a forestry context to choose between strategies of preservation and development (Albers et al., 1996 a; Albers, 1996 b; Bosetti et al., 2004), to derive optimal harvest strategies (Jacobsen & Thorsen, 2003; Malchow-Møller et al., 2004) and forest stand regeneration policies (Jacobsen, 2007). Albers (1996 b) considers a multi-plot setting, focusing on the spatial interdependence and adjacency among forest plots. Our model relates to this literature in the measure that our framework uses irreversibility, uncertainty, and closed-loop rationality in a dynamic decision process. We also consider multiple forest areas setting, but these areas can be non-adjacent, which allows us to assume they can be subject to different environmental policies. Furthermore, our approach is characterized by the choice between two alternatives which represent two exclusive real options (Abildtrup & Strange, 1999; Geltner et al., 1996; Malchow-Møller & Thorsen, 2003). That is, both alternatives are associated with uncertain future pay-offs and the possibility to adapt future management according to forthcoming information. However, we are not estimating the optimal stopping rule, as in the above mentioned studies, but define the optimal strategy at a given decision point accounting for future information and the possibility to adapt management according to this information. The aim of our paper is to shed light on the reflections about the arbitrage between different strategies in the multifunctional forest management of multiple forest areas. By multifunctional, we mean a management ensuring both economic function (timber commercialization) and ecological function (biodiversity, water quality or other amenities) of the forest. Basically, we consider the two strategies, also analyzed by Vincent & Binkley (1994) : 1) multifunctionality in every forest area and 2) specialization of forest areas such that the forest area as a whole is multifunctional. However, we focus on identifying the optimal management strategy in a dynamic setting with uncertainty about future demand for forest services and where the different management strategies may have different degrees of irreversibility. We formulate a simple model which is used to derive general decision rules which can be applied in a given situation defined by the production function and uncertainty about future changes in policy. Such changes could be explained by new biological information or changes in the climate or in the preferences for different uses of the forest.

The starting point is the following : we consider a forest owner/manager of two forest areas. We suppose that this manager is constrained by society to implement a multifunctional management of their total forest area. An immediate problem arises about the best option for the manager in order to maintain multifunctionality for their total forest area. Indeed, a first option consists of specializing one forest area through a *Clear-cutting* management while the other forest area is specialized for full *Preservation* of the ecological function, without any timber harvesting. The other option is to implement a mix of the previous strategies in each forest areas (*Mixed regime* in all the forest areas) where timber is harvested but part of the ecological value is preserved. This two forest areas framework is the simplest one to embed the problem of the arbitrage between the two alternatives for multifucntional forest management (mixed or specialized). For the general and more realistic multiple forest areas case, one can think of our two alternatives as : 1) all forest areas managed under mixed regime *versus* 2) a proportion close to 50% of clear-cut areas while the rest is preserved (for the specialization option).

The choice between a mixed regime and specialization is a fundamental decision problem for many forest managers. For example, in France the government encourages forest owners to increase their harvest of timber because it is considered that forests are exploited less than socially optimal[1]. This concerns, among others, forests in mountain regions. If timber should be harvested in hitherto unexploited mountain forests the manager will have to choose an exploitation strategy subject to the imperative protection functions of mountain forests. The relevant alternatives may include : 1) selective harvesting in all forests (mixed regime in all forests) and 2) clear cutting of some forests and letting other forest areas under protection (specialization). However, the forest manager will also have to account for future changes in the demand for the different uses of the forests. The demand may change due to changes in preferences of the population or climate change may imply that the relative value of different functions changes, and these changes in demand may be site specific. Our model considers how uncertainty and irreversibility influence this decision problem.

A first key issue that conditions the choice of forest management in each forest area is the relationships of irreversibility among management regimes. *Clear-cutting* a forest area is more irreversible than implementing *Mixed regime*, which is more irreversible than full *Preservation*. A second key issue is the uncertainty about the future environmental policy, coming from society and defined for each forest area, when the initial forest management decisions for each forest area take place.

In order to examine the implications of irreversibility and uncertainty in managing multifunctionality, we propose a simple framework where there are only two forest areas in a two-period decision process with two initial choices for multifunctionality (specialization and mixed regime in both forest areas). In the first period the manager makes management choices for each forest area, taking into account the economic and ecological values of the management regimes, the irreversibility relationships among regimes and the priors on the information to come from society about the environmental policies affecting each forest area. In the second period, the information is revealed and the manager chooses again a management regime for each forest area, being constrained by the choices made in the first period. The time elapse between the two periods is supposed to be long enough to have an uncertainty about the forthcoming environmental policy but short enough for the irreversibility relationships to be relevant (no regeneration of the forest).

Our objective is to compare these two management options and determine the conditions for choosing one or the other. Due to the difficulty of applying a standardized valuation method for multifunctional management, our method consists of building a model around assumptions describing the "worst" (reasonable) case for the option of mixed regime in all forest areas. In this way we obtain a benchmark scenario where this option is favored as little as possible, in order to determine the conditions for choosing it, within the corresponding set of restrictive assumptions. The idea is that if the mixed regime option is optimal within conditions in the worst case scenario, then, in reality, this option must be even more suitable. In order to do so, we intentionally make the following assumptions :

(i) Although society is supposed to target multifunctionality, we assume that once the information on the environmental policy is revealed, it can only be a *Preservation* or a *Clear-cutting* policy. In this case, multifunctionality is achieved by society thanks to a mix of specialization incentives.

---

[1] See the Grenelle Environment Round Table, www.legrenelle-environnement.fr

(ii) The mixed regime is modeled as an intermediate management regime, whose value is always inferior to the management that perfectly matches the environmental policy. More precisely, from an economic point of view, the mixed regime is supposed to have less value than the clear-cutting regime but more value than the preservation regime. In a similar way, from an ecological point of view, the mixed regime is supposed to have more value than the clear-cutting regime and less value than the preservation regime.

(iii) Finally, we work in a risk neutral environment. In this case, there is no difference in terms of expected value between betting on a perfect match with the environmental policy through specialization (thus taking the risk of mismatch), and ensuring an intermediate benefit in any case *via* a mixed regime in all forest areas.

Relaxing any of the restrictive assumptions above would favor the initial choice of mixed regime in all forest areas since the corresponding conditions for choosing this initial option would be less restrictive. Indeed, if we relax assumptions we see that :

(i') Considering that society could call for a mixed regime in a particular area would be anticipated by the manager and favor the initial choice of mixed regime in all forest areas. There are cases where society may call for a mixed regime (van Rensburg et al., 2002).

(ii') Considering that the mixed regime has a greater value than the specialization regimes would trivially make the choice of mixed regime optimal in every management period. In real life, in some particular cases, the mixed regime may have a greater economic or ecological value than the extreme management regimes. Biodiversity can be greater when encouraged by an appropriated mixed management than in total preservation. For instance, some bird species such as the red-cockaded woodpecker requires low density older trees that are maintained today with active forest management. In a region where such bird species would be in danger of extinction, this would add ecological value to the mixed regime compared to a full preservation. Another example is that recreation might have more value in a forest that is managed in a sustainable way than in a preserved one that could be more difficult to visit.

(iii') The interest of choosing mixed regime in all forest areas would be greater in a risk aversion framework, since this option ensures the best adaptability to the forthcoming environmental policy. Let us consider, as it can be the case in reality, that the forest owner is risk averse. We have modeled the mixed regime so that it has the same expected value than betting on one the two extreme values of preservation and clear-cutting (see *(ii)*), but mixed regime still represents a less risky initial choice. For instance if a specialization choice in period 1 for the two forest areas happened to be opposite to what society calls for at period 2 (complete mismatch of initial choice and society's preferences), then the forest owner would get the worst possible payoff. This worst payoff would never be obtained when choosing a mixed regime in period 1 (although the best possible payoff would not be reachable either, since the only way to obtain it is to bet on a specialization strategy in period 1 that result in a perfect match in period 2 with society's preferences). A risk averse forest owner would take this into account and choose the less risky initial option of mixed strategy, sacrificing the perfect match payoff, but avoiding the mismatch payoff.

In the benchmark scenario resulting from *(i)*, *(ii)* and *(iii)*, we thus eliminate from the mixed regime all the advantages that are not directly related to the irreversibility among regimes and the adaptability to the second period information about the adopted environmental policy.

2. The model

We consider two forest areas (area 1 and area 2) that are identical *ex-ante*. As we will see they may differ *ex-post* in terms of environmental policies. For each forest area, there exist three management regimes P, M and C. Regime P (*Preservation*) is assumed to guarantee a greater ecological value of the forest (greater biodiversity, for instance). To simplify, we assume that in this regime the forest remains un-cut. Regime C (*Clear-cutting*) leads to greater monetary incomes through intensive timber commercialization. We assume clear-cutting management. We choose extreme scenarios for the management regimes P and C as this is sufficient to generate the stylized facts we wish to bring to the fore. Between these two extreme regimes, M corresponds to an intermediate management regime to which we will refer to as *Mixed regime*. Regime M ensures both ecological and economic functions of the forest. It is assumed to be ecologically less good than P and economically less good than C, but it is ecologically better than C and economically better than P, since only part of the timber is harvested.

Irreversibility. Because trees have a long production period, there exists a natural relationship of irreversibility among the management regimes. This irreversibility is described in Assumption A1: $P \rightarrow M \rightarrow C$, which means that P can lead to any of the three regimes and M can only lead to M or C, while C is a dead-end.

In other words, clear-cutting is the completely irreversible option, while preservation is the more flexible one. M is modeled as an intermediate regime that has an intermediate flexibility. These irreversibility relationships are linked to an underlying assumption about the considered time frame. For given forest manager's time horizon and forest regeneration cycle there exists always a time frame in which these relationships apply. We restrict our analysis to this time frame because our model focuses on irreversibility issues. It is evident that over this time frame our irreversibility relationships may no longer be valid since, for instance, a clear-cut forest area can grow and even become a preserved forest area. In short, the time frame between periods 1 and 2 is long enough for the information revealed at period 2 to be uncertain at period 1, but too short for the forest to have time to regenerate (irreversibility among regimes play a role).

Options for multifunctionality. We assume that the manager targets a multifunctional management of the forest as a whole (area 1 plus area 2) and that the questions of the forest owner's incentives to participate in a multifunctional management are solved. Such multifunctionality can only be implemented in two ways. First, the manager can adopt a specialization strategy in which one area is preserved and the other is clear-cut. Without loss of generality, we can assume in this case that area 1 is preserved, so that this option is denoted PC. Second, the manager can choose a mixed regime in both forest areas (denoted MM). In short, only options PC and MM are considered. Both regimes are multifunctional as they maintain both ecological and economic characteristics of the forest as a whole. Our goal is to compare these two management options and to determine the conditions for choosing one or the other. To solve this arbitrage, it is necessary to observe the consequence of the irreversibility relationships (Assumption A1) on the 2 forest areas. More precisely, option PC at $t = 0$ can lead to 3 regimes at $t = 1$: PC, MC, CC. Option MM can lead to 4 regimes at $t = 1$: MM, MC, CM, CC.

Information about environmental policy. We consider a two-period management framework. As mentioned before, at $t = 0$ the manager chooses a strategy in $\{PC, MM\}$. We assume that at $t = 0$ there is no information on whether it is beneficial to preserve the forest areas or if they have no great ecological value and it is better to clear-cut. We assume that at $t = 1$

society reveals information $I \in \{P,C\} \times \{P,C\}$ concerning the environmental policy (P or C) that is suitable for each forest area. More precisely, the information that is revealed at $t = 1$ can be written $I = (i_1, i_2)$ with i_1 and i_2 in $\{P,C\}$. At society's level, we assume that a multifunctional management of the forest is targeted. Indeed, nowadays most forest countries have taken consciousness that it is not suitable to only consider forests as a source of timber, nor to renounce on timber commercialization by integrally preserving the whole forest area. Because there are many forest areas, multifunctionality can be achieved at a country level with some forest areas being preserved for their ecological value and others being used for timber harvesting, depending on the knowledge about the presence or absence of ecological interest of each particular forest region. We suppose that when having enough information on the ecological value of a particular forest region, society gives a signal to the corresponding private forest managers by revealing which management is suitable in this region. This results in a combination of environmental policies (information $I = P$ or $I = C$). If, for instance, preservation is targeted, society may set taxes for the clear-cut forests areas and subsidies for the protected forests areas inside the region. This environmental policy would change the economic values of the management options in managers' eyes. Forest managers have to choose their individual strategies in the forest areas they manage, depending on their priors about these environmental policies (forthcoming information).

The assumption according to which information I belongs to $\{P,C\} \times \{P,C\}$ means that we assume that society, even though targeting multifunctionality at an aggregated level, never calls for M as a suitable management in a given forest area. As a result, M is modeled as an intermediate regime between P and C, and not as an optimal regime (see Assumption A2). Allowing the environmental policy to be M would favor the initial choice MM. Given the information at $t = 1$, the manager chooses a new management strategy under constraints of irreversibility. Interestingly, the two-areas framework embeds two different kinds of flexibility. One is linked to the irreversibility relationships among regimes : C is more irreversible than M which is more irreversible than P. The other corresponds to an adaptability to the environmental policies revealed at $t = 1$. This adaptability comes from the fact that when starting from option MM the manager can choose the more profitable regime between MC or CM, given the revealed information. There is no such adaptability if PC is chosen at $t = 0$, because the clear-cutting implemented in area 2 is irreversible. The interest of considering a two-areas framework resides in the arbitrage between these two kinds of flexibility.

Manager's priors. We assume that the manager is risk neutral. Let $v_t(X)$ be the payment corresponding to regime $X \in \{P,M,C\} \times \{P,M,C\}$ at time $t \in \{0,1\}$. At time $t = 0$, $v_0(X)$ is deterministic and known by the manager whereas $v_1(X)$ is a random variable, which distribution is assumed to be known and corresponding to the manager's priors on the forthcoming information on the environmental policy. The priors at $t = 0$ on this information $I \in \{P,C\} \times \{P,C\}$ are described by the probabilities μ_I, verifying $\mu_{PP} + \mu_{PC} + \mu_{CP} + \mu_{CC} = 1$. At time $t = 1$, the value $v_1(X)$ is revealed to the manager since the environmental policy is then known for each forest area. To put aside the impact of the payments of the first period, we assume $v_0(PC) = v_0(MM)$. Therefore we can normalize these initial values to zero and simplify the notation at $t = 1$, $v_1(X) = v(X)$. This assumption is consistent with the fact that the manager wonders about the best alternative between PC and MM, because giving them different initial values would favor one or the other of these alternatives. In this way we only focus on the flexibility of the options and the value of the information revealed at $t = 1$.

By only considering the available options at $t = 1$ (given the irreversibility relationships), we can denote by $v^I(X)$ the benefit from implementing a management $X \in \{PC, MM, MC, CM, CC\}$ on the forest areas under information $I \in \{P, C\} \times \{P, C\}$. This 2-areas benefit is simply defined as the sum of the benefits of the two forest areas. More precisely, for forest area k (with $k \in \{1,2\}$) we denote by $v^{i_k}(x_k)$ the value associated with implementing a management $x_k \in \{P, M, C\}$ under information $i_k \in \{P, C\}$, then for $X = (x_1, x_2)$ and $I = (i_1, i_2)$, we have $v^I(X) = v^{i_1}(x_1) + v^{i_2}(x_2)$.

Hierarchy among regimes. The values of the 2-areas management regimes in each information configuration are assumed to be known, and all the uncertainty is concentrated on the fact that the manager does not know at $t = 0$ the information configuration at $t = 1$. We assume there is a hierarchy among the management regimes under the information i, in terms of value : Assumption A2 : If the manager is informed about a preservation environmental policy ($i = P$) for a particular forest area, then, whatever the management of the other area is, $v^P(P) > v^P(M) > v^P(C)$. On the contrary, if ($i = C$), then $v^C(C) > v^C(M) > v^C(P)$.

This very intuitive assumption just means that it is suitable to choose the management regime that matches the revealed information about the environmental policy. This hierarchy is justified economically by the fact that an *ex-post* perfect match to the environmental policy can lead to subsidies and a mismatch to environmental taxes. Of course, the irreversibility constraints might not always allow a perfect adaptation of the regime to the policy. Also note that mixed regime M never gives the best value, whatever the information is.

Assumption A2 has an immediate consequence on the 2-areas regime valuation given information $I \in \{P, C\} \times \{P, C\}$. Given the first period decisions (PC or MM) and the irreversibility relationships determining the available regimes at $t = 1$, the optimal 2-areas regime choices are obtained thanks to :

$$\max_X [v^I(X)] = \max_{x_1}[v^{i_1}(x_1)] + \max_{x_2}[v^{i_2}(x_2)]$$

As a result, if regime PC is chosen at $t = 0$, the optimal regime at $t = 1$ depending on the revealed information I is :

$$PC \text{ if } I = PP$$
$$PC \text{ if } I = PC$$
$$CC \text{ if } I = CP$$
$$CC \text{ if } I = CC$$

If regime MM is chosen at $t = 0$, the optimal regime at $t = 1$ is :

$$MM \text{ if } I = PP$$
$$MC \text{ if } I = PC$$
$$CM \text{ if } I = CP$$
$$CC \text{ if } I = CC$$

The optimal decision exists and is unique for both alternatives PC or MM. Note that regime MC which is available starting from PC is never chosen at $t = 1$. The decision tree depicted in figure 1 takes into account the irreversibility relationships and summarizes the optimal regime choices depending on the information revealed.

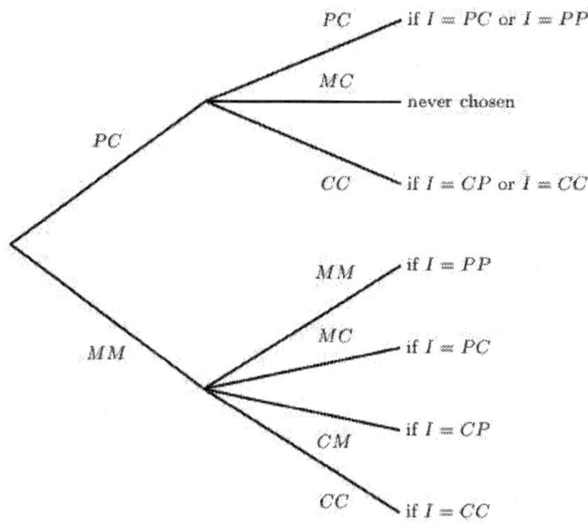

Fig. 1. Regime decisions depending on environmental policies

3. Result

We have seen that if the information is known, the manager can choose the best option at $t = 1$. However, when the initial decision takes place at $t = 0$, information I is not available. We must compare options PC and MM at $t = 0$ in terms of expected value. We assume that the manager uses the following rationality :

$$\hat{V}(PC) = E[\max\{v(PC), v(MC), v(CC)\}]$$

and

$$\hat{V}(MM) = E[\max\{v(MM), v(MC), v(CM), v(CC)\}]$$

where $\hat{V}$ gives the expected value at $t = 0$ of the initial decision in a *closed-loop* information structure[2], *i.e.* by taking into account that information is forthcoming at $t = 1$. Given the optimal strategies at $t = 1$ we can write the expressions of the expected values at $t = 0$ for each option :

$$\hat{V}(PC) = \mu_{PP}v^{PP}(PC) + \mu_{PC}v^{PC}(PC) + \mu_{CP}v^{CP}(CC) + \mu_{CC}v^{CC}(CC)$$

and

$$\hat{V}(MM) = \mu_{PP}v^{PP}(MM) + \mu_{PC}v^{PC}(MC) + \mu_{CP}v^{CP}(CM) + \mu_{CC}v^{CC}(CC)$$

By using $v^I(X) = v^{i_1}(x_1) + v^{i_2}(x_2)$, we finally obtain condition C :

$$\hat{V}(PC) \geqslant \hat{V}(MM)$$

$$\Leftrightarrow$$

---

[2] The closed-loop information structure differs from the open-loop one, in which the agent does not consider that the information is forthcoming. In an open-loop information structure, we would have for the option PC that $V^*(PC) = \max\{E[v(PC)], E[v(MC)], E[v(CC)]\}$. The *quasi-option value* corresponds to the difference $\hat{V} - V^*$, see Arrow & Fischer (1974); Henry (1974).

$$(\mu_{PP} + \mu_{PC})[v^P(P) - v^P(M)] \geqslant (\mu_{PP} + \mu_{CP})[v^P(M) - v^P(C)]$$

3.1 Analysis of the results

Let us analyze the choice condition of the previous section. Condition C determining the initial choice of PC or MM depends on the relative value of regime M under $I = P$, and on the priors of the manager on the information to come.

Concerning the relative value of M, we know (Assumption A2) that under information $I = P$, $v^P(P) > v^P(M) > v^P(C)$. Condition C shows that a value of M close to the value of C (respectively P) weights in favor of choosing PC (respectively MM) at $t = 0$, but that this is not sufficient to determine this choice because of the impact of the probabilities of occurrence of the environmental policies. The impact of the relative value of M is only avoided for $v^P(M) = (v^P(P) + v^P(C))/2 = \tilde{v}_M$. In this case, $[v^P(P) - v^P(M)] = [v^P(M) - v^P(C)]$ and condition C becomes :

$$\hat{V}(PC) \geqslant \hat{V}(MM) \Leftrightarrow \mu_{PC} \geqslant \mu_{CP}$$

Concerning the impact of the priors, let us first deal with some extreme cases. If $\mu_{CC} = 1$, then $\mu_{PP} = \mu_{PC} = \mu_{CP} = 0$, and condition C shows that the manager is indifferent between MM and PC. Indeed, if information CC is certain, management CC is always possible at $t = 1$, no matter what the initial choice is. Interesting cases are those for which $\mu_{CC} < 1$.

If we now consider that the manager has no idea of the forest area that is more likely to be subject to one or the other environmental policy at $t = 1$, that is to say if $\mu_{PC} = \mu_{CP}$, we then remark that the probabilities do not play anymore in the arbitrage, but only the relative value of M does, since condition C becomes :

$$\hat{V}(PC) \geqslant \hat{V}(MM) \Leftrightarrow v^P(M) \leqslant \tilde{v}_M$$

This is in particular the case when the forest areas are adjacent in terms of location. It is rational to think that such forest areas will be subject to the same environmental policy, so that $\mu_{PC} = \mu_{CP} = 0$.

In the light of these extreme cases, we can rewrite condition C by introducing the center of the interval of possible values of $v^P(M)$:

$$\tilde{v}_M = \frac{v^P(P) + v^P(C)}{2}$$

and the ratio γ of probabilities characterizing the asymmetry between priors :

$$\gamma = \frac{\mu_{PP} + \mu_{PC}}{\mu_{PP} + \mu_{CP}}$$

We have that :

$$\hat{V}(PC) \geqslant \hat{V}(MM)$$

$$\Leftrightarrow$$

$$\gamma[v^P(P) - v^P(M)] \geqslant [v^P(M) - v^P(C)]$$

$$\Leftrightarrow$$

$$(\gamma + 1)v^P(M) \leqslant \gamma v^P(P) + v^P(C)$$

$$\Leftrightarrow$$

$$v^P(M) \leqslant \tilde{v}_M - \frac{v^P(P) + v^P(C)}{2} + \frac{\gamma v^P(P) + v^P(C)}{\gamma + 1}$$

And we finally obtain a new expression for condition C :

$$\hat{V}(PC) \geqslant \hat{V}(MM)$$

$$\Leftrightarrow$$

$$v^P(M) \leqslant v_M^* = \tilde{v}_M + \frac{\gamma - 1}{2(\gamma + 1)}[v^P(P) - v^P(C)]$$

If $\mu_{PC} = \mu_{CP}$, then $\gamma = 1$ and the threshold over which mixed regime in both forest areas is preferred over specialization is $\tilde{v}_M$. This means that half of the range of values (from the lowest $v^P(C)$ to the highest $v^P(P)$) leads to the choice of the mixed regime in both forest areas. Over the threshold the expected value of regime M (coming from timber harvesting and the subvention rewarding the environmental quality of this regime) is big enough to make it attractive to choose this regime for both forest areas at $t = 0$. Under the threshold, it is specialization the is optimal at $t = 0$.

However if the priors are not symmetric anymore, say if $\mu_{PC} > \mu_{CP}$, then $\gamma > 1$ and the threshold v_M^* is greater than $\tilde{v}_M$. The interval of values of M that is favorable to specialization is increased in a proportion $(\gamma - 1)/2(\gamma + 1)$ of the length of the interval of the possible values of M under information P, i.e. $[v^P(P) - v^P(C)]$. This can be interpreted as the added value of the information about which area is more likely to be subject to one or the other environmental policy. For example for a system of priors such that $\mu_{PP} = \mu_{CC} = 0$ and $\mu_{PC} = 2\mu_{CP} = 2/3$, then $\gamma = 2$ and $(\gamma - 1)/2(\gamma + 1) = 1/6$. In this case $1/2 + 1/6 = 2/3$ of the interval of possible values of M lead to choosing specialization at $t = 0$.

Conversely, if $\mu_{PC} < \mu_{CP}$ then $\gamma < 1$ and $(\gamma - 1)/2(\gamma + 1) < 0$. This time the range of values of M that is favorable to specialization decreases ($v_M^* < \tilde{v}_M$). However, this situation can only occur in a situation in which the manager has asymmetric priors but cannot adapt, because of technical constraints, the first period specialization strategy in order to match these priors. In this case, although information $I = CP$ is more probable than $I = PC$, the available strategies at $t = 0$ are only PC and MM. This of course favors option MM since choosing PC could lead to a mismatch with the environmental policies at $t = 1$. If there is no such technical constraints, having priors such that $\mu_{PC} < \mu_{CP}$ should rather lead to compare CP and MM at $t = 0$. The arbitrage is then solved like in the case where PC and MM are compared by renaming forest area 1 into forest area 2 and *vice versa*, and we find again that $v_M^* > \tilde{v}_M$, that is to say a situation that is more favorable to specialization than the symmetric case $\mu_{PC} = \mu_{CP}$.

4. Concluding remarks

We have considered that when a manager of two forest areas targets a multifunctional use of the forest in the context of irreversibility among regimes and uncertainty about the forthcoming information on environmental policy affecting the forest areas, two alternatives must be compared : *Specialization* of the areas and implementation of *Mixed regime* in both forest areas. Added to the flexibility of management regimes, considering 2 areas gives rise to another kind of flexibility : the better adaptability to forthcoming information.

Our model shows that the choice of mixed regime in both forest areas depends upon the relative expected value of a forest area in a mixed regime compared to a preserved one : mixed regime should not adversely affect the ecological value of the forest too much. More precisely, the value of a mixed regime should be at least greater than the average value of preservation and clear-cutting regimes for a forest area that is subject to a preservation environmental policy. In the case where the manager has priors on which forest area is more likely to be subject to one or the other environmental policy, this favors the choice of specialization if we consider that the manager can choose the management matching these priors. The interval of expected values of the mixed regime that ensures the choice of mixed regime in both forest areas in the first period is therefore reduced. We give an expression of the resulting variation of the threshold in function of the priors of the manager. This variation can be interpreted as the added value of the information about the asymmetry in the probabilities of occurrence of the environmental policies for each forest area.

Let us underline that the range of expected values of the mixed regime that is favorable to implementing mixed regime in all forest areas has been obtained within a framework designed in order to not favor the mixed regime (worst case framework). This was aimed at finding the minimal conditions of emergence of such forest management (see assumptions i, ii, and iii in the introduction). Relaxing any of these assumptions in our framework would favor the choice of mixed regime in both forest areas over specialization. Our model, thus, provides minimal requirements for the mixed regime to be an attractive solution for owners/managers, which could be helpful to have in mind when setting minimal subventions levels.

Even though the model is theoretical it provides relatively simple decision rules which applies to a range of empirical problems. The choice between clear cutting and selective harvest strategies in hitherto un-managed forest areas is one example of a decision where the presented framework applies. However, it applies more generally to choice situations where increased intensity of land use may have an irreversible impact on the supply of ecological services and where there is uncertainty about the future location-specific demand for these services. Determining whether *Specialization* or *Mixed regime* will be the optimal strategy will, of course, depend on an assessment of the joint production function of the different ecosystem services, i.e. assessing to which degree there are synergies or conflicts in joint production of the considered services or goods.

In the present paper we considered the choice of a private forest owner facing uncertainty about future forest policy (e.g. payments for ecological services). The model applies also to situations with state forests where the forest managers are constrained by the population's current demand for ecosystem services but also consider the uncertainty about the future demand for ecosystem services.

5. References

Abildtrup, J. & Strange, N. (1999). The option value of non-contaminated forest watershed. *Forest Policy and Economics* 1: 115-125.

Albers, H.J., Fischer, A.C. & Hanemann, M.W. (1996). Valuation and management of tropical forests, implications of uncertainty and irreversibility. *Environmental and Resource Economics* 8: 39-61.

Albers, H.J. (1996). Modeling ecological constraints on tropical forest management: Spatial interdependence, irreversibility, and uncertainty. *Journal of Environmental Economics and Management* 30: 73-94.

Andersson, M., Sallnas, O. & Carlsson, M. (2005). A landscape perspective on differentiated management for production of timber and natura conservation values. *Forest Policy and Economics* 9: 153-161.

Arrow, K.J. & Fischer, A.C. (1974). Environmental preservation, uncertainty, and irreversibility. Quarterly Journal of Economics 88: 312-319.

Boscolo, M. & Vincent, J.R. (2003). Nonconvexities in the production of timber, biodiversity, and carbon sequestration. *Journal of Environmental Economics and Management* 46: 251âĂŞ268.

Bosetti, V., Conrad, J.M. & Messina, E. (2004). The value of flexibility : Preservation, Remediation, or Development for Ginostra? *Environmental and Resource Economics* 29: 219-229.

Bowes, M.D. & Krutilla, J.V. (1989). *Multiple-Use management: The Economics of Public Forestlands*. Resources for the Future, Washington D.C.

Engel, S., Pagiola, S. & Wunder, S. (2008). Designing payments for environmental services in theory and practice : an overview of the issues. *Ecological Economics* 65: 663-674.

Franklin, J. (1989). Towards a New Forestry. *American Forestry* 95: 37-44.

Geltner, D., Riddiough, T. & Stojanovic, S. (1996). Insights on the Effect on Land Use Choice: The Perpetual Option on the Best of Two Underlying Assets. *Journal of Urban Economics* 39: 20-50.

Gregory, R.G. (1955). An economic approach to multiple use. *Forest Science* 1: 6-13.

Hartman, R. (1976). The harvesting decision when a standing forest has value. *Economic Inquiry* 14: 52-58.

Helfand, G.E. & Whitney, M.D. (1994). Efficient multiple-use forestry may require land-use specialization: comment. *Land Economics* 70: 391-95.

Henry, C. (1974). Investment decisions under uncertainty: The irreversibility effect. *American Economic Review* 64: 1006-1012.

Jacobsen, J. B. & Thorsen, B. J. (2003). A Danish example of optimal thinning strategies in mixed-species forest under changing growth conditions caused by climate change. *Forest Ecology and Management* 180: 375-388.

Jacobsen, J.B. (2007). The regeneration decision: a sequential two-option approach. *Canadian Journal of Forest Research* 37: 439-448.

Katila M. & Puustjarvi, E. (2004). Des marchés pour les services forestiers environnementaux : réalité et potentiel. *Unasylva* 219: 53-58.

Malchow-Møller, N. & Thorsen, B.J. (2003). Optimal stopping with two exclusive real options.*CEBR Discussion Paper* 2003-33, Centre for Economic and Business Research, Copenhagen.

Malchow-Møller, N., Strange, N. & Thorsen, B.J. (2004). Real option aspects of adjacency constraints. *Forest Policy and Economics* 6: 261-270.

Mill, G.A., van Rensburg, T.M., Hynes, S. & Dooley, C. (2007). Preferences for multiple-use forest management in Ireland : Citizen and consumer perspectives. Ecological Economics 60: 642-653.

OECD (2001). *Multifunctionality - Towards an analytical framework*. OECD Publication Service.

Swallow, S., Talukdar, P. & Wear, D. (1997). Spatial and temporal specialization in forest ecosystem management under sole ownership. *American Journal of Agricultural Economics* 79: 311-326.

van Rensburg, T.M., Mill, G.A., Common, M. & Lovett, J. (2002). Preferences and multiple use forest management. *Ecological Economics* 43: 231-244.

Vincent, J.R. & Binkley, C.S. 1993. Efficient multiple use forestry may require land-use specialization. *Land Economics* 69: 370-376.

Vincent, J.R. & Binkley, C.S. (1994). Efficient multiple use forestry may require land-use specialization : reply. *Land Economics* 70: 396-97.

17

Ecoefficient Timber Forwarding on Lowland Soft Soils

Tomislav Poršinsky, Tibor Pentek, Andreja Bosner and Igor Stankić
Forestry Faculty of Zagreb University
Croatia

1. Introduction

Environmental acceptability is one of the criteria for assessing work efficiency of sustainable forest management. Environmentally acceptable timber harvesting is determined by procedures involving different machines and tools and adequate ways of timber processing, after which the damage to habitat (soil, water) and stand (standing trees, seedlings) are as low as possible. Due to an increasing influence of the public opinion on the current forest environment, the aesthetic appearance of the ongoing forest work site should also be taken into account as well as its appearance after the works have been completed.

Timber harvesting in main felling of lowland even-aged forests of Croatia is based on felling and processing timber by chain saws and forwarding timber processed by cut-to-length method. For this purpose, medium-weight (12–16 t) and heavy (>16 t) forwarders are used, whose mass with the load ranges between 25 and 40 t, and it is distributed to three or four axles. A three-axle (six-wheel) forwarder is equipped with two larger wheels on the front axle and four smaller wheels on the rear axle, which is constructed as a tandem, bogie axle. Four-axle (eight-wheel) forwarder is equipped with wheels of the same dimensions on the front axle and on the rear bogie axle of the vehicle. Bogie axle, with wheels in the so-called tandem distribution, increases the mobility and stability of the forwarder in forest off-road operations.

The soils of lowland forests in Croatia are of heavy mechanical content, and under conditions of frequently excessive moisture (underground water, precipitation, flood or sink water) during the whole year their bearing capacity decreases and these sites are classified as sensitive forest habitats. According to Ward and Lyons (2000), forest habitats where it is necessary to modify the common procedures of wood harvesting so as to avoid damaging effects to ecological, economic and social functions of forests, are sensitive.

The decrease of the soil bearing capacity restricts the mobility and lowers the productivity of forwarders (Poršinsky & Stankić, 2006a), and also increases the level of soil disturbance (Poršinsky & Stankić, 2006b), which can be seen in the form of soil compaction and rutting (Fig. 1). It can be concluded from the above that vehicles with the least possible contact pressure will be the most suitable for off-road wood transport in the Croatian lowland forests. From the point of view of economic use, the Croatian forestry requires the forwarder with the load capacity of 14 t and lifting torque of the hydraulic crane of 100 kNm, which can provide loading and extracting of large logs from main felling sites (Horvat et al., 2004).

Fig. 1. Forwarder mobility and soil disturbance in main felling of lowland even-aged forests of Croatia

The paradox between the application of machine work and site disturbance as side effect of its usage determines an ecoefficient mechanized timber harvesting, involving: 1) efficiency (productivity and costs) of machine work and 2) decreased impact on habitat of machines used in the system of timber harvesting (Owende et al., 2002; Akay et al., 2007; Pentek et al, 2008).

The aim of this paper is to show based on the example of a medium-weight forwarder: 1) the impact of load decrease on productivity and unit costs of timber forwarding, 2) the impact of load mass on forwarder wheel pressure, 3) the impact of tire width and use of tracks depending on load mass on nominal ground pressure. The results can be used to guide decision making in purchasing forwarders, and such approach would provide an efficient and environmentally friendly timber forwarding under conditions of limited bearing capacity of gley soils in main felling sites of lowland forests.

2. Scope

In assessing the environmental soundness of forest vehicles, whose contact with the soil is likely to cause soil damage, the main criteria are trafficking and soil compaction (Poršinsky & Horvat, 2005). Trafficking causes compaction of the soil surface due to moving of forest machines (MacDonald et al., 2002), and it depends on secondary openness of the felling unit and the highest distance of timber reach (Pentek et al., 2010) of the catching device (hydraulic crane, pulling rope of winch) of a certain means of work (forwarder, skidder).

Soil compaction, or rutting, is the consequence of the vehicle off-road travel due to short effect of contact pressures and slippage of drive wheels as well as pulled load (Horn et al., 2004). Soil compaction causes breaking of structural aggregates which decreases inter-aggregate space as well as pore quantity and soil volume (Poršinsky, 2005). Consequently the soil thermal regime is disturbed, water-air relationship in the soil changes and conditions for feeding plants are lowered to a certain extent (Arnup, 1999), for instance microbiological activity is decreased as the soil is brought into anaerobic conditions (Frey et al., 2009). Compaction primarily results in the decrease of the quantity of non-capillary pores and soil permeability to water (Halvorson et al., 2003), which accelerates surface water drainage on slopes covered with a network of vehicle ruts and eventually causes erosion (Owende et al., 2002).

Sensitivity of forest soil to compaction is determined by the following factors: value of vehicle contact pressures, soil texture, soil moisture during timber forwarding, proportion of skeleton and sand particles in the soil, soil structure, bulk density and soil porosity as well as thickness of humus accumulation layer (Arnup, 1999).

Under conditions of restricted ground bearing capacity, the wheels of the forwarder tandem (bogie) axle are equipped with semi-tracks, by which multiple benefits are achieved: 1) soil protection from damage, primarily against compaction and movement of soil layers caused by the increase of the contact area, or decrease of the contact pressure (Bygdén et al., 2004; Gerasimov & Katarov, 2010), 2) vehicle mobility by decrease of wheel slippage, as well as rut depth and vehicle rolling resistance (Bygdén et al., 2004; Bygdén & Wästerlund, 2007; Suvinen, 2006), 3) efficient timber forwarding because of the possibility of use of the vehicle payload, and also increase of the vehicle speed (Poršinsky & Stankić, 2006a), 4) reduction of fuel consumption due to lower wheel slippage (Suvinen, 2006), 5) increase of the forwarder lateral stability during timer loading and unloading, and also during vehicle travel especially when working on slopes (Sutherland, 2003).

In order to reduce the damage of soils of restricted bearing capacity, apart from semi-tracks, additional measures are taken aimed at decreasing the forwarder contact pressures such as the use of multi-wheel vehicles (Nugent et al., 2003; Partington & Ryans, 2010), wheel doubling (Ireland, 2006; Owende et al., 2002), use of wide tires (Saarilahti, 2002b), and also the regulation of tire air pressure (Eliasson, 2005; Sakai et al., 2008), use of chains on the vehicle front wheels (Suvinen, 2006), reduction of the quantity of loaded timber (Poršinsky, 2005), planning the time of work operations (Saarilahti, 2002a).

Apart from the above measures, the researchers have also dealt with the idea of improving the conditions of the soil bearing capacity by covering skid trails by sawmill slabs or pallets (Owende et al., 2002), or by the ever more present cover made of forest residues (Poršinsky & Stankić, 2006b; Eliasson & Wästerlund, 2007; Ampoorter et al., 2007; Gerasimov & Katarov, 2010), which is still treated as waste in cutting and processing timber.

2.1 Ground bearing capacity

Soil bearing capacity (strength, trafficability) is the capability of the soil to resist to external forces (action of the vehicle wheels and tracks), and it is determined by soil settling (rut depth) under external load. In forestry, the soil bearing capacity is determined as the maximum allowed contact pressure of the vehicle wheel (Saarilahti, 2002b) not causing damage to soil, which depends on the type and texture content of the soil, proportion of humus and skeleton particles (constant soil parameters) and a variable parameter – current moisture (Poršinsky, 2005).

The last classification of terrains for harvesting operations was made within the EcoWood project, and as it paid special attention to the ecoefficient wood harvesting on sensitive sites, it categorized the soil strength into four classes and prescribed the maximum contact pressure for each class (Fig. 2). This descriptive classification of the bearing strength of forest soil also recommends the use of the equation nominal ground pressure of the vehicle

(Mellgren, 1980) for determining the suitability of individual types of vehicles for timber harvesting depending on the limit contact pressure on the soil of individual strength classes (Ward et al., 2003).

Soil Strength Classes	General Description of the Soil Types	Cone index CI, kPa	Young's modulus E, MPa	Shear Strength τ, kPa	Ground Bearing Capacity GBC, kPa
1 – Strong soil	Dry sands and gravels, Firm mineral soils	>500	>60	>60	>80
2 – Average soil	Soft mineral or iron-pan soils	300–500	20–60	20–60	60–80
3 – Soft soil	Wet gleys and peaty soils	<300	<20	<20	40–60
4 – Very soft soil	Wet peats	<<300	<<20	<<20	<40

Source: Owende et al. (2002)

Photo: Poršinsky (2005)

Fig. 2. Soil strength classes

2.2 Nominal ground pressure

The vehicle contact pressure is the ratio between the weight and contact surface of the vehicle with the ground (soil), and it expresses the environmental suitability of a specific forest vehicle. The problem in calculating the vehicle contact pressures for forest off-road travel is the dependence of the tire and soil contact area on: 1) elastic deformations of the loaded wheel (tire characteristics, air pressure) and 2) plastic-elastic soil deformations (granulometric content, moisture).

Wishing to standardize the way of calculation of contact pressures of forest vehicles, primarily for providing comparison of vehicles (or different equipment levels of individual vehicles) used for forest off-road timber extraction, Mellgren (1980) introduced nominal ground pressure (Fig. 3). Nominal ground pressure is static pressure (vehicle at standstill), and theoretically it occurs in case of a rigid wheel on plastic-elastic ground where the wheel-soil contact area is calculated as the product of multiplication of the wheel semi-diameter and tire width. When identifying the contact length of the wheel and plastic ground with the wheel semi-diameter, it is important to assume that 15% of the wheel diameter sinks into the soil (wheel rut), by which full contact between the wheel tire and soil is provided

(Partington & Ryans, 2010). In case of lower wheel sinking into the soil (depending on soil bearing strength) the contact area decreases and the contact pressure of vehicles increases, and it is higher with respect to the nominal ground pressure. Actually, the nominal ground pressure is the lowest pressure realized by the vehicle under conditions of reduced soil bearing strength and hence it cannot be used for comparing the suitability of two different wheels under different soil conditions.

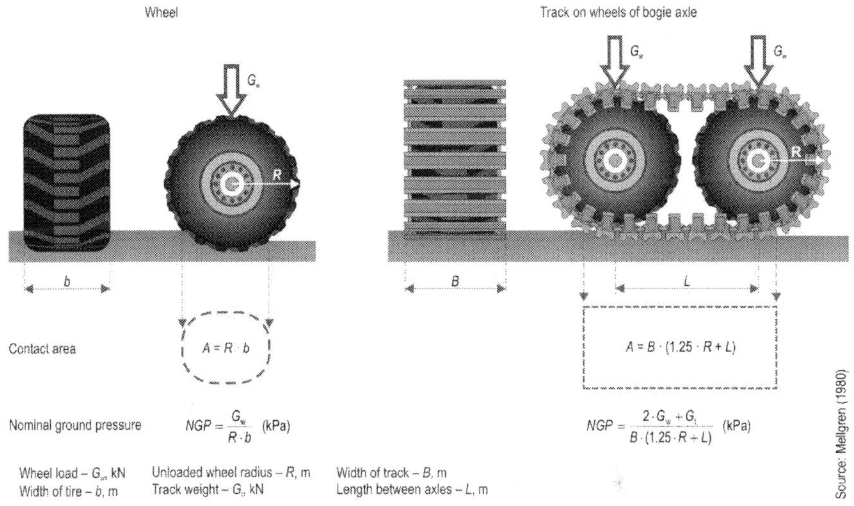

Fig. 3. Calculation of Nominal Ground Pressure

Simplification of the calculation of the contact area, i.e. approximation of the contact length of the loaded wheel with the rut depth of 15% of the wheel diameter, theoretically limits a wide use of this model. The basic objection to approximation of the wheel-soil contact length, with the wheel semi-diameter, is that it applies only in case when the angle between the beginning and end of the wheel/ground contact is 1 radian ($\approx$57.3°), meaning that the model is geometrically sustainable only in certain cases (Poršinsky & Horvat, 2005).

The advantage of the nominal ground pressure is its simple calculation, and the deficiencies are neglecting the impact of tire deflection of the loaded wheel during movement, tire air pressure, independence on soil characteristics and overestimation of use of wide tires (Saarilahti, 2002b).

3. Materials and methods

The analysis of efficiency and environmental soundness of timber forwarding was carried out on the example of a medium-weight six-wheel forwarder Valmet 840.2 with nominal payload of 12 t, whose dimensions and load distribution of unloaded vehicle are shown in Fig. 4. The surface of the cross-cut of the bunk area is 4.1 m², and 4 m in length. The vehicle is driven by a six-cylinder diesel engine with pre-charging of the nominal power of 125 kW at 2200 min-1 and 670 Nm of the maximum torque at 1400 min-1. The forwarder is equipped with hydraulic crane Cranab CFR7C, with the lifting force of 7.1 kN at the maximum range of 9.1 m.

The effect of load reduction (4 t, 8 t related to 12 t of the vehicle payload) on the forwarder efficiency with respect to the distance of timber forwarding is expressed in accordance with the multi-criteria planning model of productivity of these vehicles (Stankić, 2010). This model takes into account: 1) forwarder class, 2) soil bearing strength, 3) forwarder equipment with tracks, 4) felling density, 5) volume of the mean felling tree, and 6) distance of timber forwarding. Unit cost of timber forwarding is calculated according to machine rate made by the company »Hrvatske šume« Ltd Zagreb for the Valmet 840.2 forwarder amounting to 58.15 EUR/PMH.

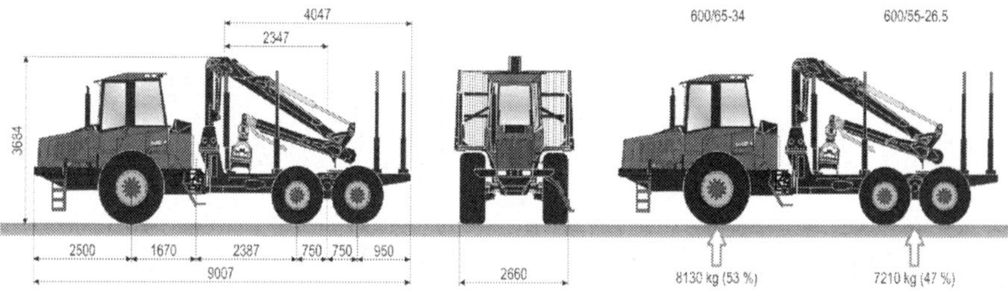

Fig. 4. Valmet 840.2 Forwarder

For calculating the nominal ground pressure, a theoretical model of axle load distribution was used, the case of vehicle at standstill on level ground, depending on mass and length of loaded logs in the forwarder load area (Poršinsky & Horvat, 2005). The analysis of axle load distribution was based on an average length of logs (4 m) made by cut-to-length method in the area of the Croatian lowland forests (Stankić, 2010), and the mass of 1800 kg of a pair of semi-tracks in case when the wheels of the rear (bogie) axle are equipped with them. The wheel load assumed even load distribution of axle load by pertaining wheels. The contact surface between wheels (semi-tracks) and soil was calculated according to Mellgren (1980), for narrow (front – 600/65-34, rear – 600/55-26.5) and wide (front – 710/55-34, rear – 710/45-26.5) tires recommended by the manufacturer of this forwarder.

The analysis of the forwarder environmental soundness was based on: 1) values of nominal ground pressure of front and rear wheels of the vehicle depending on the load mass, and equipment of the vehicle with narrow and wide tires, i.e. equipment of the wheels of the rear axle with tracks, and 2) the upper limit value of the allowed ground pressure (<60 kPa) of the limited bearing strength (Fig. 2 – class 3, soft soil), which prevails at the time of main felling in the Croatian lowland forests.

4. Results and discussion

In accordance with the objectives of the study, the results of soundness of timber forwarding, under conditions of limited soil bearing strength of the Croatian lowland forests carried out by medium-weight forwarders, are presented with respect to: 1) the impact of load decrease on forwarder efficiency, and 2) nominal ground pressure as the measure of environmental soundness considering the vehicle equipment and load mass.

4.1 Forwarder efficiency

In the Croatian forestry, six-wheel forwarders prevail, mostly equipped with tires characterized by deeper and sparser tread pattern (so-called aggressive tread). Such form of tread pattern reduces wheel slippage, but increases damage to soil and tree roots (Sutherland, 2003). The use of tires with shallower and denser tread pattern (so-called non-aggressive tread), which reduces soil damage but increases wheel slippage, and is more suitable for the use of tracks on the wheels of the rear (bogie) axle of forwarders, is more an exception than a rule (Poršinsky, 2005). The same is applicable for the improvement of conditions of soil bearing capacity on skid trails by forming a cover of branches or 3–4 m long fuelwood.

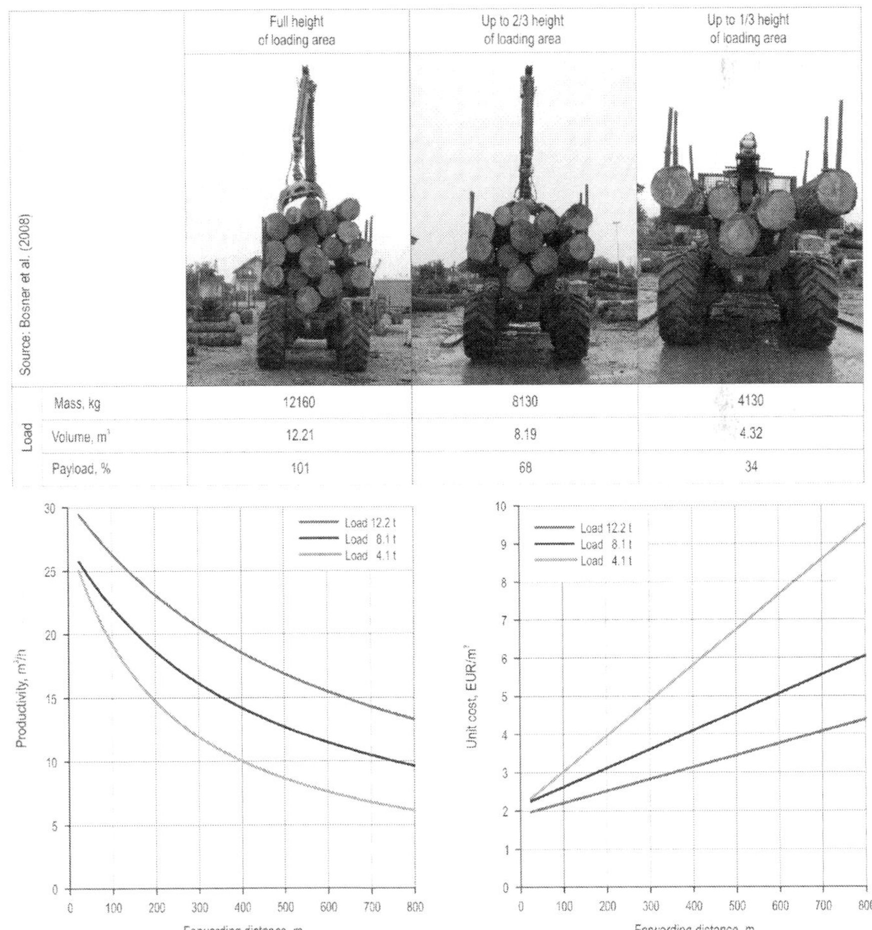

Fig. 5. Impact of load reduction on efficiency of Valmet 840.2 Forwarder

The most frequent form of providing forwarder mobility under conditions of limited soil bearing strength of the Croatian lowland forests is the reduction of the quantity of the loaded timber (load mass, volume), which has an adverse effect on the forwarder efficiency. The impact of the reduction of load volume on the efficiency of Valmet 840.2 forwarder is shown in Fig. 5 with respect to timber loading up to: 1) the full height of the load area (load – 12.2 t, 12.2 m³), 2) 2/3 height of the load area (load – 8.1 t, 8.2 m³), and 3) 1/3 height of the load area (load – 4.1 t, 4.3 m³). It should be emphasized that the load volume is expressed based on the measurement of the length and diameter with bark on the thicker end, in the middle and on the thinner end of each log in the load of the forwarder, and the volume is estimated by Riecke-Newton equation (Köhl et al., 2006).

Load reduction, up to 2/3 height of the load area (68% of the vehicle payload), resulted in the decrease of productivity ranging from 16% (distance of 100 m) to 28% (distance of 800 m) and increase of unit costs from 19% (100 m) to 38% (800 m) with respect to the nominally loaded forwarder (12 t load).

Additional load reduction up to 1/3 height of the load area (34% of the vehicle payload) resulted in the decrease of productivity ranging from 27% (distance of 100 m) to 54% (distance of 800 m) and increase of unit costs from 37% (100 m) to 117% (800 m) with respect to the nominally loaded forwarder (12 t).

Such wide range of productivity decrease and increase of the forwarder unit costs due to higher distance of timber forwarding are the consequence of interaction between time consumption of vehicle travel and loading and unloading timber or load volume (Poršinsky & Stankić, 2006a; Stankić, 2010). Obviously the decrease of the loaded timber highly affects the forwarder efficiency (especially with the increase of the forwarding distance) and therefore, from the economic point of view the method for providing vehicle mobility as well as environmental soundness of timber forwarding under conditions of limited soil bearing strength of the Croatian lowland forests is absolutely not acceptable.

4.2 Environmental soundness

Nominal ground pressure of vehicles is based on interaction between the load of vehicle wheels and its contact area, by which the methodological applicability is only restricted to the case of equal wheel tire dimensions and equal load distribution by the vehicle wheels. In case of different dimensions of the front and rear wheels, i.e. unequal load distribution between front and rear axle of the vehicle, (transfer from wheel–soil system into vehicle–terrain system), Saarilahti (2002a) uses the so-called »reference wheel« (the wheel with the highest contact pressure) or the contact ground pressure is expressed separately depending on the wheels of the front and rear axle, respectively (Poršinsky & Horvat, 2005).

Depending on mass (0–12 t) of roundwood loaded in the forwarder load area (Fig. 6a), the total vehicle mass increases, and there is a considerable increase of the load on rear wheels (1.8–4.6 t) and a relatively insignificant increase of load on front wheel (4.1–4.5 t). With mass increase of the loaded timber <11.5 t, the wheels of the front axle are the reference wheels, after which (just before reaching the vehicle payload) the rear wheels take over this role. The Bavarian federal forests have developed a special approach to the assessment of the environmental soundness of vehicles used in timber harvesting aimed at protecting soil from compaction, which is based on 4 classes of wheel load (Fig. 6a), i.e. in using wide (≥700 mm) tires (Wolf, 2010). According to the Bavarian guidelines for the whole carrying capacity range (<12 t), the wheels of the front axle of the tested forwarder are in the area of »acceptable« wheel load, while the rear wheels are in the area of »optimal« load <9.5 t of the load mass.

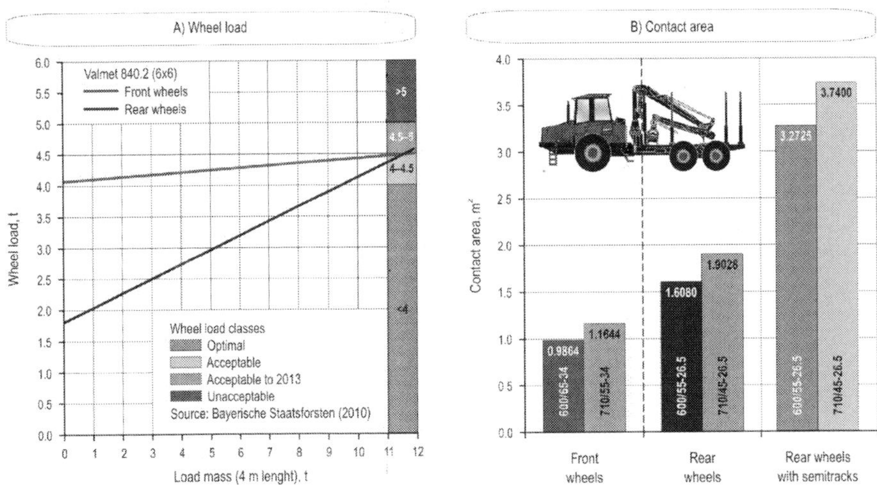

Fig. 6. Wheel load and contact area – Valmet 840.2 (6x6) Forwarder

The analysis of the vehicle-soil contact area (Fig. 6b) showed that the contact area increased by 18% (under the front wheels, but also under the rear wheels) when using wide tires (710 mm) compared to narrow tires (600 mm). The use of semi-tracks on rear wheels of the bogie axle resulted in an almost double increase of the contact area.

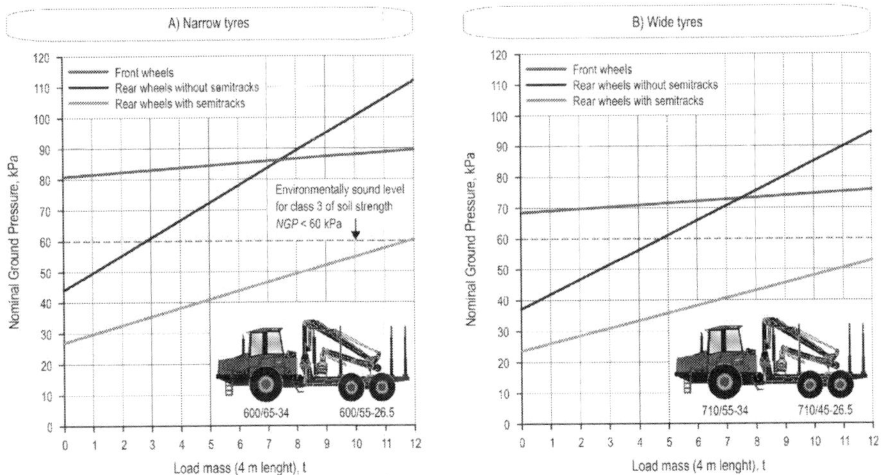

Fig. 7. Nominal Ground Pressure of front and rear wheels (track) vs. load mass

The analysis of the impact of the mass of roundwood loaded into the forwarder load area on the value of the nominal ground pressure under the front and rear wheels of the vehicle, or under the wheel semi-tracks of the rear tandem axle is shown in Fig. 7a for narrow tires (600 mm) and in Fig. 7b for wide tires (710 mm).

Regardless of the use of narrow or wide tires, and also of the load mass, the nominal ground pressure under the forwarder front wheels exceeds the allowed load of the soil of limited bearing strength (<60 kPa). With narrow tires (600/65-34), for the carrying capacity range of the vehicle (<12 t), the nominal pressure is 35–49% higher than allowed. The advantage of using wide tires (710/55-34) on the front wheels of the vehicle can be seen in the decrease of the exceeded nominal pressure, which ranges between 14 and 26% with respect to the allowed value. From the aspect of environmental soundness, the solution of the overloaded front axle of the six-wheel forwarder is an eight-wheel forwarder.

The nominal ground pressure under the forwarder rear wheels equipped with narrow tires (600/55-26.5) exceeds the allowed load of the soil of limited bearing strength in loading timber of the mass >3 t, and when using wide tires (710/45-26.5) in loading timber of the mass >5 t, which is extremely unfavorable from the aspect of timber forwarding (Fig. 5).

By using semi-tracks on rear wheels of the tandem swinging axle with narrow tires, the nominal ground pressure is lower by 55–2 % (depending on load mass) compared to the allowed value. Equipping the forwarder with semi-tracks on wide tires is additionally suitable for the environmental soundness of timber forwarding due to the additional decrease of the nominal ground pressure, which is lower ranging from 61% (unloaded vehicle) to 12% (loaded vehicle) compared to the allowed contact pressure of the soil of limited bearing capacity.

5. Conclusion

Under conditions of limited soil bearing capacity of gley soils due to increased moisture, the decrease of forwarder load, as a measure that provides vehicle mobility and also reduces the level of damage to forest soil, is highly unacceptable from the point of view of timber forwarding efficiency.

The analysis of the nominal ground pressure under the wheels (tracks) of the front and rear axle gave the following guidelines for efficient and environmentally acceptable timber forwarding, under conditions of limited soil bearing capacity:

- due to higher nominal pressure under the front wheels of the three-axle forwarder with respect to the allowed load on the soil of limited bearing capacity (<60 kPa), the use of four-axle (eight-wheel) forwarders is recommended,
- the use of wide tires (710 mm) still provides unsatisfactory increase of load mass with respect to the vehicle equipped with narrow tires (600 mm), provided that the allowed soil load is not exceeded,
- in order to provide adequate mobility, full use of the vehicle payload that assures forwarding efficiency, but also environmental soundness, the use of semi-tracks on wheels of tandem front and rear axle of an eight or ten wheeled forwarder is recommended (Fig.8).

These results should be used as guidelines for future purchasing of forwarders, by which efficient and environmentally sound timber forwarding would be provided under conditions of limited bearing capacity of gley soils in main felling sites of lowland forests in Croatia and elsewhere.

Source: www.ponsee.fi

Fig. 8. Eight and ten wheeled forwarder with semi-tracks

6. Acknowledgment

The research was carried out within the scientific and research project of the Ministry of Science, Education and Sport of the Republic of Croatia »Environmentally Sound Timber Harvesting (068-0682111-0390)«.

7. References

Akay, A.E.; Sessions, J. & Aruga, K. (2007). Designing a forwarder operation considering tolerable soil disturbance and minimum total cost. *Journal of Terramechanics*, Vol.44, No.2, (April 2007), pp. 187-195, ISSN 0022-4898

Ampoorter, E.; Goris, R.; Cornelis, W.M. & Verheyen, K. (2007). Impact of mechanized logging on compaction status of sandy forest soils. *Forest Ecology and Management*, Vol.241, No.1-3, (March 2007), pp. 162-174, ISSN 0378-1127

Arnup, R.W. (1998). The Extent, Effect and Management of Forestry-related Soil Disturbance, with Reference to Implications for the Clay Belt: A Literature Review. Ontario Ministry of Natural Resources, Northeast Science & Technology, TR-37, pp. 1-30., ISBN 0-7778-7453-9

Bosner, A.; Nikolić, S.; Pandur, Z. & Benić, D. (2008). Development and Calibration of Mobile Measuring System of Vehicle Axle Mass – Measurements on Forwarder. *Nova mehanizacija šumarstva*, Vol.29, December 2008, pp. 1-15, ISSN 1845-8815

Bygdén, G.; Eliasson, L. & Wästerlund, I. (2004) Rut depth, soil compaction and rolling resistance when using bogie tracks. *Journal of Terramechanics*, Vol.40, No.3, (Jully 2003), pp. 179-190, ISSN 0022-4898

Bygdén, G. & Wästerlund, I. (2007). Rutting and soil disturbance minimized by planning and using bogie tracks. *Forestry Studies*, Vol.46, pp. 5-12, ISSN 1406-9954

Eliasson, L. (2005). Effects of forwarder tyre pressure on rut formation and soil compaction. *Silva Fennica*, Vol.39, No.4, (September 2005), pp. 549-557, ISSN 0037-5330

Eliasson, L. & Wästerlund, I. (2007). Effects of slash reinforcement of strip roads on rutting and soil compaction on a moist fine-grained soil. *Forest Ecology and Management*, Vol.252, No.1-3, November 2007, pp. 118-123, ISSN 0378-1127

Frey, B.; Kremer, J.; Rüdt, A.; Sciacca, S.; Matthies, D. & Lüscher, P. (2009). Compaction of forest soils with heavy logging machinery affects soil bacterial community structure. *European Journal of Soil Biology*, Vol.45, No.4, (July-August 2009), pp. 312-320, ISSN 1164-5563

Gerasimov, J. & Katarov, V. (2010). Effect of Bogie Track and Slash Reinforcement on Sinkage and Soil Compaction in Soft Terrains. *Croatian Journal of Forest Engineering*, Vol.31, No.1, (Juny 2010), pp. 35-45, ISSN 1845-5719

Halvorson, J.J.; Gatto, L.W. & McCool, D. K. (2003). Owerwinter changes to near-surface bulk density, penetration resistance and infiltration rates in compacted soil. *Journal of Terramechanics*, Vol.40, No.1, (January 2003), pp. 1-24, ISSN 0022-4898

Horn, R.; Vossbrink, J. & Becker, S. (2004). Modern forestry vehicles and their impacts on soil physical properties. *Soil and Tillage Research*, Vol.79, No.2, (December 2004), pp. 207-219, ISSN 0167-1987

Horvat, D.; Poršinsky, T.; Krpan, A.; Pentek, T. & Šušnjar, M. (2004). Suitability Evaluation of Forwarders Based on Morphological Analysis. *Strojarstvo*, Vol.46, No.4-6, (July-December 2004), pp. 149-160, ISSN 0562-1887

Ireland, D. (2006). Traction Aids in Forestry. *Forestry Commission*, Technical Note 13, pp. 1-8, ISBN 0-85538-700-9, Edinburgh, UK

Köhl, M.; Magnussen, S. S. & Marchetti, M. (2006). *Sampling Methods, Remote Sensing and GIS Multiresource Forest Inventory*, Springer-Verlag, ISBN 3-540-32571-9, Berlin, Germany

Mellgren, P.G. (1980). Terrain Classification for Canadian Forestry. *Canadian Pulp and Paper Association*, 1-13

McDonald, T.P.; Carter, E.A. & Taylor, S.E. (2002). Using the global positioning system to map disturbance patterns of forest harvesting machinery. *Canadian Journal of Forest Research*, Vol.32, No.2, (February 2002), pp. 310-319, ISSN 0045-5067

Nugent, C.; Canali, C.; Owende, P.M.O.; Nieuwenhuis, M. & Ward, S. (2003). Characteristic site disturbance due to harvesting and extraction machinery traffic on sensitive forest sites with peat soils. *Forest Ecology and Management*, Vol.180, No.1-3, (July 2003), pp. 85-98, ISSN 0378-1127

Owende, P.M.O.; Lyons, J.; Haarlaa, R.; Peltola, A.; Spinelli, R.; Molano, J. & Ward, S.M. (2002). Operations protocol for Eco-efficient Wood Harvesting on Sensitive Sites. Project ECOWOOD, Funded under the EU 5th Framework Project (Quality of Life and Management of Living Resources) Contract No. QLK5-1999-00991 (1999-2002), pp. 1-74.

Partington, M. & Ryans, M. (2010). Understanding the nominal ground pressure of forestry equipment. *FPInnovations*, Vol.12, No.5, (July 2010), pp. 1-8, ISSN 1493-3381

Pentek, T.; Poršinsky, T.; Šušnjar, M.; Stankić, I.; Nevečerel, H. & Šporčić, M. (2008). Environmentally Sound Harvesting Technologies in Commercial Forests in the

Area of Northern Velebit – Functional Terrain Classification. *Periodicum Biologorum*, Vol.110, No.2, (Juny 2008), pp. 127-135, ISSN 0031-5362

Pentek, T.; Nevečerel, H.; Dasović, K.; Poršinsky, T.; Šušnjar, M. & Potočnik, I. (2010). Analysis of Secondary Relative Opennnes in Hilly Areas as a Basis for Selection of Winch Rope Length. *Šumarski list*, Vol.134, No.5-6, (Juny 2010), pp. 241-248, ISSN 0373-1332

Poršinsky, T. (2005). Efficiency and Environmental Evaluation of Timberjack 1710B Forwarder on Roundwood Extraction from Croatian Lowland Forests. PhD Thessis, Forestry Faculty of Zagreb University, pp. 1-170.

Poršinsky, T. & Horvat, D. (2005). Wheel Numeric as Parameter for Assessing Environmental Acceptability of Vehicles for Timber Extraction. *Nova mehanizacija šumarstva*, Vol.26, December 2005, pp. 25-38, ISSN 1845-8815

Poršinsky, T. & Stankić, I. (2006a). Efficiency of Timberjack 1710B Forwarder on Roundwood Extraction from Croatian Lowland Forests. *Glasnik za šumske pokuse*, Special issue 5: pp. 573-587, ISSN 0352-3861

Poršinsky, T. & Stankić, I. (2006b). Environmental Evaluation of Timberjack 1710B Forwarder on Roundwood Extraction from Croatian Lowland Forests. *Glasnik za šumske pokuse*, Special issue 5: pp. 589-600, ISSN 0352-3861

Saarilahti, M. (2002a). Soil interaction model. Project deliverable D2 (Work package No. 1) of the Development of a Protocol for Ecoefficient Wood Harvesting on Sensitive Sites (ECOWOOD). EU 5th Framework Project (Quality of Life and Management of Living Resources) Contract No. QLK5-1999-00991 (1999-2002), pp. 1-87.

Saarilahti, M. (2002b). Modelling of the wheel and tyre, 1. Tyre and soil contact – Survey on tyre contact area and ground pressure models for studying the mobility of forest tractors. Soil interaction model, Appendix Report No.5, 1-43.

Sakai, H.; Nordfjell, T.; Suadicani, K.; Talbot, B. & Bøllehuus, E. (2008). Soil Compaction on Forest Soils from Different Kinds of Tires and tracks and Possibility of Accurate Estimate. *Croatian Journal of Forest Engineering*, Vol.29, No.1, (Juny 2008), pp. 15-27, ISSN 1845-5719

Stankić, I. (2010). Multicriterial Planning of Timber Forwarding in Croatian Lowland Forests. PhD Thessis, Forestry Faculty of Zagreb University, pp. 1-123.

Sutherland, B.J. (2003). Preventing Soil Compaction and Rutting in the Boreal Forest of Western Canada: A Practical Guide to Operating Timber-Harvesting Equipment. *FERIC Advantage*, Vol.4. No.7, (March 2003), pp. 1-52, ISSN 1493-3381

Suvinen, A. (2006). Economic Comparison of the Use of Tyres, Wheel Chains and Bogie Tracks for Timber Extraction. *Croatian Journal of Forest Engineering*, Vol.27, No.2, (December 2006), pp. 181-102, ISSN 1845-5719

Ward, S.M. & Lyons, J. (2000). The development of an operations protocol for wood harvesting on sensitive sites. Proceedings of International conference »Thinnings: A valuable forest management tool«, September 9-14, 2001, IUFRO Unit 3.09.00 & FERIC & Natural Resources Canada & Canadian Forest Service, CD, 1-12.

Ward, S.M. & Owende, P.M.O. (2003). Development of a protocol for eco-efficient wood harvesting on sensitive sites. Proceedings of the 2nd International Scientific Conference »Forest and Wood-Processing Technology vs. Environment –

Fortechenvi Brno 2003«, May 26–30, 2003, Brno, Chech Republic, Mendel University of Agriculture and Forestry Brno & IUFRO WG 3.11.00, pp. 473-482.

Wolf, M. (2010). Bodenschutz bei den Bayerischen Staatsforsten. Bayerische Staatsforsten Aör, pp. 1-8.

18

Comparison of Timber Consumption in U.S. and China

Lanhui Wang[1] and Huiwen Wang[2]
[1]Beijing Forestry University
[2]Beihang University
China

1. Introduction

Timber consumption is now a global concern from environmental protection and sustainable development prospective in the past decades. The nature of the environmental stress depends upon the quality and level of difference in the trends and patterns of natural resources consumption across countries (Parikh *et al.,*1991).

Forest resources supply timber, but that is not forest's only function. People are more concerned with its other functions like air-cleaning, and slowing down climate change. So the timber supply function is no longer the first choice of forest resource. Therefore, the large amount of timber consumption arose the claim that timber production is destroying the forests of the world, countries which consume large amount of timber are the focus of the controversy, especially the big countries such as China and U.S. As a matter of fact, there are different patterns of timber production and consumption in different countries because of different preference of using timber. So it is better to make a deep insight to the phenomena rather than believe the claims simply. On the other hand, China imports logs and processes them then exports large amount of forest products to the world. The processing brings more contamination to China as a longterm sacrifice as well as the contribution to the world. It is kind of "Smile Curve with two ends out" as Japan does in forest sector.

Though the possession quantity per capita is above world level, US is still a net timber import country. Given the strong consumption drive, imagine how much area of forest needs to be logged to meet such a huge demand. According to some statistics from forest service (James, 2007), the population of US increased by 52.70% from 194 million to 296 million during 1965-2005, meanwhile, the consumption of timber increased by 58.89% from 377 million m³ to 599 million m³. The dependence on import timber of US increased from 2% in 1991, up to 9% in 1996. This increased to 16% in 2002, and it is predicted to be 19% in 2050 if the current policy remains (State Forestry Administration, 2008). It is true that the timber consumption is dependent on population and consumption per capita, while the consumption per capita is a comprehensive factor and it changes over time because of the variation of the recycling rate, technical change, as well as people's disposable income.. What are the factors affecting timber consumption according to evidence from the past? Are there any policies affecting timber consumption in U.S. or not, when it comes to the sub-prime loan crisis on the housing sector? This brings out the need to examine the factors affecting the consumption of different products from the evidence of history data. Such an exercise is important to know what are the variables or policy instruments imposing impacts to different products.

On China side, due to the vast application of timber in industries like furniture, paper and pulp, housing construction and decoration, there are a great number of factors affecting timber consumption. For example, population, industry demand, and international trade. The research on this issue is important to know what are the variables or policy instruments imposing impacts to timber products.

2. Forest resources in China and U.S.

According to the seventh national forest resource survey (State Forestry Administration, 2004-2008), the forest area in mainland of China is 193.33 million ha, the forest coverage is 20.36% of total territory, the forest stock in terms of forest is 133.63 million m^3. the area of natural forest is 119.69 million ha with the forest stock of 11402 million m^3, the plantation area is 61.69 million ha with volume of 1961 million m^3 , which rank the first place in the world. Because of the relatively young age structure of the plantation, China still needs large amount of timber from outside. On the other hand, China, a country with 19 percent population of the world has about 5 percent forest of the world, consumes about 25 percent industrial wood of the world. It is not easy for China to be self-sufficient in timber in a short run, but in a long run, the plantation will be the back up for China to be domestic-sufficient in forest sector.

While according to the forest facts (USDA, 2009), the forest area in U.S. is 303.93 million ha, the forest coverage is about 33% while the forest stock is 26393 million m^3. Forestry issues are of considerable significance to the United States, which has 5 percent of the world population and consumes 27 percent of the world's industrial wood products. Although domestic timber inventory is only 8 percent of the world total, 76 percent of U.S. consumption of industrial wood comes from domestic supplies.

Fig.1 gives the comparison of U.S. and China in population and forest resources, which shows the contrast of U.S. and China in forest resource. With almost the same territory with China, U.S. is more abundant in forest resource.

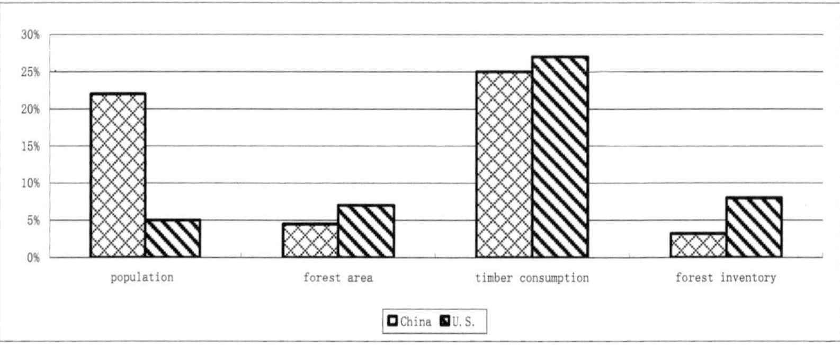

Fig. 1. Comparison of U.S. and China in social and forestry sector

2.1 Timber consumption in China and U.S.

In a global context, United States and China are both among the biggest timber consumers. The timber consumption per capita of U.S is 2.01 m^3 on average, almost 3.2 times of the average level in the world (State Forestry Administration, 2008). Because the large population, though the per capita level of China is relatively low about 0.24 m^3, China consumes lots of timber(see Fig. 2 and Fig. 3). Also the historical trends indicate that China's timber consumption is in rapid increase, while in the U.S. the consumption in the last two decades is in fluctuation mode, a bit increase compared with past.

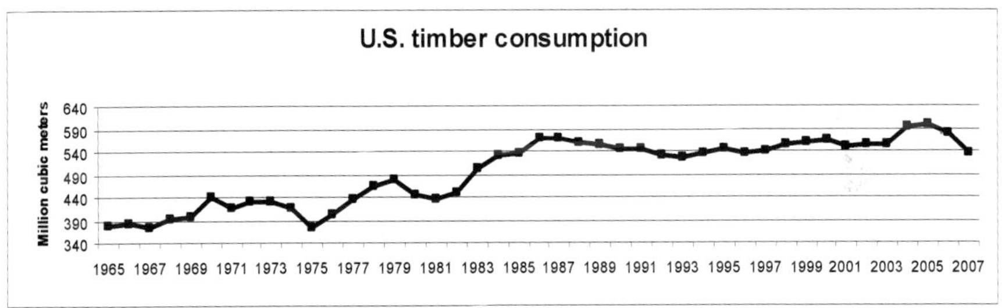

Fig. 2. U.S timber consumption

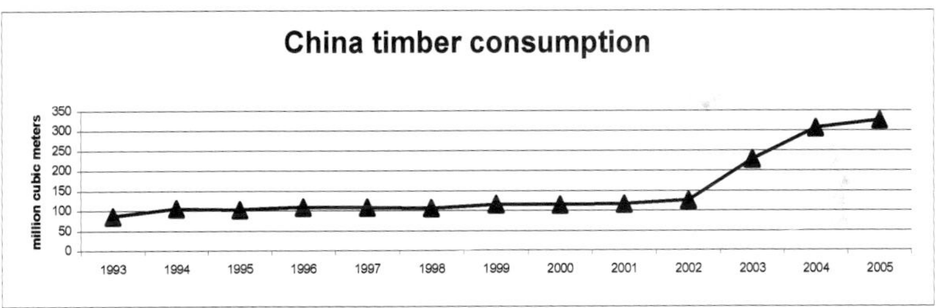

Fig. 3. China timber consumption

Furniture manufacturing, paper making and house construction are the three most timber-consuming industries in China. Furniture accounts for about one third of the total consumption. The rapid development of furniture industry enables China to be the No. 1 wood furniture export country since 2005. The quick increase of new house starts contribute to the timber consumption and log import increase. The average annual increase of timber consumption in newly-built house construction and interior decoration is over 20% since 2000. Paper making is on the rapid increase as well, the timber consumption by roundwood equivalent increase from 31million m^3 in 2000 to nearly 80 million m^3 in 2007.

Fig. 4 shows the dynamic share of the main industries in timber consumption in China.

New housing construction accounts for more than one third of U. S. softwood lumber and structural panels consumed about 10% volumes of other softwood and hardwood products, strengthened considerably in 2005, but declined soon after 2005. Housing starts of single family units led the increase and multi-family housing (see Fig.5). New housing and repair and remodeling drive the wood product demand, unfortunately it did not maintain the strong trend after 2005. However, investment in residential repair and remodeling rose to 226,359 million dollars, increasing by about 6.12% in 2006, while almost stabilized in 2007. The data collected by United States Census shows the average size of new single family home increase considerably. In the last 50 years, the size of new single-family home doubled.

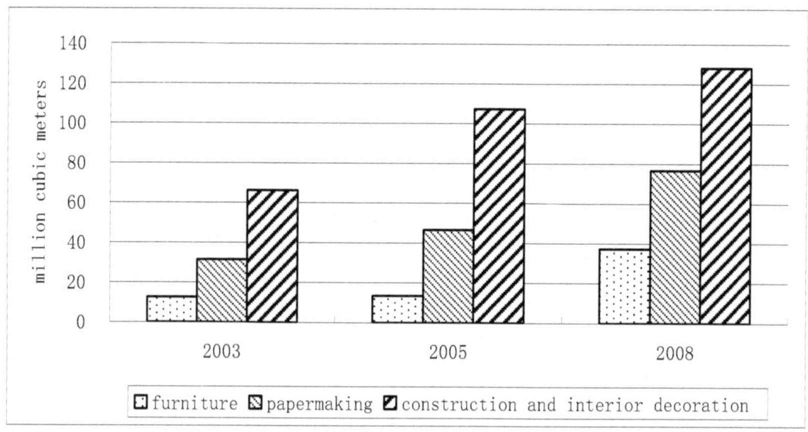

Fig. 4. Timber consumption of main wood related industries in China

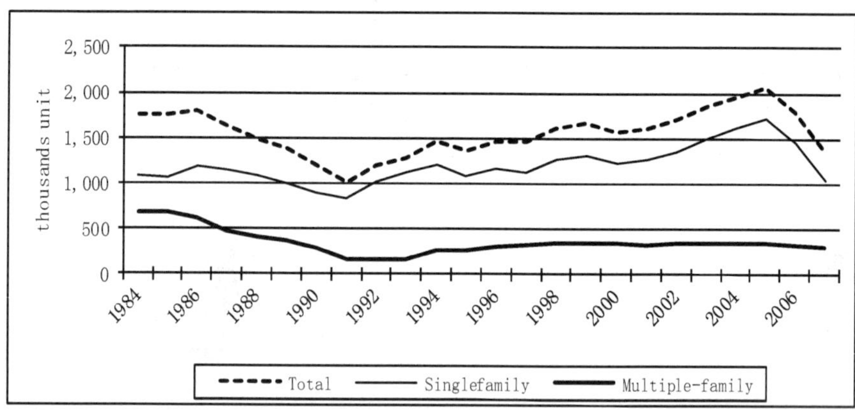

Data source: U.S. Department of Agriculture

Fig. 5. Housing starts in U.S. by type 1984-2007

The timber consumption patterns are quite different in China and U.S. The former is more industry-oriented and the latter is more domestic construction related. The following work in section 3 and 4 will take a deep insight to the question what indeed are the factors affecting the timber consumption in both countries.

2.2 Previous research

In the literature, quantitative analysis and forecasting of forest product markets started from 1950s, based mostly on pure time-series analysis. Considerable improvements have been made in the theoretical issues of models, statistical estimation methods and the coverage of datasets. Some of previous studies tried to estimate the elasticity of different forest products like round-wood, pulpwood and newsprint. For a particular country or groups of countries (Hetemaki *et al.* 1992 ; Nebebe *et al*,1992) carried a study on Nirgeria and applied international consumption function to predict the demand and supply requirements (disaggregated product types) by using cross-section data of different countries, taking income and population as exogenous variables. Few studies (Rao, 1990; Rai, 1988) forecasted the demand of forest products in India using time series analysis. Although the above studies helped decision makers in estimating the demand for timber products, they ignored some of the important factors (population growth, economic condition, technical change, substitute materials) that affected the overall consumption and hence had limited utility in projecting the future requirements and the model result is hard to interpret (Samprit, 2000).

Few studies paid attention to variable selection and the interaction between variables. Since the initial set of independent variables is usually quite large selected subjectively and a number of coefficients will most likely have been estimated by Ordinary Least Squares methods (Forward, Backward or Stepwise approaches provided by most standard statistical programs). Although partial least squares regression (PLS) and principal component regression (PCR) provide good results, the problems still cannot be solved efficiently (Johan *et al*, 1998). More recently, several modified PCR and PLS methods have been proposed, for example a hierarchical PCA (HPCA), a consensus PCA (CPCA，a hierarchical PLS (HPLS) and a multi-block PLS (MBPLS) (Svante *et al*,1996).These methods divide the set of independent variables into multiple blocks according their physical context and try to explain the dependent variables from different dimensions, in order to solve regression by multi-block independent variables. However, these methods usually synthesize the information by extracting components from each block, then explain the dependent variables by components, but do not provide a variable selection in each block in advance. The initially selected independent variables differ in explanatory power and there may be severe collinearity existing in between. Therefore, other studies (e.g. (Wang Huiwen 2008, Lanhui Wang & Huiwen Wang, 2008) have proposed a solution which selects the main independent variables in each block, and then implement PLS regression on these variables. The present study attempts to apply this approach to timber sector in both countries and tries to find the statistical evidence between consumption of timber products and socioeconomic variables grouped into multiple blocks in order to see the difference in timber consumption pattern.

3. Methodology

3.1 The Schmidt orthogonal transformation and PLS method: Our theoretical basis

To begin, we introduce the Schmidt orthogonal transformation method to explore whether the Gauss_Markov assumption can be rejected if any group of independent variables were transformed by the Schmidt orthogonal transformation. After that, we present the PLS method.

3.1.1 The Schmidt orthogonal transformation and its reverse transformation

The proof of the following theorems is provided by Hao (2009).

Theorem 1: Any set of nonlinearly correlated vectors $x_1, x_2, \cdots, x_s$ can be transformed into a set of orthogonal vectors by a Schmidt transformation.

$$z_1 = x_1$$

$$z_2 = x_2 - \frac{x_2' z_1}{z_1' z_1} z_1$$

$$z_3 = x_3 - \frac{x_3' z_1}{z_1' z_1} z_1 - \frac{x_3' z_2}{z_2' z_2} z_2 \tag{1}$$

$$\cdots\cdots$$

$$z_s = x_s - \sum_{k=1}^{s-1} \frac{x_s' z_k}{z_k' z_k} z_k$$

Based on the theorem 1 and theorem 2, we prove the corollary 1 and get the reversal transformation of of Schmidt transformation.

Corollary1: For any variable set $x_1, x_2, \cdots, x_p$, with rank s where $s \leq p$ the set $z_1, z_2, \cdots, z_p$ can be obtained. Of this set, there are s orthogonal vectors, while the remaining p-s vectors are zero vectors after the Schmidt transformation. For clarity and convenience in later discussion, we called z_j the Schmidt variable and the variables corresponding with z_j, i.e., x_j are called variables related with z_j. There are two functions of a Schmidt orthogonal transformation: the first is an orthogonal decomposition of the information of the variable set and the second function is to exclude those variables not related with the Schmidt variables (these are the p-s vectors that changed into zero vectors).

Proof: Inverse Proof . If corollary 1 is not true, for any variable set $x_1, x_2, \cdots, x_p$ let

$$z_1 = x_1$$

$$z_j = x_j - \sum_{k=1}^{j-1} \frac{x_j' z_k}{z_k' z_k} z_k, \quad j = 2, \cdots, p \tag{2}$$

If there is $s + 1$ non-zero vectors, it is easy to prove that they are orthogonal in between, while since the rank of is s, here comes the contradiction. Therefore, the corollary 1 is proved.

Theorem 2: For any variable set $x_1, x_2, \cdots, x_p$ with rank s · where $s \leq p$ after the Schmidt transformation, there is a non-zero orthogonal vector $Z = (z_1, z_2, \cdots, z_s)$. In general, we assume the variables related to $z_1, z_2, \cdots, z_s$ are the set $x_1, x_2, \cdots, x_s$, denoted as $X = (x_1, x_2, \cdots, x_s)$. As well, we define

$$r_{jk} = \frac{x_j' z_k}{z_k' z_k}, \quad j = 2, \cdots s ; k = 1, \cdots, s-1$$

i.e. the transformation matrix of X and Z is R.

$$\tilde{R} = \begin{pmatrix} 1 & r_{21} & r_{31} & \cdots & r_{s1} \\ 0 & 1 & r_{32} & \cdots & r_{s2} \\ 0 & 0 & 1 & \cdots & r_{s3} \\ \vdots & \vdots & \vdots & \ddots & \vdots \\ 0 & 0 & 0 & \cdots & 1 \end{pmatrix}_{s\times s}$$

The reversal transformation of Schmidt transformation is as follows:

$$Z = X\tilde{R}^{-1} \tag{3}$$

Prove: neglect those zero variable vectors after Schmidt transformation $x_{s+1}, x_{s+2}, \cdots, x_p$, the matrix form of Schmidt transformation is:

$$(z_1, z_2, \cdots, z_s) = (x_1, x_2, \cdots, x_s)$$
$$- (z_1, z_2, \cdots, z_s) \begin{bmatrix} 0 & r_{21} & \cdots & r_{s1} \\ 0 & 0 & \cdots & r_{s2} \\ \vdots & \vdots & \ddots & \vdots \\ 0 & 0 & \cdots & 0 \end{bmatrix} \tag{4}$$

From (4), it can be deduced that

$$(z_1, z_2, \cdots, z_s) \begin{bmatrix} 1 & r_{21} & \cdots & r_{s1} \\ 0 & 1 & \cdots & r_{s2} \\ \vdots & \vdots & \ddots & \vdots \\ 0 & 0 & \cdots & 1 \end{bmatrix} = (x_1, x_2, \cdots, x_s) \tag{5}$$

Obviously, $\tilde{R}$ is a reversible matrix. Therefore the conclusion of theorem 2 can be obtained

According to theorem 2, if carrying out regression on y Schmidt variables $Z = (z_1, z_2, \cdots, z_s)$, the model will be:

$$\hat{Y} = Z\hat{\beta} \tag{6}$$

Then if reverse the model to the regression y on $X = (x_1, x_2, \cdots, x_s)$, it will be

$$\hat{Y} = X\tilde{R}^{-1}\hat{\beta} \tag{7}$$

3.1.2 The Schmidt orthogonal transformation and the Gauss-Markov condition in OLS

In classical multiple linear regression, y is denoted as the dependent variable and $x_1, x_2, \cdots, x_p$ as the set of independent variables and the regression model is

$$y = \beta_0 + \beta_1 x_1 + \cdots + \beta_p x_p + \varepsilon \tag{8}$$

We take the variables transformed by the Schmidt transformation as our independent variables and then regress y on this set. It is easily proven that in model (8), if we regress y on $z_1, z_2, \cdots, z_s$, the Gauss-Markov condition can be satisfied.

From this discussion, the parameter test and model evaluation method can be applied directly in the regression of y on $z_1, z_2, \cdots, z_s$.

3.1.3 Introduction of PLS regression method

Assume there is one dependent variable denoted as y_1 and p independent variables denoted as $\{x_1, x_2, \cdots, x_p\}$ for a set of n observations, which provides us with the two following matrices: $X = [x_1, x_2, \cdots, x_p]_{n \times p}$ and $Y = [y_1]_{n \times 1}$. PLS extracts the component t_1 from X, i.e., t_1 is a linear combination of $x_1, x_2, \cdots, x_p$. Two requirements should be met during extraction:

1. t_1 should carry as much as possible the variance information on variation of each matrix;
2. the correlation of t_1 and y_1 should be maximum.

These two requirements state that t_1 should represent most the information of matrix X, while t_1 should possess the greatest power of explanation towards y.

After the first component is extracted, PLS will implement the regression of X on t_1 and the regression of Y on t_1. If the regression model has reached a satisfactory level of precision, the algorism terminates. Otherwise, the residuals, after the extraction of X and Y, are regressed on t_1 for a second time. This step is repeated until a required level of precision is reached. If there are m components, i.e., $t_1, t_2, \cdots, t_m$ extracted from X, then y_1 is regressed on these m components. The model is then expressed in terms of y_1 and $x_1, x_2 \cdots x_p$.

3.2 PLS regression of multi-block variables based on the Schmidt transformation

Given the theory introduced in the last section, we now present the PLS regression method of multi-block variables, based on the Schmidt transformation.

We begin by standardizing the variables and denote the standardized dependent variables as y and the independent variables as x_{ij}, $i = 1, 2 \cdots m$, where m is the number of independent variables in the set and $j = 1, 2 \ldots p_i$, is the number of variables in set i. The specific steps are given as follows:

Step 1. Regress y on each variable x_{1j} ($j = 1, 2 \ldots p_i$,) in set 1, we choose (Or the algorithm chooses) the Schmidt variable with largest t-value into the model. Assume $z_{11} = x_{11}$, then carry out the Schmidt orthogonal transformation of the other with z_{11} and let :

$$z_{1j}^2 = x_{1j} - \frac{x_{1j}' z_{11}}{z_{11}' z_{11}} z_{11}, \quad j = 2, 3, \cdots, p_1 \tag{9}$$

Step 2. Carry out a bi-variate regression of y on z_{11} and z_{1j}^2, $j = 2, \cdots, p_1$, choose the Schmidt variable with the largest t-value into the model . Let the related variable with z_{12} be x_{12}.

Step 3.   Let $z_{1j}^3 = x_{1j} - \dfrac{x_{1j}' z_{11}}{z_{11}' z_{11}} z_{11} - \dfrac{x_{1j}' z_{12}}{z_{12}' z_{12}} z_{12}, \quad j = 3, 4, \cdots, p_1$ (10)

Do a tri-variate regression of y on z_{11}, z_{12} and $z_{1j}^3, j = 3, 4 \cdots p_1$ and select the Schmidt variable with the largest t-value for introduction into the model.

Step 4. Repeat steps 1 to 3, until no more variables outside the model can pass the t test. Then the first subset of regressions s_1 can be obtained, i.e., those variables related to the Schmidt variables such as z_{11}, z_{12}). s_1 is called the first main variable subset of the first block of variables

Step 5. For subset i，repeat steps 1 to 4 and make a variable selection for each block of variables. Then the final main subset can be obtained as $s = s_1 \cup s_2 \cup \cdots \cup s_m$.

Step 6. mplement PLS regression of y on the variables included in s, then the regression model is as follows：

$$\hat{y}_i = \hat{\beta}_0 + \hat{\beta}_1 x_1 + \hat{\beta}_2 x_2 + \cdots + \hat{\beta}_k x_k \tag{11}$$

Step 7. Reverse the set of standardized data into the original their original scale. Then the final result can be obtained.

4. Dataset and result

Table 1 and table 2 give the variables for each country. The data used here is time series data, for U.S 1965-2007; for china 1993-2005 because the dataset is incomplete when compared with U.S side and the two countries have different national forest survey timing. The independent variables are grouped into three blocks related with macro economy, industry and international trade which are the main factors affecting timber demand and supply.

Block	Independent variables included	Symbol and unit
s_1 Macro-economy	GDP per capita	x_{11} , billion in 1996 dollars
	Population	x_{12} , million persons
	Personal disposable income	x_{13} , billion in 1996 dollars
s_2 Industry	Expenditure on housing maintenance	x_{21} , billion in 1996 dollars
	Expenditure on new housing	x_{22} , billion in 1996 dollars
	construction	x_{23}, percentage
	Mortgage rate	x_{24} , unit
	Housing stars	x_{25} , square feet
	One family house average size	x_{26} , square feet
	Multi-family house average size	x_{27} , percentage
	industry index	
s_3 International Trade	Exchange rate with Yuan	x_{31} , ratio
	Exchange rate with Euro	x_{32} , ratio
	Exchange rate with Canadian dollar	x_{33} , ratio (before 1999, European Currency Unit is used)

Table 1. Independent variables for US timber consumption

By using the method proposed above (Lanhui,Wang&Huiwen,Wang, 2008), following the detailed steps, the main factors affecting the U.S. and China were captured. For U.S timber sector, the factors affecting timber consumption are: GDP per capita(x_{11}), population (x_{12}), expenditure in house maintenance (x_{21}), among them, population is the most powerful drive. The results are shown as follows:

U.S.: $$\hat{y} = 2013.48 + +19.06 x_{11} + 1029.5 x_{12} + 42.18 x_{21} \tag{12}$$

China: $$\hat{y} = 1356.76 + 0.21 x_{13} + 2.15 x_{23} \tag{13}$$

Block	Independent variables included	Symbol and unit
s_1 Macro-economy	GDP	x_{11} , billion in 1990 RMB
	Output of manufacturing industry	x_{12} , million Yuan in 1990 RMB
	Output of construction industry	x_{13} ,million Yuan in 1990 RMB
	Out put of secondary industry	x_{14} ,million Yuan in 1990 RMB
	Personal disposable income	x_{15} ,Yuan/year
s_2 Industry	Paper and paper board Production	x_{21} , million tons
	Floor area for new housing	x_{22} , million square meter
	Wood panel production	x_{23} , million cubic meters
s_3 International Trade	Exchange rate with US dollar	x_{31} , ratio
	Exchange rate with Euro	x_{32} , ratio (before 1999, European Currency Unit is used)

Table 2. Independent variables for China timber consumption

The fitting charts can be seen in Fig.6 .

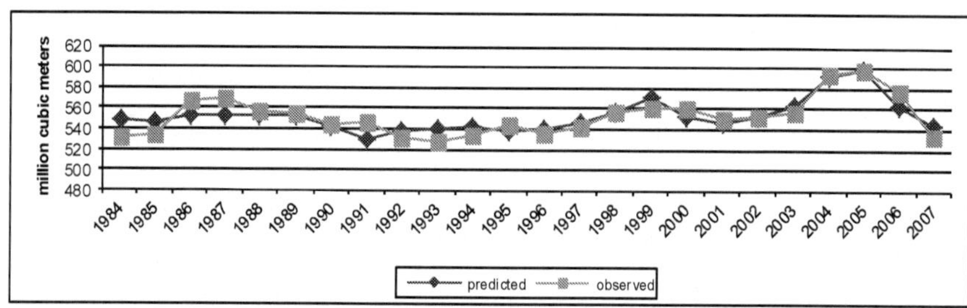

Fig. 6. Fitting chart of U.S timber consumption

For China , the determinants are output value of manufacturing industry (x_{12}) and paper and paper board production (x_{21}) and they have a positive effect on timber consumption. The fitting chart can be seen in Fig.7.

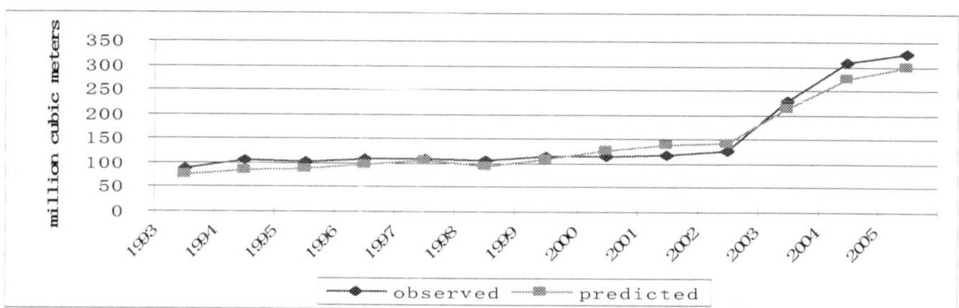

Fig. 7. Fitting chart of China timber consumption

5re the main factors affecting timber demand and supply.bundant in forest resource. The results show different timber consumption pattern in U.S. and China. China is more industry driven but U.S. is more housing construction driven. Different driving forces explain the different consumption patterns in the two countries. Furthermore, it is foreseeable that the long term the factors affecting timber consumption will not change the increase trend abruptly so the timber consumption patterns will continue for a relative long period respectively in the long run in both countries. Given the growing environment concerns, China will confront more difficult timber trade situation to satisfy her own increasing domestic demand. U.S. will be mostly self-sufficient in timber supply.

5. Discussion

China is more industry-oriented in timber consumption and the main factors are from processing-industry. While U.S. is mainly domestic consumption-oriented, the drives come from the GDP and population and housing sector. The result coincides with the "China-factory phenomena" in many timber-related industries. U.S. is more economy and population oriented in timber industries. This multi-block PLS method is effective for selecting the important variables. In a large dataset, the variable selection based method-Multiple Block Partial Least Squares is a good choice.

From the view point of forest resource, the different consumption patterns propose different prospective for both countries. China is a country lack of forest resources; while U.S. has abundant forests to support its domestic-consumption, though U.S. relies on import to satisfy its demand to some extent in order to meet the environmental concerns. The gap between production and consumption of China to a large degree was filled by the international trade. In view of the security of timber resources and promotion of environmental consciousness, many countries tightened or are trying to tighten the timber trade policy. For example, Russia increased its log export tariff to 80% since January 2010, this policy affects many countries in the world especially its neighboring country China, whose nearly 70% log import is from Russia. It is hard for China to source the timber overseas . So the fast growth plantation is a feasible choice. So far it seems underperforming in terms of quantity and quality, it doesn't follow the original ambitious goal about 13 million ha by 2015 with supply capability of 130 million m³(China report, 2002). Compared with U.S, China's timber industry needs to enlarge the forest investment to guarantee sustainable timber supply. The pressure of forest resource in both countries are quite different. On the U.S. side, it can rely on its own forest resource for timber and environmental system in a long run. The feasible way for China is to follow the steps of U.S. to be self-sufficient and environmentally-friendly consumption mode in the future.

6. Acknowledgment

This work is funded by the China Fundamental Research Funds for the Central Universities RW2011-13, BLRW2010-04 ,China Nature Science Fund No.71003006 and China Social Science Fund of Ministry of Education No.09YJA910001.

7. References

Hao Zhifeng. Linear Algebra. Higher education press, Beijing, 2009,pp23-35
G.S.Haripriya, Jyoto K.Parikh.(1998). Socioeconomic development and demand for timber products, Global Environmental Change, Vol8 ,No.3, pp 249-262

Hetemaki, L and Kuuluvainen, J (1992).Incorporating data and theory in roundwood supply and demand estimation. American Journal of Agricultural Economics 74, pp1010-1018

James. L. Howard. U.S. Timber Production, Trade, Consumption and Price Statistics 1965 to 2005(2008). Research Paper FPL-RP-637. Madison, WI: U.S. Department of Agriculture, Forest Service, Forest Products Laboratory, pp91

Johan A. Westerhuis, Theodora Kourti and John F. Macgregorj. (1998). Analysis of multiblock and Hierachyical PCA and PLS models, Journal of Chemometrics,vol 12, pp301–321

Lan-hui WANG, Hui-wen WANG.(2008). A PLS Method based on Schmidt Transformation of Multi-block Independent Variables: an Application，Recent Advance in Statistics Application and Related Areas,Part2, Conference Proceedings of 2008 International Institute of Applied Statistics Studies,Yantai,China,Aussino academic Publishing house

Nebebe, F. and Kira, D. (1992).Estimating elasticities in pulp and paper industry: An econometric analysis of Canadian regions, Journal of Quantitative Economics, pp 451-466

Parikh, Jyoti K., Gokarn, S, J.P.Painuly.(1991). Consumption Patterns: the Driving Force of Environmental Stress, India Gandi Institute of Development Research Discussion paper No 59,1-3,India Gandhi Institute of Development Research, Bombay.

Rai, K.N., Shri Niwas and Kharkar, R.K. (1988). Demand and supply analysis of forest products in India, Indian Journal of Agricultural Economics 39, pp300-308

Rao, D.N. (1991),Forest economics in India: Issues and Perspectives. Center for Economic Studies and Planning Discussion Paper, Jawahar Lal Nehru University, New Delhi

Samprit Chatterjee, Ali S. Hadi, Bettram Price (2000).Regression Analysis by Example, John Wiley & Sons. Inc. 2000, pp230~250

State Forestry Administration.(2008). American Expert discusses sustainable forest management. Retrieved from
<Http://www.forestry.gov.cn/distribution/2008/04/02/n_sjly-2008-04-02-395.html>

Svante Wold, Nouna Kettaneh, Kjell Tjessem.(1996). Hierarchical multiblock PLS and PC models for easier model Interpretation and as an alternative to variable selection, Journal of Chemometrics, 1996 (10), pp463-482

United State Department of Agriculture. (2009).U.S. Forest resource facts and historical trends, FS-801, 2009, pp2-17

Wang Huiwen, Chen Meiling, Saporta (2008). Multivariate Regression Variable Selection Method based on Schmidt orthogonal transformation(in Chinese), The Journal of Beijing University of Aeronautics and Astronautics, 2008,34(6), pp729-733.

China report. Fast plantation project launched in China, Retrieved from
<Http://www.forestry.gov.cn/distribution/2008/04/02/n_sjly-2008-04-02-395.html>